Automation of Finite Element Methods

Jože Korelc · Peter Wriggers

Automation of Finite Element Methods

Springer

Jože Korelc
Faculty of Civil and Geodetic Engineering
University of Ljubljana
Ljubljana
Slovenia

Peter Wriggers
Leibniz University Hannover
Hannover
Germany

ISBN 978-3-319-39003-1 ISBN 978-3-319-39005-5 (eBook)
DOI 10.1007/978-3-319-39005-5

Library of Congress Control Number: 2016940112

Printed on acid-free paper

This Springer imprint is published by Springer Nature
The registered company is Springer International Publishing AG Switzerland

The original version of the bookfrontmatter was revised: For detailed information please see Erratum. The Erratum to this chapter is available at 10.1007/978-3-319-39005-5_9

Preface

Finite elements analysis (FEM) is one of the most powerful tools for the numerical simulation of complex industrial problems. Still new formulations and discretization schemes have to be developed that account for the needs of advanced applications in engineering.

The book is intended for students and engineers who want to develop new finite element formulations, especially for nonlinear problems. The derivations of the finite element matrices and vectors needed for an efficient treatment of nonlinear applications within a finite element environment can become extremely complex and error-prone. Due to the power of symbolic computations it is nowadays possible to automatically generate efficient and highly compressed code for linear and nonlinear problems. By this, tedious hand calculations can be avoided leading to more accurate implementations and huge time savings in the development phase, e.g., of new finite elements or material models.

Based on the symbolic system *AceGen* fast and reliable code can be created with a minimum of effort and immediately be tested and verified by using the associated finite element program *AceFEM*. *AceGen* is a package within *Mathematica* and produces source code for different finite element environments. The use of *AceGen* is described within this book for applications in solid mechanics. For that the basic equations of continuum mechanics are summarized and the input for the symbolic systems is added in order to provide a guide to apply *AceGen* for nonlinear problems of three-dimensional solids such as hyperelasticity, finite deformation plasticity, and sensitivity analysis. In addition, element formulations for structural elements like nonlinear truss-, beam-, and shell structures will be developed.

We would like to mention the German Science Council (DFG) which supported the first author when he visited the Leibniz Universität Hannover as Mercator Professor in 2015. The DFG also funded many grants of the second author for different projects on nonlinear finite element methods over the years. The results of this work can be found at many different places throughout the book. Last but not least, we would like to thank the Springer Verlag for the pleasant collaboration during the past years.

Ljubljana
Hannover
March 2016

Jože Korelc
Peter Wriggers

Contents

Notation

In many ways the presented notation departs from the standard notation adopted in many monographs about the finite element approach to the problems in computational mechanics. The standard notation in many ways still predates the era of computers and thus represents the obstacle for the true automation of computational modeling. The newly introduced notation reflects the need for the unification of the classical mathematical notation of computational mechanics models and the actual computer implementation. This unification has general consequences on the way how the computational models are formulated and solved and has a potential to shape the future of the field of computing.

The tables below are composed of three columns. In the first column the standard notation is given that will be used throughout the text of the book. In the second column the corresponding notation is given for the program codes included in the book. A short description of the quantity is given in the third column. The naming of the quantities follows the well-established rules as far as possible. The changes are made to adjust the name for the specifics of the syntax of symbolic system *Mathematica* and computer languages in general. In particular, the accented characters, subscripts, and superscripts are a well-established way to distinguish between the quantities; however in most of the programing languages they cannot be used as a part of the names of variables. In the program code the accented characters are transformed into a sequence of two characters (e.g., $\bar{u}$ is transformed into `ub`, $\tilde{u}$ is transformed into `ut`) and the subscripts and superscripts are added at the and of the name (e.g., h_e^j is transformed into `hej`). The bold characters that are used to indicate the vectors and matrices in standard notation are in programing notation transformed to blackboard characters (e.g., $\mathbf{R}_e$ is transformed into $\mathbb{R}_\mathrm{e}$). The reason is that *Mathematica* does not distinguish between the plain and bold characters in the names of variables, however it does distinguish between the plain and blackboard characters. The slanted bold characters that are used in standard notation to indicate tensorial quantities are also transformed into blackboard characters in programing notation (e.g., $\boldsymbol{X}$ is transformed into $\mathbb{X}$). Quantities that are input/output parameters of the subroutines are indicated by the suffix `IO`.

General	
$\boldsymbol{A} \cdot \boldsymbol{B} = \mathrm{tr}(\boldsymbol{A}\boldsymbol{B}^T)$	Scalar product
$(\partial a/\partial \boldsymbol{B})_{ij} = \partial a/\partial B_{ij}$	Derivative of a scalar function with respect to tensor argument
$(\partial \boldsymbol{a}/\partial \boldsymbol{B})_{ijk} = \partial a_i/\partial B_{jk}$	Derivative of a vector function with respect to tensor argument
$((\partial \boldsymbol{A}/\partial \boldsymbol{B})_{ijkl} = \partial A_{ij}/\partial B_{kl}$	Derivative of a tensor function with respect to tensor argument
$(\partial \boldsymbol{A}/\partial \boldsymbol{b})_{ijk} = \partial A_{ij}/\partial b_k$	Derivative of a tensor function with respect to vector argument
$(\boldsymbol{A}^T)_{ijk} = A_{jik}$	Transpose of a third-order tensor
$(A[\mathbf{B}])_{ij} = \sum_k \sum_l A_{ijkl} b_{kl}$	Inner product of fourth (A) and second ($\mathbf{B}$) order tensor
$\mathbf{M} = \mathsf{A}_{e=1}^{n_e} \mathbf{M}_e$	A is an operator that assembles global matrix $\mathbf{M}$ from individual element contributions $\mathbf{M}_e$
$\mathrm{vec}(\mathbf{a})$	Vectorization operator that converts matrix $\mathbf{a}$ into column vector while accounting for symmetry of $\mathbf{a}$ and skipping the zero entries

Computational Derivatives		
$\frac{\hat{\delta}(\Box)}{\hat{\delta}(\Box)}$		Computational derivative
$\left.\frac{\hat{\delta}(\Box)}{\hat{\delta}(\Box)}\right\|_{\frac{D\mathrm{b}}{D\mathrm{a}}=\mathrm{M}}$		Computational derivative with the total derivatives of variables $\mathbf{b}$ with respect to variables $\mathbf{a}$ set to be equal to arbitrary matrix $\mathbf{M}$
$\widehat{abc}$	abcIO	abc is considered from the algorithmic point of view an independent variable (e.g., part of the input/output data of the subroutine)
$\widehat{D_{x_j} y_i}$	DxjDyiIO	Independent variable that holds total derivative of variable y_i with respect to variable x_j
$\widehat{D_{\mathrm{x}} \mathrm{y}}$	DyDxIO	Matrix of independent variables that hold total derivatives of vector of variables $\mathbf{y}$ with respect to vector of variables $\mathbf{x}$

Quantities at the Global Level of the FE Problem		
n_e		Number of elements
n_{tn}		Total number of nodes
$\mathbf{X}^J$		Spatial coordinates of the Jth global node
n_{tp}		Total number of global unknowns
n_{dim}	ndim	Number of spatial dimensions
n_{step}		Total number of time (load) steps (also index of the terminal time step)
n		Index of the last converged time step (for the quantities without the index it is assumed that they have the index $n+1$)

(continued)

(continued)

$n+1$		Index of the current time step ($\mathbf{p} \equiv \mathbf{p}_{n+1}$)
n_{tg}		Total number of integration points of the problem
B		Body (domain of the problem)
$B^h = \bigcup_{I=1}^{n_B} B_I^h = \bigcup_{e=1}^{n_e} \Omega_e$	discretized body (domain)	
n_B		Number of subdomains of the problem
$\mathbf{d}_B$	dB	Domain specific data (e.g., material parameters)
n_{d_B}	`ndB`	Length of domain specific data vector (e.g., number of material parameters)
B_I^h		Discretized subdomain of the problem
∂B^h		Boundary of discretized body
$\partial B_{\text{int}}^h$		Interior element boundaries
$\mathbf{p}$		Set of global variables with unknown value ($\mathbf{p} = \mathbf{p}_{n+1} = \cup \hat{\mathbf{p}}$)
$\mathbf{p}_n$		Set of global variables with unknown value at time t_n
$\mathbf{p}_1, \mathbf{p}_2, \ldots, \mathbf{p}_n, \mathbf{p}_{n+1}, \ldots, \mathbf{p}_{n_{\text{step}}}$		time domain discretized into the finite number of time steps
$\hat{\mathbf{p}}$		Nodal-wise ordered nested set of global DOF
$\bar{\mathbf{p}}$		Set of global DOF with the imposed homogeneous essential (Dirichlet) boundary condition $\bar{\mathbf{p}} \cup \mathbf{p} = \cup \hat{\mathbf{p}}$
$\mathbf{K}, \mathbf{R}$		Global tangent matrix and residual
t	`t`	Parameter time
$\bar{t}$	`tb`	Terminal time
λ	λ	Parameter (load level)
$\bar{\lambda}$	λb	Terminal value of parameter (total load level)

Element Level Quantities		
Ω_e		Element domain
$\partial\Omega_e$		Element boundary
n_{en}	`nen`	Number of element nodes
n_p	`np`	Total number of element nodal DOF
$\mathbf{R}_e$	*R*`el`	Element residual
$\mathbf{K}_e$	*K*`e`	Element tangent matrix
$\mathbf{p}_e$	*P*`e`	Set of element nodal DOF at time t_{n+1}
$\mathbf{p}_{e\,n}$	*P*`en`	Set of element nodal DOF at time t_n
$n_{p\,I}$	`npiDOF`	Number of element nodal DOF in I^{th} node
$\hat{\mathbf{p}}_e$	*P*`eIO`	Nodal-wise ordered nested set of element DOF at time t_{n+1}

(continued)

(continued)

$\hat{\mathbf{p}}_{e\,n}$	*Pen*IO	nodal-wise ordered nested set of element DOF at time t_n
$\bar{\mathbf{p}}_e$	*P*be	Set of element DOF with imposed homogeneous essential (Dirichlet) boundary condition $\bar{\mathbf{p}}_e \subset \mathbf{p}_e$
$\hat{\mathbf{p}}_e$	*P*ce	Set of true element unknowns $\hat{\mathbf{p}}_e = \mathbf{p}_e \backslash \bar{\mathbf{p}}_e$
l_{dhe}	ldhe	Length of history data vector per element
$\mathbf{d}_{he}$	*D*heIO	All history data per element at time t_{n+1}
$\mathbf{d}_{he\,n}$	*D*henIO	all history data per element at time t_n
l_{dae}	ldae	Number of arbitrary real values per element (history independent)
$\mathbf{d}_{ae}$	*D*aeIO	Vector of arbitrary real values per element

Coordinate Systems and Shape Functions		
$\boldsymbol{X} = \{X_e, Y_e, Z_e\} = \{X_1, X_2, X_3\}$	*X* {Xe, Ye, Ze} {X1, X2, X3}	Initial (material)coordinates
$\boldsymbol{\Xi} = \{\xi, \eta, \zeta\} = \{\Xi_1, \Xi_2, \Xi_3\}$	$\boldsymbol{\Xi}$ $\{\xi, \eta, \zeta\}$ $\{\Xi_1, \Xi_2, \Xi_3\}$	Coordinates of the reference element
$\mathbf{X}$	*X*IO	Nodal-wise ordered nested set of all coordinates of all element nodes
$\mathbf{X}_c$	*X*c	Set of X coordinates of all element nodes
$\mathbf{Y}_c$	*Y*c	Set of Y coordinates of all element nodes
$\mathbf{Z}_c$	*Z*c	Set of Z coordinates of all element nodes
$\mathbf{J}_e$	*J*e	Jacobi matrix $\frac{\partial \boldsymbol{X}}{\partial \Xi}$
J_e	Jed	Jacobi determinant
$\mathbf{J}_0$	*J*0	Jacobi matrix evaluated at the center of the element
J_0	J0d	Jacobi determinant evaluated at the center of the element
t_ζ	tζ	Thickness in ζ direction (for 2D and plate/shell formulations)
$\mathbf{N}$	*N*h	Set of global shape functions
$\bar{\mathbf{N}}$	*N*b	Set of shape functions local to the element domain
$\tilde{\mathbf{N}}$	*N*t	All shape functions $\tilde{\mathbf{N}} = \mathbf{N} \cup \bar{\mathbf{N}}$

Integration Point Level Quantities		
n_g	ng	Number of integration points of element
I_g	Ig	integration point index

(continued)

(continued)

w_{gp}	wgp	integration point weight
$\mathbf{R}_g$	*R*g	integration point contribution to the element residual
$\mathbf{K}_g$	*K*g	integration point contribution to the element tangent matrix

Coupled Problems		
$\mathbf{h}$	*h*	Set of all coupled unknowns at time t_{n+1}
$\mathbf{h}_n$	*h*n	Set of all coupled unknowns at time t_n
j_{NR}	jNR	Iteration index of local Newton iterative loop

Element Level Coupled Problem		
$\mathbf{Q}_e$	*Q*e	System of equations local to the element $\mathbf{Q}_e(\mathbf{p}_e, \mathbf{h}_\mathbf{e})$
n_{he}	nhe	Number of unknowns local to the element
$\mathbf{h}_e$	*h*eIO	Unknowns local to the element at time t_{n+1} $(\mathbf{h} = \bigcup_e \mathbf{h}_e)$
$\mathbf{h}_{e\,n}$	*h*enIO	Unknowns local to the element at time t_n $(\mathbf{h}_n = \bigcup_e \mathbf{h}_{e\,n})$
$\mathbf{h}_e^{(j)}$	*h*ej	Local unknowns at j^{th} iteration of local iterative loop
$\mathbf{A}_e$	*A*e	Local tangent matrix
I_{eh}	Ieh	Position of element related data within the element history data vector $\mathbf{d}_e$
$\breve{\mathbf{p}}_e$	*P*ae	Set of nodal and locally coupled element DOF at time t_{n+1} $(\breve{\mathbf{p}}_e = \mathbf{p}_e \cup \mathbf{h}_e)$
$\breve{\mathbf{p}}_{e\,n}$	*P*aen	Set of nodal and locally coupled element DOF at time t_n
$\breve{\mathbf{R}}_e$	*R*ae	Extended element residual
$\breve{\mathbf{K}}_e$	*K*ae	Extended element tangent matrix
$\breve{\mathbf{R}}_g$	*R*ag	Extended integration point residual
$\breve{\mathbf{K}}_g$	*K*ag	Extended integration point tangent matrix

Integration Point Level Coupled Problem		
$\mathbf{Q}_g$	*Q*g	Integration point system of equations $\mathbf{Q}_g(\mathbf{p}_e, \mathbf{h}_g)$
n_{hg}	nhg	Number of integration point unknowns
$\mathbf{h}_g$	*h*gIO	Integration point unknowns at time t_{n+1}
$\mathbf{h}_{g\,n}$	*h*gnIO	Integration point unknowns at timet_n
$\mathbf{h}_g^{(j)}$	*h*gb	Integration point unknowns at *j*th iteration of local iterative loop
$\mathbf{A}_g$	*A*g	Integration point tangent matrix
$\mathbf{r}_g$	*r*g	set of intermediate variables such that $\mathbf{Q}_g = \mathbf{Q}_g(\mathbf{r}_g(\mathbf{p}_e), \mathbf{h}_g)$
$\frac{D\mathbf{h}_g}{D\mathbf{r}_g}$	D*h*D*r*	Intermediate tangent matrix

(continued)

(continued)

I_{hg}	Ihg	Position of integration point data within the element history data vector $\mathbf{d}_{he}$ or $\mathbf{d}_{hen}$
l_{hg}	lhg	Length of history data vector per integration point
$\mathbf{d}_{hg}$	*d*hgIO	All history data per integration point
$\mathbf{d}_{hgn}$	*d*hgnIO	All history data per integration point at time t_n

Continuum Mechanics		
General		
ρ_0	ρ0	Mass density in initial or reference configuration
ρ	ρ	Mass density in current configuration
Π		Variational potential
Π^*	Πs	Variational potential per unit of initial volume
K	K	Kinetic energy per unit volume
ψ	ψ	Potential
W	W	Strain energy per unit of initial volume ($W = \rho_0 \psi$)
$\bar{\boldsymbol{b}}$	*B*b	Force per unit mass ($\rho_0 \bar{\boldsymbol{b}}$ = force per unit of initial volume)
$\bar{\boldsymbol{t}}$	*T*b	Surface traction in initial configuration
W^{ext}	Wext	Potential of external forces per unit volume
W^P	W	Pseudo-potential
$\delta \mathbf{p}_e$ or $\boldsymbol{\eta}_e$		variation of element nodal DOF
Kinematics		
φ		Configuration mapping $\varphi : B \rightarrow E^3$
$\varphi(B)$		Current configuration of body B
$\boldsymbol{x} = \{x_e, x_e, x_e\} = \{x_1, x_2, x_3\}$	x {xe, ye, ze} {x1, y2, z3}	Current (spatial) coordinates $\boldsymbol{x} = \boldsymbol{\varphi}(\boldsymbol{X}, \boldsymbol{t})$
$\boldsymbol{u} = \{u_e, v_e, w_e\} = \{u_1, v_2, w_3\}$	u {ue, ve, we} {u1, v2, w3}	Interpolation of displacements at time t_{n+1}
$\boldsymbol{u}_n = \{u_{en}, v_{en}, w_{en}\} = \{u_{1n}, v_{2n}, w_{3n}\}$	un {uen, ven, wen} {u1n, v2n, w3n}	Interpolation of displacements at time t_n
$\mathbf{u}$	*u*IO	Displacement DOF of all element nodes represented by nodal-wise ordered nested set
$\mathbf{u}_n$	*u*nIO	Displacement DOF of all element nodes at time t_n

(continued)

(continued)

$\mathbf{u}_c$	*u*c	Set of displacements in X direction of all element nodes
$\mathbf{v}_c$	*v*c	Set of displacements in Y direction of all element nodes
$\mathbf{w}_c$	*w*c	Set of displacements in Z direction of all element nodes
$\boldsymbol{H}$	*H*	Displacement gradient
$\boldsymbol{F}$	*F*	Deformation gradient $F = \text{Grad}\varphi(\boldsymbol{X}, \boldsymbol{t})$
$\hat{\boldsymbol{F}}$	*F*ISO	Isochoric part of deformation gradient
J_F	JF	Determinant of $\boldsymbol{F}$
$\bar{\boldsymbol{H}}$	*H*b	Enhancement of displacement gradient
$\bar{\boldsymbol{F}}$	*F*b	Enhancement of deformation gradient

Strain Measures		
$\boldsymbol{C}$	*C*t	Left Cauchy–Green deformation tensor
$\hat{C}$	*C*ISO	Isochoric part of left Cauchy–Green deformation tensor
$\boldsymbol{b}$	*b*	Right Cauchy–Green deformation tensor
$\boldsymbol{E}$	*E*g	Green–Lagrange strain tensor
$\boldsymbol{e}$	*e*	Almansi strain tensor
ε	ϵ	Small strain tensor
$\boldsymbol{e_D}$	*e*	Deviatoric small strain tensor
Stress measures		
σ	σ	Small stress tensor (also Cauchy stress tensor)
τ	τ	Kirchhoff stress tensor
$\boldsymbol{S}$	*S*	2nd Piola–Kirchhoff stress tensor
s	*s*	Deviatoric stress tensor
Elasticity		
E	E	Modulus of elasticity
K	K	Bulk modulus
G	G	Shear modulus
ν	ν	Poisson's ratio
Λ, μ	Λ, μ	Lamé constants ($\mu = G$)
C		Elasticity tensor

Plasticity		
ε^p	εp	Pastic small strain tensor
ε^e	εe	Elastic small strain tensor
W^e	W	Elastic strain energy per unit of reference volume
W^α	Wα	Potential for hardening variables per unit of reference volume
α	α	Set of hardening variables

(continued)

(continued)

f	fg	Flow condition / yield criteria
$\hat{\sigma}$	σh	Von Mises stress
$\lambda,\dot{\gamma}$	λg,γg	Plastic multiplier
$\boldsymbol{q}$	q	Back stress (kinematic hardening)
$\boldsymbol{r}$, $\boldsymbol{h}$	r, h	Flow direction and change of hardening for non-associative plasticity
$\boldsymbol{n}$	n	Flow direction for associative plasticity
$\hat{\alpha}$	αh	Equivalent accumulated plastic strain
Y_0	Y0	Initial yield stress
$\hat{H}$	Hh	Isotropic linear hardening modulus
H	Hk	Kinematic hardening modulus

Sensitivity Analysis and Optimization		
F	F	Response functional
$\boldsymbol{\phi}$	ϕ	Global set of all sensitivity parameters
ϕ_I	ϕ	I^{th} sensitivity parameter
I_ϕ	Iϕ	Sensitivity parameter index
n_ϕ	nϕ	Number of sensitivity parameters
${}^I\tilde{\mathbf{R}}_g$	Rtg	Integration point contribution to independent sensitivity pseudo-load vector
${}^I\tilde{\mathbf{R}}_e$	Rte	Element contribution to independent sensitivity pseudo-load vector

Design Velocity Fields		
n_ψ		Number of input data scalar fields
$\boldsymbol{\psi}$		Global set of input data fields ($\boldsymbol{\psi} = \boldsymbol{\psi}(\boldsymbol{\phi})$)
$\boldsymbol{\psi}_L$		Lth input data field
$\boldsymbol{\psi}^J$		Value of input data field at Jth node
$D_{\boldsymbol{\phi}}\boldsymbol{\psi}$		Design velocity matrix
$D_{\boldsymbol{\phi}}\boldsymbol{\psi}_L$		Lth design velocity field
$D_{\boldsymbol{\phi}}\boldsymbol{\psi}^J$		Value of design velocity matrix at Jth node
$\boldsymbol{\psi}_e$	$\boldsymbol{\psi}$e	Set of general input data fields for eth element (e.g., spatial coordinates, material parameters, except nodal unknowns)
$n_{\psi e}$	n$\boldsymbol{\psi}_e$	Number of general input data fields for eth element
$D_{\phi_I}\boldsymbol{\psi}_e$	D$\boldsymbol{\psi}$eDϕIO	General design velocity matrix for eth element

Sensitivity of Element DOF		
$\frac{D\mathbf{p}_e}{D\phi_I}$		Sensitivity of a set of element DOF at time t_{n+1}
$\frac{D\hat{\mathbf{p}}_e}{D\phi_I}$	`DpeD`ϕ	Sensitivity of a set of element DOF represented by nodal-wise ordered nested set at time t_{n+1}
$\frac{D\hat{\mathbf{p}}_{e\,n}}{D\phi_I}$	`DpenD`ϕ	Sensitivity of a set of element DOF represented by nodal-wise ordered nested set at time t_n
$D_{\phi_I}\bar{\mathbf{p}}_e$	`DpbeD`ϕ`IO`	Element essential boundary condition velocity field data structure
$D_{\phi_I}\bar{\mathbf{p}}_{e\,n}$	`DpbenD`ϕ`IO`	Element essential boundary condition velocity field data structure at time t_n
$D_{\phi_I}\check{\mathbf{p}}_e$	`DpceD`ϕ`IO`	Sensitivities of true element unknowns $(\mathbf{p}_e\backslash\bar{\mathbf{p}}_e)$ represented by nodal-wise ordered nested set
$D_{\phi_I}\check{\mathbf{p}}_{e\,n}$	`DpcenD`ϕ`IO`	Sensitivities of true element DOF $(\mathbf{p}_{e\,n}\backslash\bar{\mathbf{p}}_{e\,n})$ at time t_n
$D_{\phi_I}\hat{\mathbf{p}}_e$	`DpeD`ϕ`IO`	Sensitivities of a set of nodal DOF (with and without prescribed boundary condition) represented by the nodal-wise ordered nested set at time t_{n+1}
$D_{\phi_I}\hat{\mathbf{p}}_{e\,n}$	`DpenD`ϕ`IO`	Sensitivities of a set of nodal DOF at time t_n

Coupled Problems Related Sensitivity Data		
$\frac{D\mathbf{h}_g}{D\phi_I}$	`DhgD`ϕ	Sensitivity of a set of integration point history variables at time t_{n+1}
I_{sg}	`Isg`	Pointer to the position of $\frac{D\mathbf{h}_g}{D\phi_I}$ or $\frac{D\mathbf{h}_{g\,n}}{D\phi_I}$ fields within the history data sets $\mathbf{d}_{he}$ or $\mathbf{d}_{he\,n}$
$\frac{D\mathbf{h}_{g\,n}}{D\phi_I}$	`DhgnD`ϕ	Sensitivity of a set of integration point history variables at time t_n
$D_{\phi_I}\mathbf{h}_e$	`DheD`ϕ`IO`	Set of sensitivities of unknowns local to the element
$D_{\phi_I}\mathbf{h}_{e\,n}$	`DhenD`ϕ`IO`	Set of sensitivities of unknowns local to the element at time t_n
$D_{\phi_I}\mathbf{h}_g$	`DhgD`ϕ`IO`	Set of sensitivities of integration point unknowns
$D_{\phi_I}\mathbf{h}_{g\,n}$	`DhgnD`ϕ`IO`	Set of sensitivities of integration unknowns variables at time t_n
${}^I\mathbf{Z}_g$	`zg`	Auxiliary sensitivity quantity

Introduction

Finite element simulations can be used as predictive tools in many complex processes in engineering, medicine, or physics. Finite element methods are a general method for the solution of partial differential equations. They provide the power for the virtual description of, e.g., new materials, structures, implants, and biological processes and thus are a key in the emerging area of predictive science.

Analysts working in many different application areas demand efficient and reliable software for their investigations. To date this software was handwritten based on formulations that were derived by scientists and software engineers. The related process is—especially in nonlinear applications—slow and can take more than several weeks for a new finite element when the underlying theory is involved and derivations of complex tensor fields to obtain residuals and tangent matrices are prone to error. Thus the demand for an automation of these processes is evident. Several attempts to derive finite element formulations automatically can be found in the literature. To reduce the effort of developing related new source code, symbolic code generation was developed over the past decade. It, at a stage where the automatically generated source code is as small as handwritten code, is efficient and reliable. In this book we like to describe a general approach that can be applied to many different applications in engineering and science. It is based on the variational formulations of the problem but by using weak formulations it could be adapted to arbitrary partial differential equations. The main advantage of using symbolic code development is that the development time, especially for complex materials or elements, reduces by orders of magnitude. This book is written to show the merits of the automatic code development and the ease of use once the software developer is acquainted with the basic underlying concepts and ideas. The book will mainly concentrate on solid and structural mechanics problems; however the general potential of the automatic code generation goes far beyond these engineering applications.

Engineers and scientists rely on a number of all-purpose finite element codes that can be applied for the solution of many different problems. Such applications require large numerical finite element models with several thousands up to several million degrees of freedom. Thus besides the correct formulations of the problem in the continuum mechanics setting it is also necessary to provide efficient and robust methods for the solution. Often new material models but also new finite element formulations and related algorithms have to be added to these codes. These provide the necessary user interfaces for which also the automatically generated finite have to be developed such that they can be immediately used to solve large scale problems.

In solid mechanics there are different types of nonlinearities

- **Geometrical nonlinearity** is related to problems that depict large displacements and rotations. Applications related to that are structural elements like cables, frames, membranes, or shells. In these cases the strains are still small.
- **Finite deformations** are related to problems that exhibit large (finite) strains. Thus not only the displacements are large but so are the strains. Applications are metal forming or tyre mechanics.
- **Physical nonlinearity** is associated with material response that is nonlinear. The material behavior is characterized by a nonlinear response function between stresses and strains or by a set of evolution equations.
- **Stability problems** can be subdivided into geometrical and material instability. Geometrical instability includes bifurcations like buckling or other singularities like the snap-through behavior of a structure. Examples for material instabilities are necking or shear bands in metals.
- **Nonlinear boundary conditions** characterize nonlinearities stemming from the boundary. Examples are contact constraints or deformation dependent loads.

More nonlinearities can occur when different interacting fields that describe solids, heat conduction in solids or fluids are coupled. Examples are thermo-mechanical coupling, fluid–structure interaction or problems in which chemical reactions, heat generation, and conduction and mechanical stresses are coupled. The latter formulation describes an engineering process, like the virtual design of a new material. Here nonlinearities in each of the different field equations have to be considered.

Goal of modern simulation-based engineering analysis is the better understanding of processes and thus design of methods for nonlinear problems which are robust, accurate, and efficient is necessary. With the achievement of these goals finite element methods can be applied more safely to nonlinear problems in engineering. A large step for the software development is the use of the symbolic approach which leads to automated finite element analysis.

Linearizations are needed within the algorithmic treatment of the solution process for the nonlinear boundary value problems. This is the case for finite element methods where Newton–Raphson algorithms are employed to solve the nonlinear algebraic equation systems. With the use of automated finite element method, it is no longer necessary to compute the linearizations of models and algorithms by hand. This is all done by the system *AceGen*. It should be emphasized that the linearization of a continuous problem does not coincide in all cases with the linearization of the discrete problem which results from a finite element discretization.

Thus a clear consequence is to apply linearizations within the automated FEM approach only at discretization level.

This book shows how automated FEM could be efficiently applied to problems of solid mechanics. For this the book includes the basic relations needed for the mathematical modeling of an engineering problem in solid mechanics and the algorithmic treatment used for its numerical simulation. Each of the different cases like elastic or elasto-plastic analysis is accompanied by a description on how such problem could be solved using the automated finite element system *AceGen* . Furthermore there will be investigations with respect to robustness, accuracy, and efficiency of the automatically generated finite elements.

The book is subdivided into eight chapters which include the following subject areas:

- Notation contains a list of abbreviations that is used throughout the book.
- The introduction to the major idea and the contents of the book can be found in Introduction.
- The foundation of nonlinear continuum mechanics is summarized in Chap. 1 to establish a basis for a unified treatment of finite element formulations. Different strain measures for finite deformations are introduced, and the associated stress tensors are presented together with the general balance laws and related weak forms.
- In order to get started using the symbolic system Chap. 2, provides the general idea related to automatic differentiation and its application to the finite element method.
- Chapter 3 discusses the automation process when problems are formulated in primal variables. For this a classification of the problems is needed as well as solution procedures.
- Chapter 4 is related to the automation of discretization techniques and different possibilities how to discretize potentials and weak forms.
- Constitutive equations for elastic and inelastic response for small and finite strain primal analysis are provided in Chap. 5. Theoretical formulation of the models is provided as well as its algorithmic treatment. For all models the input needed for *AceGen* is provided.
- Continuum elements for different two- and three-dimensional applications are developed in Chap. 6. After a general discussion of solid finite element discretizations, first standard displacement elements are considered followed by mixed and enhanced elements.
- Special two- and three-dimensional structural elements such as trusses, beams and shells are formulated in Chap. 7. These are constructed especially for finite deformations and large rotations and deflections. All elements are formulated in the initial configuration.
- The application of automation to sensitivity analysis is provided in Chap. 8. This necessary technique for different types of problems is easily incorporated in the automation process.

Within Chaps. 3–8 different formulations are derived based on the finite element method. Within all chapters the symbolic code is provided for the described formulations that are needed to automatically generate the associated source code.

Chapter 1
Basic Equations of Continuum Mechanics

This chapter contains a summary of the continuum mechanics background that is needed for the finite element formulation of solid mechanics and structural problems. In detail the kinematical relations and the balance laws with their weak forms are described in this chapter.

Kinematical relations will be formulated for the current and the referential description of motion. Based on that, strain measures will be introduced. Variational formulations will be derived which are basis for nonlinear finite element methods.

Since this book is devoted to nonlinear finite element formulations the underlying theory of continuum mechanics cannot be treated in the necessary depth. This chapter summarizes main results. Extensive derivations are not presented but at such point the relevant literature will be cited. For a more in depth treatment of continuum mechanics the reader is referred to standard books, e.g. Truesdell and Toupin (1960), Truesdell and Noll (1965), Eringen (1967), Malvern (1969), Chadwick (1999) or Holzapfel (2000) for the basics on continuum mechanics, Ogden (1984) for the theory of elasticity and Marsden and Hughes (1983) or Ciarlet (1988) for the mathematical background of the theory of elasticity.

1.1 Kinematics

The kinematical relations concern the description of the deformation and motion of a body, the derivation of strain measures and the time derivatives of kinematical quantities. All kinematical relations are needed within the constitutive equations and the weak formulation of balance laws.

1.1.1 Motion and Deformation Gradient

In this section the motion and deformation of homogeneous bodies are considered. Here the continuum approach is applied in which a body B is described in a formal

J. Korelc and P. Wriggers, *Automation of Finite Element Methods*,
DOI 10.1007/978-3-319-39005-5_1

way by a set of continuously distributed points $P \in B$, also called particles or material points, which occupy a region within the Euclidean point space $\mathbb{E}^3$. The configuration of a body B is a one-to-one mapping $\varphi\colon B \longrightarrow \mathbb{E}^3$, which places the particles of B in $\mathbb{E}^3$. With this definition the location of a particle X from B is given for the configuration φ as $\boldsymbol{x} = \varphi(X)$. Thus the placement of a body B is described by $\varphi(B) = \{\varphi(X) \mid X \in B\}$ and called configuration $\varphi(B)$ of body B.

The motion of body B is then given as a one-parametric series of configurations $\varphi_t\colon B \rightarrow \mathbb{E}^3$. The location of a particle X at time $t \in \mathbb{R}^+$ yields

$$\boldsymbol{x} = \varphi_t(X) = \varphi(X, t). \tag{1.1}$$

This equation describes a curve in $\mathbb{E}^3$ for a particle X. $\boldsymbol{X} = \varphi_0(X)$ defines the reference configuration of body B, with $\boldsymbol{X}$ being the location of the particle X for this configuration. Thus from (1.1)

$$\boldsymbol{x} = \varphi(\varphi_0^{-1}(\boldsymbol{X}), t) \tag{1.2}$$

can be deduced.[1]

Usually it is not necessary to distinguish between $\boldsymbol{X}$ and X. Then the notation simplifies and instead of (1.2) the relation

$$\boldsymbol{x} = \varphi(\boldsymbol{X}, t) \tag{1.3}$$

is obtained, where $\boldsymbol{X}$ represents particle X in the reference configuration B. Based on this the placements $\boldsymbol{x}$ and $\boldsymbol{X}$ can be formulated as position vectors in $\mathbb{E}^3$ with respect the origin $\boldsymbol{O}$, see Fig. 1.1. Point X is defined in the reference configuration by the position vector $\boldsymbol{X} = X_a \boldsymbol{E}_a$. $\boldsymbol{E}_a$ defines an orthogonal base system in the reference configuration with origin $\boldsymbol{O}$.

If the motion is characterized with respect to the material coordinates $\{X_1, X_2, X_3\}$ this is called material or referential description. In the material description—often also referred to as Lagrangian description of motion—one follows the movement of a particle of body B in time.

In the following capital letters are used as indices for components of vectors and tensors with respect to the basis $\boldsymbol{E}_a$ of the reference configuration where X_a[2] are the Lagrangian coordinates of the particle X.

Another possibility is the use of the spatial coordinates $\{x_1, x_2, x_3\}$ when the motion of body B has to be described. In this formulation attention is paid to a point

[1] It is not necessary that the body assumes the reference configuration at any time. Since the reference configuration can be chosen arbitrarily it is often assumed for practical purposes that the configuration of body B at the beginning of the deformation (initial configuration) is equivalent to the reference configuration. However there are applications like isoparametric interpolation functions within finite element formulations for which reference configurations will be defined which are purely fictitious.

[2] In the following we will use indices a, b, c, ... to refer in index notation to the referential description while indices $i, j, k, \ldots$ refer to the current or spatial configuration.

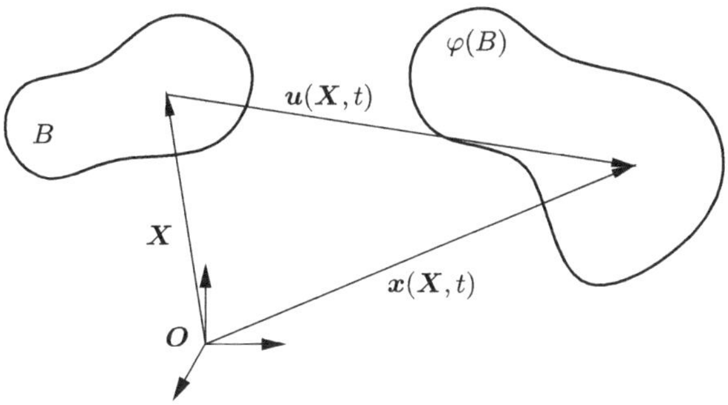

Fig. 1.1 Motion of body B

in space and the change of the motion with time t at this point. This description is called current, spatial or Eulerian description of motion. Small letters are used for indices of vectors and tensors which are related to the basis $\boldsymbol{e}_i$ of the current or spatial configuration. The quantities x_i are the spatial coordinates of X.

For simplicity an orthogonal cartesian basis will be assumed in the following. This is in accordance with the formulation of numerical methods based on FEM in which often isoparametric interpolations are used which rely on an orthogonal base system. A more general description using curvilinear coordinates is just technical but leads eventually to a significantly more complex formulation.

The equations of continuum mechanics can be formulated with respect to the deformed or undeformed configurations of a body. From the theoretical point of view there is no difference or preference whether the equations are related to the initial or current configuration. Thus the configuration can be chosen freely. However physical implications like in the theory of plasticity have to be taken into account, see e.g. (Lubliner 1990, p. 453) when selecting a certain description. Additionally this selection can have consequences regarding the selected numerical method. Here differences with respect to efficiency can be observed which will be discussed in later chapters.

Since it is not clear from the outset which formulation is preferable, the following strain measures will be derived for the reference configuration B as well as for the current configuration $\varphi(B)$.

To describe the deformation process locally a tensor $\boldsymbol{F}$ is introduced which relates tangent vectors of initial and current configuration to each other. It maps a material line element of the initial configuration $\boldsymbol{dX}$ in B, to a line element $\boldsymbol{dx}$ of the current configuration $\varphi(B)$.

$$\boxed{\boldsymbol{dx} = \boldsymbol{F}\,\boldsymbol{dX} \quad \text{or} \quad dx_i = F_{ia}\,dX_a} \tag{1.4}$$

By the structure of this equation it is clear that $\boldsymbol{F}$ represents a gradient. Hence $\boldsymbol{F}$ is called deformation gradient. From the symbolic form $\boldsymbol{F} = \partial \boldsymbol{x} \,/\, \partial \boldsymbol{X}$ follow the components of the deformation gradient as partial derivatives $\partial x_i \,/\, \partial X_a = x_{i,a}$. With (1.3)

$$\boxed{\boldsymbol{F} = \text{Grad } \varphi(\boldsymbol{X}, t) = F_{ia}\, \boldsymbol{e}_i \otimes \boldsymbol{E}_a = \frac{\partial x_i}{\partial X_a}\, \boldsymbol{e}_i \otimes \boldsymbol{E}_a} \tag{1.5}$$

is obtained. The matrix formulation of $\boldsymbol{F}$ yields

$$[F_{ia}] = \begin{bmatrix} x_{1,1} & x_{1,2} & x_{1,3} \\ x_{2,1} & x_{2,2} & x_{2,3} \\ x_{3,1} & x_{3,2} & x_{3,3} \end{bmatrix}. \tag{1.6}$$

Since the gradient in (1.5) is a linear operator also the local transformation in (1.4) is linear. To maintain the connection of B during the deformation process the mapping (1.4) has to be one-to-one which excludes a singularity of $\boldsymbol{F}$. The latter condition can be recast in the form

$$J_F = \det \boldsymbol{F} \neq 0, \tag{1.7}$$

where J_F defines a determinant named after Jacobi. Furthermore—to exclude a self penetration of the body—the following constraint has to be fulfilled by the deformation gradient: $J_F > 0$. Since $\boldsymbol{F}$ cannot be singular the inverse $\boldsymbol{F}^{-1}$ exists, which can be applied to invert relation (1.4)

$$\boldsymbol{dX} = \boldsymbol{F}^{-1}\, \boldsymbol{dx}. \tag{1.8}$$

Knowing the deformation gradient allows to express further transformations of differential quantities between B and $\varphi(B)$. The transformation of surface area elements between B and $\varphi(B)$ is given by the formula of Nanson (see e.g. Ogden 1984, p. 88)

$$\boxed{\boldsymbol{da} = \boldsymbol{n}\, da = J_F\, \boldsymbol{F}^{-T}\, \boldsymbol{N}\, dA = J_F\, \boldsymbol{F}^{-T}\, \boldsymbol{dA}} \tag{1.9}$$

In this equation $\boldsymbol{n}$ is the normal vector of the surface of the deformed body $\varphi(B)$ and $\boldsymbol{N}$ is the normal vector in B, see Fig. 1.2. Often the notion of the cofactor of $\boldsymbol{F}$ is introduced as $\text{Cof}\boldsymbol{F} = J_F\, \boldsymbol{F}^{-T}$. Hence the surface area transformation is then given as $\boldsymbol{n}\, da = \text{Cof}\boldsymbol{F}\, \boldsymbol{N}\, dA$. J_F is the Jacobi determinant, defined in (1.7) and da and dA are the area elements of the associated configurations, respectively.

The transformation between volume elements of initial and current configuration is provided by the relation

$$\boxed{dv = J_F\, dV} \tag{1.10}$$

By introducing a displacement vector $\boldsymbol{u}(\boldsymbol{X}, t)$ as difference between the position vectors of current and initial configuration

$$\boldsymbol{u}(\boldsymbol{X}, t) = \varphi(\boldsymbol{X}, t) - \boldsymbol{X} \tag{1.11}$$

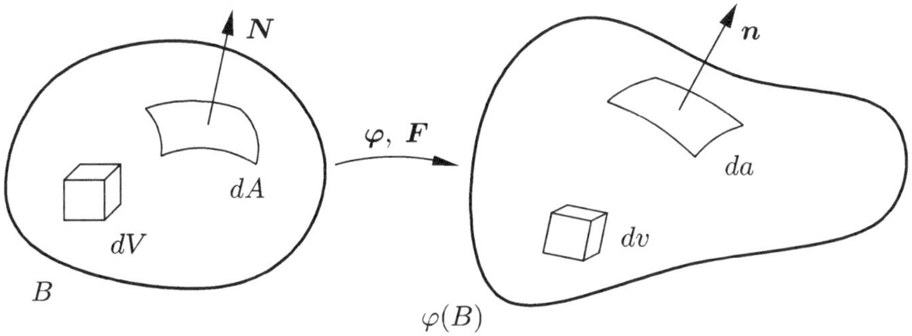

Fig. 1.2 Transformation of differential elements

the deformation gradient (1.5) can be written as follows

$$\boldsymbol{F} = \operatorname{Grad}\left[\,\boldsymbol{X} + \boldsymbol{u}(\boldsymbol{X},t)\,\right] = \mathbf{1} + \operatorname{Grad}\boldsymbol{u} = \mathbf{1} + \boldsymbol{H} \tag{1.12}$$

where $\boldsymbol{H}$ is the displacement gradient.

1.1.2 Strain Measures

In this section different strain measures are discussed which are used in forthcoming formulations. The first strain tensor is referred to the initial configuration B is defined by

$$\boxed{\boldsymbol{E} = \frac{1}{2}\,(\boldsymbol{F}^T\boldsymbol{F} - \mathbf{1}\,) = \frac{1}{2}\,(\,\boldsymbol{C} - \mathbf{1}\,)} \tag{1.13}$$

and called Green–Lagrange strain tensor.[3] The tensor $\boldsymbol{C} = \boldsymbol{F}^T\,\boldsymbol{F}$ in (1.13) is the right Cauchy–Green tensor which expresses the square of the line element $\boldsymbol{dx}$ by the material line element $\boldsymbol{dX}$: $\boldsymbol{dx}\cdot\boldsymbol{dx} = \boldsymbol{dX}\cdot(\boldsymbol{C}\,\boldsymbol{dX})$. Hence the strain $\boldsymbol{E}$ describes the change of the square of the line elements from B to $\varphi(B)$.

A generalization of (1.13) can be found e.g. in Ogden (1984) and is defined by

$$\boldsymbol{E}^{(\alpha)} = \frac{1}{\alpha}\,(\,\boldsymbol{U}^{\alpha} - \mathbf{1}\,), \quad \alpha \in \mathbb{R} \quad \text{for } \alpha = 0 \rightarrow \boldsymbol{E}^{(0)} = \ln\boldsymbol{U}. \tag{1.14}$$

This generalized strain tensor is referred to the initial configuration B.

[3]The Green–Lagrange strain measure is often used in nonlinear structural engineering applications. Mostly this strain measure is applied for problems with large displacements but small strains (e.g. within beam or shell theory) since it can describe arbitrary rigid body motions correctly.

The poplar decomposition of the deformation gradient was used within the definition of the strain measures (1.14). This splits the deformation gradient in a multiplicative way in a proper orthogonal rotation tensor $\boldsymbol{R}$ (with $\boldsymbol{R}^{-1} = \boldsymbol{R}^T$) and the symmetrical stretch tensors $\boldsymbol{U}$ and $\boldsymbol{V}$, see e.g. (Ogden 1984, p. 92):

$$\boxed{\boldsymbol{F} = \boldsymbol{R}\,\boldsymbol{U} = \boldsymbol{V}\,\boldsymbol{R}} \tag{1.15}$$

Due to the orthogonality of $\boldsymbol{R}$ the right Cauchy–Green tensor can be written as $\boldsymbol{C} = \boldsymbol{F}^T\,\boldsymbol{F} = \boldsymbol{U}^T\,\boldsymbol{R}^T\,\boldsymbol{R}\,\boldsymbol{U} = \boldsymbol{U}^T\,\boldsymbol{U} = \boldsymbol{U}^2$. The last result follows from the symmetry of $\boldsymbol{U}$. Hence the Green–Lagrange strain tensor (1.13) can be written as $\boldsymbol{E} = \frac{1}{2}\,(\boldsymbol{U}^2 - \boldsymbol{1})$ which is included in (1.14) for the special case of $\alpha = 2$.

A special strain tensor is due to Hencky

$$\ln\,\boldsymbol{V} = \frac{1}{2}\,\ln\boldsymbol{b}, \tag{1.16}$$

see the work of Hencky (1933).

Another special case is the so called Almansi strain tensor

$$\boldsymbol{e} = \frac{1}{2}\,(\,\boldsymbol{1} - \boldsymbol{V}^{-2}\,) = \frac{1}{2}\,(\,\boldsymbol{1} - \boldsymbol{b}^{-1}\,) = \frac{1}{2}\,(\,\boldsymbol{1} - \boldsymbol{F}^{-T}\,\boldsymbol{F}^{-1}\,). \tag{1.17}$$

In this equation the left Cauchy–Green tensor

$$\boxed{\boldsymbol{b} = \boldsymbol{F}\,\boldsymbol{F}^T = \boldsymbol{V}\,\boldsymbol{R}\,\boldsymbol{R}^T\,\boldsymbol{V}^T = \boldsymbol{V}^2} \tag{1.18}$$

was introduced. The left Cauchy–Green tensor is related to the spatial configuration and will be of significance in later chapters in which constitutive equations are formulated and implemented in a numerical scheme.

We observe that the evaluation of matrix functions operating over tensors that are essential part of formulation of different strain measures, see e.g. (1.14) and (1.16). In Hudobivnik and Korelc (2016) a method was presented that is able to automatically derive numerically efficient closed-form representation of an arbitrary matrix function and its first and second derivatives for 3×3 matrices with real eigenvalues. The method offers an unique solution to the standard problem of ill-conditioning in the vicinity of multiple equal eigenvalues which is characteristic for all closed-form representations. A compiled library of subroutines with closed-form representation of most commonly used matrix functions along with their first and second derivatives has been created and is available in *AceGen* for the use in general finite element environments. Consequently, the matrix functions can become as accurate, efficient and commonly available as their scalar counterparts, resulting in more common use of advanced strain and stress measures, such as the Hencky strain measure (1.16), that have so far been considered difficult for numerical implementation.

Incompressibility. For motions which are constraint by special conditions it is often possible to incorporate these constraint conditions directly into the kinematical relations. In case of incompressibility which plays a prominent role in rubber materials and metal plasticity the constraint condition $\det \boldsymbol{F} = J_F = 1$ has to be fulfilled. The following multiplicative decomposition of the deformation gradient

$$\boldsymbol{F} = J_F^{\frac{1}{3}}\,\hat{\boldsymbol{F}}\hat{\boldsymbol{F}} \qquad \hat{\boldsymbol{F}} = J_F^{-\frac{1}{3}}\,\boldsymbol{F} \tag{1.19}$$

was suggested in Flory (1961). It preserves a priori the volume of $\hat{\boldsymbol{F}}$ (isochoric motion), since $\det \hat{\boldsymbol{F}} \equiv 1$.

By inserting (1.19) in (1.13) a relation between the isochoric part of the right Cauchy Green deformation tensor $\hat{\boldsymbol{C}}$ and $\boldsymbol{C}$ is obtained

$$\boxed{\hat{\boldsymbol{C}} = \hat{\boldsymbol{F}}^T\,\hat{\boldsymbol{F}} = J_F^{-\frac{2}{3}}\,\boldsymbol{F}^T\,\boldsymbol{F} = J_F^{-\frac{2}{3}}\,\boldsymbol{C}} \tag{1.20}$$

The multiplicative split of $\boldsymbol{F}$ in a volume changing part (J_F) and a volume preserving part ($\hat{\boldsymbol{F}}$) in the nonlinear theory corresponds to an additive decomposition of the strain tensor in the geometrically linear theory in a deviator $\boldsymbol{e}_D$ and a volumetric part

$$\boldsymbol{\varepsilon} = \boldsymbol{e}_D + \frac{1}{3}\,\mathrm{tr}\,\boldsymbol{\varepsilon}\,\mathbf{1}. \tag{1.21}$$

1.1.3 Transformation of Vectors and Tensors

Knowledge regarding the transformation between differential quantities in the current and reference configuration is essential for many theoretical derivations and their applications in finite element methods. Tangent fields and one forms which are related to the current configuration can be expressed in terms of quantities in the reference configuration. With the notation introduced in Marsden and Hughes (1983) this is called *pull back*. Conversely a *push forward* relates tangent fields and one forms referred to the initial configuration to the current configuration $\varphi(B)$. For a detailed mathematical background see e.g. Marsden and Hughes (1983).

Tangent fields or one forms are connected to the base vectors, see Appendix B.2.4. For a covariant gradient of a scalar field $G(\boldsymbol{X}) = g(\boldsymbol{x}) = g[\varphi(\boldsymbol{X})]$ relations

$$\mathrm{Grad}\,G = \boldsymbol{F}^T\,\mathrm{grad}\,g \iff \frac{\partial G}{\partial X_a} = \frac{\partial g}{\partial x_i}\,\frac{\partial x_i}{\partial X_a}, \tag{1.22}$$

$$\mathrm{grad}\,g = \boldsymbol{F}^{-T}\,\mathrm{Grad}\,G \tag{1.23}$$

can be derived. In an analogous way the transformation for the covariant gradient of the vector field $\boldsymbol{W}(\boldsymbol{X}) = \boldsymbol{w}\,(\boldsymbol{x}) = \boldsymbol{w}\,[\varphi(\boldsymbol{X})]$ is obtained

$$\mathrm{Grad}\,\boldsymbol{W} = \mathrm{grad}\,\boldsymbol{w}\,\boldsymbol{F} \iff \mathrm{grad}\,\boldsymbol{w} = \mathrm{Grad}\,\boldsymbol{W}\,\boldsymbol{F}^{-1}. \tag{1.24}$$

As an application the deformation gradient is computed from a displacement field $\boldsymbol{u}\,[\varphi(\boldsymbol{X})]$ which is referred to the current configuration. With (1.12) and (1.24) it follows

$$\boldsymbol{F}^{-1} = \mathbf{1} - \operatorname{grad} \boldsymbol{u}. \tag{1.25}$$

Hence the inverse of the deformation gradient can be obtained directly from displacements when referred to the current configuration. This result will be applied later in formulations of the finite element method.

A typical application of a *pull back* operation to tensors is given by the transformation of the Almansi strain tensor to the Green–Lagrange strains using (1.13) and (1.17)

$$\boldsymbol{E} = \boldsymbol{F}^T \frac{1}{2} (\mathbf{1} - \boldsymbol{F}^{-T} \boldsymbol{F}^{-1}) \boldsymbol{F} = \boldsymbol{F}^T \boldsymbol{e} \boldsymbol{F}, \tag{1.26}$$

which of course does not change the physical meaning of the strain measure. It only chances the configuration.

Initial and current configuration are often parametrized in numerical methods by the introduction of convective coordinates. These can be thought as lines which are carved on the body B, see Appendix B.1.2, especially Fig. A.1. Hence the convective coordinates lines deform with the body under external loading. In this parametrization it is assumed that the cartesian coordinates $\{X_a\}$ and $\{x_i\}$ can be represented as functions of the convective coordinates $\{\Theta^j\}$. Using convective coordinates the tangent vector can be computed in each point $\boldsymbol{X}$ in B as

$$\boldsymbol{G}_j = \frac{\partial \boldsymbol{X}}{\partial \Theta^j} = \boldsymbol{X}_{,j}. \tag{1.27}$$

This is also true for a point which is described with respect to the current configuration $\varphi(\boldsymbol{X}, t)$ in $\varphi(B)$

$$\boldsymbol{g}_j = \frac{\partial \varphi(\boldsymbol{X}, t)}{\partial \Theta^j} = \varphi_{,j}. \tag{1.28}$$

Using the chain rule

$$\boldsymbol{g}_j = \frac{\partial \varphi(\boldsymbol{X}, t)}{\partial \mathbf{X}} \frac{\partial \mathbf{X}}{\partial \Theta^j} = \boldsymbol{F} \boldsymbol{G}_j \tag{1.29}$$

is derived from both previous relations. This means that tangent vectors transform like line elements $\boldsymbol{dx}$ and $\boldsymbol{dX}$, see (1.4). With Eq. (1.29) it is possible to describe the deformation gradient by the tangent vectors as follows

$$\boldsymbol{F} = \boldsymbol{g}_i \otimes \boldsymbol{G}^i. \tag{1.30}$$

The tangent vectors are covariant vectors which are connected to their contravariant counter parts (one forms) by $\boldsymbol{g}_i \cdot \boldsymbol{g}^k = \delta_i^k$. Using Eq. (1.29) transformation $\boldsymbol{g}^k = \boldsymbol{F}^{-T}\, \boldsymbol{G}^k$ is deduced.

The covariant or contravariant base vectors can serve as basis for a vector or tensor as well as the cartesian basis vectors. Once this basis is known it is relatively simple to perform *pull back* or *push forward* operations, see Appendix B.2.4.[4]

1.1.4 Time Derivatives

The dependence of the deformation $\varphi\,(\boldsymbol{X}, t)$ on time t has to be considered in non-linear problems in case that the constitutive behavior is history dependent (e.g. in plasticity or visco-elasticity) or in case that the complete process is of dynamical nature. In such applications time derivatives are needed which will be derived here for kinematical quantities.

The velocity of a material point with respect to the reference configuration is defined by the material time derivative

$$\boldsymbol{v}\,(\boldsymbol{X}, t) = \frac{D\varphi}{Dt} = \frac{\partial \varphi\,(\boldsymbol{X}, t)}{\partial t} = \dot{\varphi}\,(\boldsymbol{X}, t). \tag{1.32}$$

In the current configuration the velocity $\hat{\boldsymbol{v}}$ of a particle which assumes point $\boldsymbol{x}$ at time t in $\varphi(B)$ is given by

$$\hat{\boldsymbol{v}}\,(\boldsymbol{x}, t) = \hat{\boldsymbol{v}}\,(\varphi(\boldsymbol{X}, t), t) = \boldsymbol{v}\,(\boldsymbol{X}, t). \tag{1.33}$$

The acceleration is given in an analogous way by the second derivative with respect to time

$$\boldsymbol{a} = \ddot{\varphi}\,(\boldsymbol{X}, t) = \dot{\boldsymbol{v}}\,(\boldsymbol{X}, t). \tag{1.34}$$

Based on this definition the acceleration can be determined with reference to the current configuration. With (1.33) and the chain rule it yields

$$\hat{\boldsymbol{a}} = \dot{\hat{\boldsymbol{v}}} = \frac{\partial}{\partial t}\,[\,\hat{\boldsymbol{v}}\,(\varphi(\boldsymbol{X}, t), t)\,] = \frac{\partial \hat{\boldsymbol{v}}}{\partial t} + \operatorname{grad} \hat{\boldsymbol{v}}\, \hat{\boldsymbol{v}}. \tag{1.35}$$

---

[4]For the Green–Lagrange strain tensor $\boldsymbol{E} = \frac{1}{2}\,(\boldsymbol{F}^T\boldsymbol{F} - \boldsymbol{1})$ this leads with $\boldsymbol{F}^T\boldsymbol{F} = (\boldsymbol{G}^i \otimes \boldsymbol{g}_i)(\boldsymbol{g}_k \otimes \boldsymbol{G}^k)$ to

$$\begin{aligned} \boldsymbol{E} &= \frac{1}{2}\,(g_{ik} - \delta_{ik})\, \boldsymbol{G}^i \otimes \boldsymbol{G}^k \\ &= \frac{1}{2}\,(g_{ik} - \delta_{ik})\, \boldsymbol{F}^T\, \boldsymbol{g}^i \otimes \boldsymbol{F}^T\, \boldsymbol{g}^k = \boldsymbol{F}^T\,[\,\frac{1}{2}\,(g_{ik} - \delta_{ik})\, \boldsymbol{g}^i \otimes \boldsymbol{g}^k\,]\boldsymbol{F}. \end{aligned} \tag{1.31}$$

where g_{ik} is the so called metric tensor that defines the deformation of the solid. This result is equivalent to the *pull back* operation in (1.26).

The first term is called local part and the second term convective part of the acceleration. The local time derivative is computed holding the current position $\boldsymbol{x}$ fixed. The time derivative (1.35) is of significance in fluid mechanics.

The time derivative of the deformation gradient $\boldsymbol{F}$ yields with (1.5), (1.32) and (1.24)

$$\dot{\boldsymbol{F}} = \text{Grad}\, \dot{\boldsymbol{\varphi}}\, (\boldsymbol{X}, t) = \text{Grad}\, \boldsymbol{v} = \text{grad}\, \hat{\boldsymbol{v}}\, \boldsymbol{F}. \tag{1.36}$$

In this equation the spatial velocity gradient grad $\hat{\boldsymbol{v}}$ occurs which is often denoted by $\boldsymbol{l}$. It can be written with (1.36) as

$$\boldsymbol{l} = \dot{\boldsymbol{F}}\, \boldsymbol{F}^{-1}. \tag{1.37}$$

Equation (1.36) can now be used to compute the time derivative of the Green–Lagrange strain tensor (1.13)

$$\dot{\boldsymbol{E}} = \frac{1}{2}\, (\dot{\boldsymbol{F}}^T\, \boldsymbol{F} + \boldsymbol{F}^T\, \dot{\boldsymbol{F}}). \tag{1.38}$$

Using (1.37) in (1.36) yields for the time derivative of $\boldsymbol{E}$

$$\boxed{\dot{\boldsymbol{E}} = \boldsymbol{F}^T\, \frac{1}{2}\, (\boldsymbol{l} + \boldsymbol{l}^T)\, \boldsymbol{F} = \boldsymbol{F}^T\, \boldsymbol{d}\, \boldsymbol{F}} \tag{1.39}$$

This equation has an equivalent structure as (1.26) and hence denotes a *pull back* of the symmetric spatial velocity gradient $\boldsymbol{d} = \frac{1}{2}\, (\boldsymbol{l} + \boldsymbol{l}^T)$ to the reference configuration.

Finally the convective time derivative of a spatial tensor is considered which is also called Lie derivative. The Lie derivative is constructed such that it yields objective rates of spatial tensors. Since time derivative and variation have the same mathematical structure, this concept can also be applied to compute a spatial variation. In general the Lie derivative of a time dependent tensor field $\boldsymbol{t}$ which is related to a vector field (e.g. the velocity) $\boldsymbol{v}$ in $\varphi\,(B)$, see Marsden and Hughes (1983), is defined by

$$L_{\boldsymbol{v}}\,(\boldsymbol{t}) = \left[\frac{d}{dt} \varphi^*_{t,s}\, \boldsymbol{t}_t \right] \bigg|_{t=s} \tag{1.40}$$

where $\varphi^*_{t,s}$ is given by $\{\varphi^*_{t,s} \mid \varphi^*_{t,s} = \varphi_t \circ \varphi_s^{-1} : \varphi_s(B) \longrightarrow \varphi_t(B)\}$. It relates to the flux of the vector field $\boldsymbol{v}$. The superscript $(\ldots)^*$ denotes that a *pull back* operation has to be applied.

When this concept is applied to compute a variation the flux $\varphi^*_{\epsilon,0}$ of the vector field $\boldsymbol{u}$ is defined by $\{\varphi^*_{\epsilon,0} \mid \varphi^*_{\epsilon,0} = \varphi_\epsilon \circ \bar{\varphi}^{-1} : \bar{\varphi}(B) \longrightarrow \varphi_\epsilon(B)\}$. Then the Lie variation is given as

$$L_{\boldsymbol{u}}\,(\boldsymbol{t}) = \left[\frac{d}{d\epsilon} \varphi^*_{\epsilon,0}\, \boldsymbol{t}_\epsilon \right] \bigg|_{\epsilon=0} \tag{1.41}$$

This Lie-derivative can be formulated in a different way by inserting $\varphi_\epsilon \circ \bar{\varphi}^{-1}$ into (1.41) which simplifies the application of the concept

$$L_u\,(t) = \bar{\varphi}_* \left[\frac{d}{d\epsilon}\, \varphi_\epsilon^*\,(t) \right] \Bigg|_{\epsilon=0} . \tag{1.42}$$

This yields the "algorithm": (1) *pull back* $((\ldots)^*)$ the spatial field $\boldsymbol{t}$ to the initial configuration, (2) compute the material time derivative and finally (3) use a *push forward* $((\ldots)_*)$ operation to get the Lie-derivative in the spatial configuration.

With this concept we compute the objective variation of the Almansi strain tensor

$$L_u\,(\boldsymbol{e}) = \frac{1}{2}\,[\,L_u\,(\mathbf{1}) - L_u\,(\boldsymbol{b}^{-1})\,] = \nabla^S\,\boldsymbol{\eta} \tag{1.43}$$

Since the *pull back* of $\boldsymbol{b}^{-1}$ is $\mathbf{1}$: $L_u\,(\boldsymbol{b}^{-1}) = \varphi_{t\,*}\,(\frac{d}{dt}\,\mathbf{1}\,) = 0$. Furthermore for the spatial form of $\mathbf{1}$ it follows

1. pull back: $\boldsymbol{C} = \boldsymbol{F}^T\,\mathbf{1}\,\boldsymbol{F} = \boldsymbol{F}^T\,\boldsymbol{F}$,
2. variation in the initial configuration: $\delta\boldsymbol{C} = \delta\boldsymbol{F}^T\,\boldsymbol{F} + \boldsymbol{F}^T\,\delta\boldsymbol{F}$,
3. push forward: $\boldsymbol{F}^{-T}\,\delta\boldsymbol{F}^T\,\boldsymbol{F}\,\boldsymbol{F}^{-1} + \boldsymbol{F}^{-T}\,\boldsymbol{F}^T\,\delta\boldsymbol{F}\,\boldsymbol{F}^{-1} = \nabla\,\boldsymbol{\eta} + \nabla^T\,\boldsymbol{\eta}$

The advantage of the concept using Lie derivative is that its application yields directly objective fluxes of second order tensors.

With the above definitions the Lie derivative can be defined for a spatial tensor $\boldsymbol{g}(\boldsymbol{x}, t)$ with covariant basis by

$$\mathcal{L}_v\,\boldsymbol{g} := \boldsymbol{F} \left\{ \frac{\partial}{\partial t}\,[\,\boldsymbol{F}^{-1}\,\boldsymbol{g}\,\boldsymbol{F}^{-T}\,] \right\} \boldsymbol{F}^T . \tag{1.44}$$

This means that tensor $\boldsymbol{g}$ must be transformed first to the reference configuration by a *pull back* operation. Here the material time derivative can be computed and afterwards the resulting quantity is related to the current configuration by a *push forward* operation.

The analogous rule for the Lie derivative of a spatial tensor $\hat{\boldsymbol{g}}$ with contravariant basis is given by

$$\mathcal{L}_v\,\hat{\boldsymbol{g}} := \boldsymbol{F}^{-T} \left\{ \frac{\partial}{\partial t}\,[\,\boldsymbol{F}^T\,\hat{\boldsymbol{g}}\,\boldsymbol{F}\,] \right\} \boldsymbol{F}^{-1} . \tag{1.45}$$

Using this relation the Lie derivative of the Almansi strain tensor

$$\mathcal{L}_v\,\boldsymbol{e} = \boldsymbol{F}^{-T} \left\{ \frac{\partial}{\partial t}\,[\,\boldsymbol{F}^T\,\boldsymbol{e}\,\boldsymbol{F}\,] \right\} \boldsymbol{F}^{-1} = \boldsymbol{F}^{-T}\,\dot{\boldsymbol{E}}\,\boldsymbol{F}^{-1} \tag{1.46}$$

is obtained which can be rewritten as $\dot{\boldsymbol{E}} = \boldsymbol{F}^T\,\mathcal{L}_v\,\boldsymbol{e}\,\boldsymbol{F}$. A comparison with Eq. (1.39) shows that the Lie derivative of the Almansi strain tensor is equivalent to the symmetric spatial velocity gradient $\boldsymbol{d}$ (Table 1.1).

Table 1.1 Summary of nonlinear strain measures

Deformation/displ. gradient	$\boldsymbol{F} = \text{Grad}\,\boldsymbol{\varphi}$	$\boldsymbol{H} = \text{Grad}\,\boldsymbol{u}$
Polar decomposition	$\boldsymbol{F} = \boldsymbol{R}\,\boldsymbol{U}$	$\boldsymbol{F} = \boldsymbol{V}\,\boldsymbol{R}$
Right/left Cauchy Green tensor	$\boldsymbol{C} = \boldsymbol{F}^T\boldsymbol{F} = \boldsymbol{U}^2$	$\boldsymbol{b} = \boldsymbol{F}\,\boldsymbol{F}^T = \boldsymbol{V}^2$
Green Lagrange strains	$\boldsymbol{E} = \frac{1}{2}(\boldsymbol{C} - \boldsymbol{1})$	$\boldsymbol{E} = \frac{1}{2}(\boldsymbol{F}^T\boldsymbol{F} - \boldsymbol{1})$
Hencky strains	$\boldsymbol{e}^{(0)} = \ln\,\boldsymbol{V} = \frac{1}{2}\ln\boldsymbol{b}$	
Volumetric/deviatoric split	$\boldsymbol{F} = J_F^{\frac{1}{3}}\hat{\boldsymbol{F}}$	$\boldsymbol{C} = J_F^{\frac{2}{3}}\widehat{\boldsymbol{C}}$
Velocity gradient	$\boldsymbol{l} = \dot{\boldsymbol{F}}\,\boldsymbol{F}^{-1}$	$\boldsymbol{d} = \frac{1}{2}(\boldsymbol{l} + \boldsymbol{l}^T)$
Strain rate	$\dot{\boldsymbol{E}} = \frac{1}{2}(\dot{\boldsymbol{F}}^T\boldsymbol{F} + \boldsymbol{F}^T\dot{\boldsymbol{F}})$	$\dot{\boldsymbol{E}} = \boldsymbol{F}^T\,\boldsymbol{d}\,\boldsymbol{F}$

1.2 Balance Equations

This section contains the differential formulations which describe the local balance equations such as balance of mass, balance of linear and angular momentum as well as the first law of thermodynamics. These equations represent the fundamental relations of continuum mechanics. A detailed derivation of these equations can be found in e.g. Truesdell and Toupin (1960), Truesdell and Noll (1965), Malvern (1969, Chap. 5) and Holzapfel (2000).

1.2.1 Balance of Mass

In this section only processes are considered in which the mass of a system is conserved. This means that the change of mass has to be zero ($\dot{m} = 0$). Hence an infinitesimal mass element in initial and current configuration has to be equal which leads with $dm(\boldsymbol{X}) = \rho_0\,dV$ and $dm(\boldsymbol{x}) = \rho\,dv$ to

$$\rho\,dv = \rho_0\,dV. \tag{1.47}$$

Here ρ_0 and ρ are the densities in initial and current configuration, respectively. With Eq. (1.10) the volume elements dV and dv can be transformed leading to the Lagrangian description of the mass balance

$$\boxed{\rho_0 = J_F\,\rho} \tag{1.48}$$

For completeness the rate form of mass continuity in spatial form is presented

$$\dot{\rho}(\boldsymbol{x}, t) + \rho(\boldsymbol{x}, t) \operatorname{div} \boldsymbol{v}(\boldsymbol{x}, t) = 0 \tag{1.49}$$

which follows from the evaluation of $\dot{m} = \frac{D}{Dt} \int\limits_B \rho(\boldsymbol{x}, t)\, dv = 0$.

1.2.2 Balance of Linear and Angular Momentum

The linear momentum or the translational momentum is given in the current and initial configuration with (1.47) by

$$\boldsymbol{L} = \int\limits_{\varphi(B)} \rho\, \boldsymbol{v}\, dv = \int\limits_{B} \rho_0\, \boldsymbol{v}\, dV \tag{1.50}$$

for the continuous case. The balance of linear momentum reads: *The change of linear momentum* $\mathbf{L}$ *in time (material active) is equal to the sum of all external forces (volume and surface forces) acting on body B.*

Mathematically this statement can be expressed by

$$\dot{\boldsymbol{L}} = \int\limits_{\varphi(B)} \rho\, \bar{\boldsymbol{b}}\, dv + \int\limits_{\varphi(\partial B)} \boldsymbol{t}\, da. \tag{1.51}$$

$\rho\, \bar{\boldsymbol{b}}$ defines the volume force (e.g. gravitational force). $\boldsymbol{t}$ is the stress vector acting on the surface of the body. With Cauchy's theorem which relates the stress vector $\boldsymbol{t}$ to the surface normal $\boldsymbol{n}$ via the linear mapping

$$\boxed{\boldsymbol{t} = \boldsymbol{\sigma}\, \boldsymbol{n}, \quad t_i = \sigma_{ik}\, n_k, \quad \begin{Bmatrix} t_1 \\ t_2 \\ t_3 \end{Bmatrix} = \begin{bmatrix} \sigma_{11} & \sigma_{12} & \sigma_{13} \\ \sigma_{21} & \sigma_{22} & \sigma_{23} \\ \sigma_{31} & \sigma_{32} & \sigma_{33} \end{bmatrix} \begin{Bmatrix} n_1 \\ n_2 \\ n_3 \end{Bmatrix}} \tag{1.52}$$

(here presented in direct tensor notation, sum- and matrix formulation) the stress vector can be expressed in terms of a stress tensor $\boldsymbol{\sigma}$. Using now the divergence theorem the local balance equation of linear momentum is derived from (1.51). With reference to the current configuration $\varphi(B)$ relation

$$\boxed{\operatorname{div} \boldsymbol{\sigma} + \rho\, \bar{\boldsymbol{b}} = \rho \dot{\boldsymbol{v}}, \quad \sigma_{ik,i} + \rho\, \bar{b}_k = \rho\, \dot{v}_k} \tag{1.53}$$

is obtained. The stress tensor $\boldsymbol{\sigma}$ is called Cauchy stress tensor. $\rho \dot{\boldsymbol{v}}$ describes the inertial forces which can be neglected in case of purely static investigations.

The angular momentum with reference to a point O given by $\boldsymbol{x}_0$ is defined with respect to the current and initial configuration with (1.47) as

$$\boldsymbol{J} = \int_{\varphi(B)} (\boldsymbol{\varphi} - \boldsymbol{x}_0) \times \rho\, \boldsymbol{v}\, dv = \int_{B} (\boldsymbol{\varphi} - \boldsymbol{x}_0) \times \rho_0\, \boldsymbol{v}\, dV. \tag{1.54}$$

The balance of angular momentum can be phrased as: *The change in time (material time derivative) of angular momentum* $\boldsymbol{J}$ *with respect to a point* O *is equal to the sum of all moments stemming from external volume and surface forces with respect to point* 0.

$$\dot{\boldsymbol{J}} = \int_{\varphi(B)} (\boldsymbol{\varphi} - \boldsymbol{x}_0) \times \rho \bar{\boldsymbol{b}}\, dv + \int_{\varphi(\partial B)} (\boldsymbol{\varphi} - \boldsymbol{x}_0) \times \boldsymbol{t}\, da. \tag{1.55}$$

This equation yields after some manipulations the local balance of angular momentum which simply demands the symmetry of the Cauchy stress tensor

$$\boxed{\boldsymbol{\sigma} = \boldsymbol{\sigma}^T, \qquad \sigma_{ik} = \sigma_{ki}} \tag{1.56}$$

Observe that the balance of linear and angular momentum leads in the special case of non-existing external forces to conservation of linear and angular momentum

$$\dot{\boldsymbol{L}} = \boldsymbol{0} \Leftrightarrow \boldsymbol{L} = \text{const.}, \tag{1.57}$$

$$\dot{\boldsymbol{J}} = \boldsymbol{0} \Leftrightarrow \boldsymbol{J} = \text{const.} \tag{1.58}$$

1.2.3 First Law of Thermodynamics

Another balance law which postulates the conservation of energy in a thermodynamical process is known as the first law of thermodynamics. It reads: *the change in time (material time derivative) of the total energy* E *is equal to the sum of the mechanical power* P *of all external loads plus the heat supply* Q

$$\dot{E} = P + Q. \tag{1.59}$$

The mechanical power due to volume and surface loads is given by

$$P = \int_{\varphi(B)} \rho \bar{\boldsymbol{b}} \cdot \boldsymbol{v}\, dv + \int_{\varphi(\partial B)} \boldsymbol{t} \cdot \boldsymbol{v}\, da. \tag{1.60}$$

The heat supply

$$Q = -\int_{\varphi(\partial B)} \boldsymbol{q} \cdot \boldsymbol{n} \, da + \int_{\varphi(B)} \rho \, r \, dv \tag{1.61}$$

consists of a conduction through the surface of the body which is described by the heat flux vector $\boldsymbol{q}$ and the surface normal $\boldsymbol{n}$ and a distributed inner heat source r (specific heat supply).

The total energy is composed of the kinetic energy

$$K = \int_{\varphi(B)} \frac{1}{2} \rho \boldsymbol{v} \cdot \boldsymbol{v} \, dv \tag{1.62}$$

and the internal energy

$$U = \int_{\varphi(B)} \rho \, u \, dv. \tag{1.63}$$

u is the specific internal energy. Inserting all these relations into equation $\dot{E} = P + Q$ yields after several manipulations the local form of the first law of thermodynamics

$$\rho \, \dot{u} = \boldsymbol{\sigma} \cdot \boldsymbol{d} + \rho \, r - \operatorname{div} \boldsymbol{q}. \tag{1.64}$$

The term $\boldsymbol{\sigma} \cdot \boldsymbol{d}$ is called specific stress power.

In the framework of constitutive theory the free Helmholtz energy ψ is often introduced by the relation

$$\psi = u - \eta \, \theta. \tag{1.65}$$

Here η denotes the entropy of the system and θ is the absolute temperature. With this definition the first law of thermodynamics can be recast as

$$\boxed{\rho \, \dot{\psi} = \boldsymbol{\sigma} \cdot \boldsymbol{d} + \rho \, r - \operatorname{div} \boldsymbol{q} - \dot{\eta} \, \theta - \eta \, \dot{\theta}} \tag{1.66}$$

The special case that neither heat is supplied to an elastic body nor external forces act on the body leads to conservation of total energy

$$\dot{E} = \dot{K} + \dot{U} = 0 \Leftrightarrow E = \text{const.} \tag{1.67}$$

1.2.4 Introduction of Different Stress Tensors and Stress Rates

Equations (1.53) and (1.56) are referred to the current configuration. Often it is desirable to relate all quantities to the initial configuration B. For this purpose further stress tensors have to be introduced. Since a given stress vector does not change when referred to the current or initial configuration the following transformation can be performed using Nanson's formula (1.9) for surface elements

$$\int_{\partial\varphi(B)} \boldsymbol{\sigma}\,\boldsymbol{n}\,da = \int_{\partial B} \boldsymbol{\sigma}\,J_F\,\boldsymbol{F}^{-T}\,\boldsymbol{N}\,dA = \int_{\partial B} \boldsymbol{P}\,\boldsymbol{N}\,dA, \tag{1.68}$$

which defines the first Piola–Kirchhoff stress tensor $\boldsymbol{P}$. Observe that the first Piola–Kirchhoff stress can be written in terms of the Cauchy stress

$$\boxed{\boldsymbol{P} = J_F\,\boldsymbol{\sigma}\boldsymbol{F}^{-T}; \quad P_{ai} = J_F\,\sigma_{ik}(F_{ak})^{-1}} \tag{1.69}$$

The spatial Cauchy stress tensor $\boldsymbol{\sigma}$ in Eq. (1.69) is only from one side multiplied by $\boldsymbol{F}$ hence the tensor $\boldsymbol{P}$ is a two field tensor with one basis referred to the current and the other to the initial configuration.

Starting from the stress power in (1.66) the conjugated strain measure to the first Piola–Kirchhoff stress tensor can be computed

$$\boldsymbol{\sigma}\cdot\boldsymbol{d} = \boldsymbol{\sigma}\cdot\boldsymbol{l} = \mathrm{tr}(\boldsymbol{\sigma}\,\boldsymbol{l}^T) = \mathrm{tr}\left(\frac{1}{J_F}\boldsymbol{P}\boldsymbol{F}^T\boldsymbol{l}^T\right) = \mathrm{tr}\left(\frac{1}{J_F}\boldsymbol{P}\,\dot{\boldsymbol{F}}^T\right) = \frac{1}{J_F}\boldsymbol{P}\cdot\dot{\boldsymbol{F}} \tag{1.70}$$

Here the symmetry of the Cauchy stress tensor was used as well as relation (1.37).

Naturally it is simpler to work in the initial configuration with symmetrical stress tensors the second Piola–Kirchhoff stress was introduced. This tensor results from a complete transformation of the Cauchy stress to the initial configuration of B

$$\boxed{\boldsymbol{S} = \boldsymbol{F}^{-1}\,\boldsymbol{P} = J_F\,\boldsymbol{F}^{-1}\,\boldsymbol{\sigma}\boldsymbol{F}^{-T} \Leftrightarrow \boldsymbol{\sigma} = \frac{1}{J_F}\,\boldsymbol{F}\,\boldsymbol{S}\,\boldsymbol{F}^T} \tag{1.71}$$

which can be written in index notation as

$$S_{ab} = (F_{ia})^{-1}\,P_{bi} = J_F\,(F_{ia})^{-1}\,\sigma_{ik}(F_{bk})^{-1}. \tag{1.72}$$

$\boldsymbol{S}$ does not represent a stress which can be interpreted physically. Hence it is a pure mathematical quantity which however plays a prominent role in constitutive theory since $\boldsymbol{S}$ is work conjugated to the Green–Lagrange strain tensor (1.13).

Besides the Cauchy stress tensor $\boldsymbol{\sigma}$ often the so called Kirchhoff stress tensor $\boldsymbol{\tau}$ is introduced which results from a *push forward* of the second Piola–Kirchhoff stress tensor $\boldsymbol{S}$ to the current configuration

$$\boxed{\boldsymbol{\tau} = \boldsymbol{F}\,\boldsymbol{S}\,\boldsymbol{F}^T, \quad \boldsymbol{\tau} = J_F\,\boldsymbol{\sigma}} \tag{1.73}$$

Stress rates. The time derivative of stress tensors is of significance for the statement of incremental forms of constitutive equations. For stresses which are referred to the initial configuration (e.g. the second Piola–Kirchhoff stress tensor $\boldsymbol{S}$) the derivative with respect to time is given by the material time derivative

$$\dot{\boldsymbol{S}} = \frac{\partial \boldsymbol{S}(\boldsymbol{X}, t)}{\partial t}. \tag{1.74}$$

Time derivatives for stress tensors like the Cauchy stress tensor $\boldsymbol{\sigma}$ which are related to the current configuration are computed according to (1.35)

$$\dot{\boldsymbol{\sigma}} = \frac{\partial \boldsymbol{\sigma}}{\partial t} + \operatorname{grad} \boldsymbol{\sigma}\, \boldsymbol{v}. \tag{1.75}$$

It can easily be shown, see e.g. Truesdell and Toupin (1960), that the material time derivative of the Cauchy stress tensor is not objective, but objectivity is an inevitable prerequisite for the formulation of constitutive equations. Hence numerous time derivatives were formulated—so called objective time derivatives—which can be applied to compute stress rates. The Lie derivative of a stress tensor provides an objective stress rate, see e.g. Truesdell and Toupin (1960) or Marsden and Hughes (1983). It is given for the Kirchhoff stress tensor using (1.44) as

$$\mathcal{L}_v\, \boldsymbol{\tau} = \boldsymbol{F} \left\{ \frac{\partial}{\partial t} [\, \boldsymbol{F}^{-1} \boldsymbol{\tau} \boldsymbol{F}^{-T}] \right\} \boldsymbol{F}^T. \tag{1.76}$$

With $\dot{\boldsymbol{F}}^{-1} = -\boldsymbol{F}^{-1} \dot{\boldsymbol{F}} \boldsymbol{F}^{-1}$ and some algebraic manipulations

$$\boxed{\mathcal{L}_v\, \boldsymbol{\tau} = \dot{\boldsymbol{\tau}} - \boldsymbol{l}\, \boldsymbol{\tau} - \boldsymbol{\tau}\, \boldsymbol{l}^T = \overset{\Delta}{\boldsymbol{\tau}}} \tag{1.77}$$

can be derived using (1.37). The term $\overset{\Delta}{\boldsymbol{\tau}}$ is also called Oldroyd stress rate, see e.g. Marsden and Hughes (1983). It is equivalent to the Lie derivative of the Kirchhoff stress tensor. Observe that the Lie derivative of $\boldsymbol{\tau}$ is obtained as *push forward* of the material time derivative of the second Piola–Kirchhoff stress if Eq. (1.73) is employed in (1.76)

$$\mathcal{L}_v\, \boldsymbol{\tau} = \boldsymbol{F}\, \dot{\boldsymbol{S}}\, \boldsymbol{F}^T. \tag{1.78}$$

Another objective stress rate called the Jaumann stress rate is applied in many formulations of elasto-plastic material behaviour at finite strains. This rate is defined by

$$\boxed{\overset{\nabla}{\boldsymbol{\tau}} = \dot{\boldsymbol{\tau}} - \boldsymbol{w}\, \boldsymbol{\tau} + \boldsymbol{\tau}\, \boldsymbol{w}} \tag{1.79}$$

where $\boldsymbol{w} = \frac{1}{2}(\boldsymbol{l} - \boldsymbol{l}^T) = -\boldsymbol{w}^T$ is the skew symmetric part of the spatial velocity gradient. Since $\boldsymbol{l} = \boldsymbol{d} + \boldsymbol{w}$ is valid, the Lie derivative of $\boldsymbol{\tau}$ can be written with (1.77) as

$$\mathcal{L}_v \boldsymbol{\tau} = \overset{\nabla}{\boldsymbol{\tau}} - \boldsymbol{d}\,\boldsymbol{\tau} - \boldsymbol{\tau}\,\boldsymbol{d}. \tag{1.80}$$

This relates the Jaumann stress rate to the Lie derivative (1.76).

1.2.5 Balance Equations with Respect to Initial Configuration

With the first Piola–Kirchhoff stress the local balance of linear momentum (1.53) can be recast with respect to the initial configuration as

$$\boxed{\mathrm{DIV}\,\boldsymbol{P} + \rho_0\,\bar{\boldsymbol{b}} = \rho_0\,\dot{\boldsymbol{v}}} \tag{1.81}$$

where DIV denotes the divergence operation with respect to the initial configuration. Furthermore the use of (1.69) in the balance of angular momentum (1.56) yields

$$\boxed{\boldsymbol{P}\,\boldsymbol{F}^T = \boldsymbol{F}\,\boldsymbol{P}^T} \tag{1.82}$$

From this it is clear that the first Piola–Kirchhoff stress tensor is non-symmetric. Using (1.71), the balance of angular momentum yields the symmetry of the second Piola–Kirchhoff stress tensor: $\boldsymbol{S} = \boldsymbol{S}^T$.

Transformation of the first law of thermodynamics (1.64) to the initial configuration can be obtained with the transformation of the stress power using (1.39)

$$J_F\,\boldsymbol{\sigma} \cdot \boldsymbol{d} = \boldsymbol{F}\,\boldsymbol{S}\,\boldsymbol{F}^T \cdot \boldsymbol{F}^{-T}\,\dot{\boldsymbol{E}}\,\boldsymbol{F}^{-1} = \boldsymbol{S} \cdot \dot{\boldsymbol{E}} \tag{1.83}$$

and (1.47) as

$$\boxed{\rho_0\,\dot{u} = \boldsymbol{S} \cdot \dot{\boldsymbol{E}} - \mathrm{DIV}\,\boldsymbol{Q} + \rho_0\,R} \tag{1.84}$$

Here the heat source R and the heat flux vector $\boldsymbol{Q}$ are referred to the initial configuration. The stress power (1.83) can be written with (1.73) or (1.13) as

$$\boldsymbol{S} \cdot \dot{\boldsymbol{E}} = \frac{1}{2}\,\boldsymbol{S} \cdot \dot{\boldsymbol{C}} = \boldsymbol{\tau} \cdot \boldsymbol{d}. \tag{1.85}$$

Here the first two terms are is related to the initial configuration whereas the last term is referred to the current configuration (Table 1.2).

Table 1.2 Summary of stress tensors and work conjugated strain rates

	Stresses	Rates
Cauchy stress	$\boldsymbol{\sigma}$	$\boldsymbol{d}$
1st Piola–Kirchoff	$\boldsymbol{P} = J_F\,\boldsymbol{\sigma}\,\boldsymbol{F}^{-T}$	$\dot{\boldsymbol{F}}$
2nd Piola–Kirchoff	$\boldsymbol{S} = J_F\,\boldsymbol{F}^{-1}\boldsymbol{\sigma}\,\boldsymbol{F}^{-T}$	$\dot{\boldsymbol{E}} = \frac{1}{2}\dot{\boldsymbol{C}}$
Kirchhoff stress	$\boldsymbol{\tau} = J_F\,\boldsymbol{\sigma} = \boldsymbol{F}\,\boldsymbol{S}\,\boldsymbol{F}^T$	$\boldsymbol{d}$
Biot stress	$\boldsymbol{T}_B = \boldsymbol{R}^T\,\boldsymbol{P} = \boldsymbol{U}\,\boldsymbol{S}$	$\dot{\boldsymbol{U}}$

1.3 Weak Form of Equilibrium, Variational Principles

For the analysis of nonlinear initial boundary value problems in continuum mechanics a coupled system of partial differential equations has to be solved which consist of kinematical relations, local balance of momentum and the constitutive equations. The strong form of these equations is presented in the following for hyperelastic solids by two alternative descriptions. These are the description with respect to the initial and current configurations of the bodies. For the description with respect to the initial configuration B different stress measures can be used like the first Piola–Kirchhoff stresses or the second Piola–Kirchhoff stresses which yield two different formulations, see Sect. 1.2,

$$
\begin{array}{lll}
\text{Kinematics:} & \boldsymbol{F} & \boldsymbol{E} = \frac{1}{2}\,(\boldsymbol{F}^T\,\boldsymbol{F} - \boldsymbol{1}) \\
\text{Equilibrium:} & \operatorname{Div}\boldsymbol{P} + \rho_0\,\bar{\boldsymbol{b}} = \rho_0\,\dot{\boldsymbol{v}} & \operatorname{Div}(\boldsymbol{F}\,\boldsymbol{S}) + \rho_0\,\bar{\boldsymbol{b}} = \rho_0\,\dot{\boldsymbol{v}} \\
\text{Constitutive equation:} & \boldsymbol{P} = \dfrac{\partial W}{\partial \boldsymbol{F}} & \boldsymbol{S} = \dfrac{\partial W}{\partial \boldsymbol{E}}
\end{array}
$$

Additionally the boundary conditions for the displacements have to be prescribed on ∂B_u and boundary conditions for the tractions have to be formulated on ∂B_σ which leads to

$$
\boldsymbol{u} = \bar{\boldsymbol{u}} \quad \text{on} \quad \partial B_u \qquad \text{and} \qquad \boldsymbol{P}\,\boldsymbol{N} = \boldsymbol{F}\,\boldsymbol{S}\,\boldsymbol{N} = \bar{\boldsymbol{t}} \quad \text{on} \quad \partial B_\sigma
$$

All equations stated above can be transformed to the current configuration $\varphi(B)$ where the constitutive equations are formulated in terms of the Cauchy stress tensor, $\boldsymbol{\sigma}$, and the Kirchhoff stress tensor, $\boldsymbol{\tau}$,

$$
\begin{array}{lll}
\text{Kinematics:} & \boldsymbol{b} = \boldsymbol{F}\,\boldsymbol{F}^T & \\
\text{Equilibrium:} & \operatorname{div}\boldsymbol{\sigma} + \rho\,\bar{\boldsymbol{b}} = \rho\dot{\boldsymbol{v}} & \operatorname{div}(\frac{1}{J_F}\,\boldsymbol{\tau}) + \rho\,\bar{\boldsymbol{b}} = \rho\dot{\boldsymbol{v}} \\
\text{Constitutive equation:} & \boldsymbol{\sigma} = 2\,\rho\,\boldsymbol{b}\,\dfrac{\partial \psi}{\partial \boldsymbol{b}} & \boldsymbol{\tau} = 2\,\boldsymbol{b}\,\dfrac{\partial W}{\partial \boldsymbol{b}}
\end{array}
$$

The displacement boundary conditions are given on $\varphi(\partial B_u)$ as $\boldsymbol{u} = \bar{\boldsymbol{u}}$. For the tractions $\boldsymbol{\sigma n} = \hat{\boldsymbol{t}}$ on $\varphi(\partial B_\sigma)$ holds.

An analytical solution of these systems of nonlinear partial differential equations is only possible for a selected number of simple initial boundary value problems. Hence approximate methods like the method of finite differences or finite elements have to be applied to solve this set of equations. The use of the finite element method, which is based on a variational formulation of the equations, summarized above, expands the solution range to a broad spectrum of applications. The necessary variational formulation will be described in the following sections based on a referential and spatial description.

Several approaches can be applied to derive the variational formulation which are related to the problem at hand. In case of hyperelastic material responses a functional in the strain energy can be formulated, leading to a variational principle. For arbitrary processes the equations, summarized above, can be fulfilled in a weak sense, which yields a formulation minimizing the error of the finite element approximation for arbitrary test functions, see e.g. Johnson (1987). In the engineering literature the principle of virtual work is often basis for the derivation of the finite element approximations. It can, however, easily be shown that this formulation is equivalent to using the weak form.

In the following section several variational formulations are derived which can be applied in the context of finite elements.

1.3.1 Weak Form of Linear Momentum in the Initial Configuration

When an approximation $\boldsymbol{u}_h$ of the exact solution $\boldsymbol{u}$ is inserted in the above set of equations then an error will occur since the approximate solution is usually not equal to the exact solution. Hence the insertion of the approximate solution into the momentum balance equation $\operatorname{Div} \boldsymbol{P} + \rho_0 \bar{\boldsymbol{b}} - \rho_0 \dot{\boldsymbol{v}} = 0$ will lead to

$$\operatorname{Div} \boldsymbol{P}(\boldsymbol{u}_h) + \rho_0 \bar{\boldsymbol{b}} - \rho_0 \dot{\boldsymbol{v}}_h = \boldsymbol{R}_h$$

The residual $\boldsymbol{R}_h$, which denotes the error not fulfilling the momentum balance equation by $\boldsymbol{u}_h$, will now be reduced to zero in a weak sense by multiplying the residual by a weighting function $\boldsymbol{\eta}$ and by integrating the residual over the whole domain. The vector valued function, which has to be zero at the displacement boundaries $\boldsymbol{\eta} = \{\boldsymbol{\eta} \mid \boldsymbol{\eta} = \boldsymbol{0} \text{ on } \partial \mathrm{B_u}\}$, is often called virtual displacement or test function. This procedure leads to

$$\int_B \boldsymbol{R}_h \cdot \boldsymbol{\eta} \, dV = 0 \Longrightarrow \int_B \operatorname{Div} \boldsymbol{P}(\boldsymbol{u}_h) \cdot \boldsymbol{\eta} \, dV + \int_B \rho_0 \, (\bar{\boldsymbol{b}} - \dot{\boldsymbol{v}}_h) \cdot \boldsymbol{\eta} \, dV = 0$$

which of course has to hold for the exact solution $\boldsymbol{u}$

$$\int_B \operatorname{Div} \boldsymbol{P} \cdot \boldsymbol{\eta} \, dV + \int_B \rho_0 \, (\bar{\boldsymbol{b}} - \dot{\boldsymbol{v}}) \cdot \boldsymbol{\eta} \, dV = 0. \tag{1.86}$$

The weak form is also known as principle of virtual work in engineering. Since no further assumptions, like existence of a potential, are made, the weak form is applicable to general problems like inelastic materials, friction, non-conservative loading, etc.

By partial integration of the first term in (1.86), application of the divergence theorem and introduction of the traction boundary condition the weak form of linear momentum

$$\boxed{\int_B \boldsymbol{P} \cdot \operatorname{Grad} \boldsymbol{\eta} \, dV - \int_B \rho_0 \, (\bar{\boldsymbol{b}} - \dot{\boldsymbol{v}}) \cdot \boldsymbol{\eta} \, dV - \int_{\partial B_\sigma} \bar{\boldsymbol{t}} \cdot \boldsymbol{\eta} \, dA = 0} \tag{1.87}$$

is obtained. The gradient of the test function $\boldsymbol{\eta}$ can also be interpreted as the directional derivative of the deformation gradient $D\boldsymbol{F} \cdot \boldsymbol{\eta}$ also known as variation $\delta \boldsymbol{F}$ of the deformation gradient. In the weak form (1.87) the first Piola–Kirchhoff stress tensor can be replaced through $\boldsymbol{P} = \boldsymbol{F}\boldsymbol{S}$ by the second Piola–Kirchhoff stress tensor leading to

$$\boldsymbol{P} \cdot \operatorname{Grad} \boldsymbol{\eta} = \boldsymbol{S} \cdot \boldsymbol{F}^T \operatorname{Grad} \boldsymbol{\eta} = \boldsymbol{S} \cdot \frac{1}{2} \, (\boldsymbol{F}^T \operatorname{Grad} \boldsymbol{\eta} + \operatorname{Grad}^T \boldsymbol{\eta} \, \boldsymbol{F}) = \boldsymbol{S} \cdot \delta \boldsymbol{E}, \tag{1.88}$$

where the fact has been used that the scalar product of a symmetrical tensor (here $\boldsymbol{S}$) with an anti-symmetrical part of a tensor is zero. $\delta \boldsymbol{E}$ denotes the variation of the Green–Lagrange strain tensor which is obtained via the directional derivative

$$\begin{aligned} D\boldsymbol{E} \cdot \boldsymbol{\eta} &= \frac{d}{d\alpha} \frac{1}{2} \left[\boldsymbol{F}^T (\boldsymbol{\varphi} + \alpha \, \boldsymbol{\eta}) \, \boldsymbol{F} (\boldsymbol{\varphi} + \alpha \, \boldsymbol{\eta}) - \mathbf{1} \right] \Big|_{\alpha=0} \\ &= \frac{d}{d\alpha} \frac{1}{2} \left[[\operatorname{Grad} (\boldsymbol{\varphi} + \alpha \, \boldsymbol{\eta})]^T \operatorname{Grad} (\boldsymbol{\varphi} + \alpha \, \boldsymbol{\eta}) - \mathbf{1} \right] \Big|_{\alpha=0} \\ &= \frac{1}{2} \left[(\operatorname{Grad} \boldsymbol{\eta})^T \boldsymbol{F} + \boldsymbol{F}^T \operatorname{Grad} \boldsymbol{\eta} \right] = \delta \boldsymbol{E}. \end{aligned} \tag{1.89}$$

Using (1.88) Eq. (1.87) can be rewritten as

$$\int_B \boldsymbol{S} \cdot \delta \boldsymbol{E} \, dV - \int_B \rho_0 \, (\bar{\boldsymbol{b}} - \dot{\boldsymbol{v}}) \cdot \boldsymbol{\eta} \, dV - \int_{\partial B_\sigma} \bar{\boldsymbol{t}} \cdot \boldsymbol{\eta} \, dA = 0. \tag{1.90}$$

The first term in (1.90) denotes the internal virtual work also called stress divergence term. The last two terms describe the virtual work of the applied loading and the inertia term.

Another reformulation of the first term in (1.87) is advantageous in the automated finite element method. From

$$\boldsymbol{P} \cdot \operatorname{Grad} \boldsymbol{\eta} = \boldsymbol{P} \cdot \delta \boldsymbol{F} = \boldsymbol{P} \cdot \frac{\partial \boldsymbol{F}}{\partial \boldsymbol{\varphi}} \, \delta \boldsymbol{\varphi}$$

follows with the hyperelastic constitutive equation[5] $\boldsymbol{P} = \dfrac{\partial W}{\partial \boldsymbol{F}}$ the equivalent form for the stress divergence term

$$\boldsymbol{P} \cdot \text{Grad}\,\boldsymbol{\eta} = \frac{\partial W}{\partial \boldsymbol{F}} \cdot \frac{\partial \boldsymbol{F}}{\partial \boldsymbol{\varphi}}\,\delta\boldsymbol{\varphi} = \frac{\partial W}{\partial \boldsymbol{\varphi}} \cdot \delta\boldsymbol{\varphi} = \frac{\partial W}{\partial \boldsymbol{\varphi}} \cdot \boldsymbol{\eta}. \tag{1.91}$$

Thus the weak form (1.87) can be written as

$$\boxed{\int_B \frac{\partial W}{\partial \boldsymbol{\varphi}} \cdot \boldsymbol{\eta}\, dV - \int_B \rho_0\,(\bar{\boldsymbol{b}} - \dot{\boldsymbol{v}}) \cdot \boldsymbol{\eta}\, dV - \int_{\partial B_\sigma} \bar{\boldsymbol{t}} \cdot \boldsymbol{\eta}\, dA = 0} \tag{1.92}$$

This relation simplifies automatic differentiation since only one differentiation of the strain energy function with respect to all displacement variables is needed instead of two differentiations, first to obtain the stress tensor $\boldsymbol{P}$ and then to get the variation of the deformation gradient Grad $\boldsymbol{\eta}$, see Sect. 5.1.

1.3.2 *Weak Form of Linear Momentum in the Current Configuration*

The transformation of the weak form (1.87) to the current or spatial configuration is performed by kinematical operations in which the base vectors are *push forward* to the configuration $\varphi(B)$. With the transformation $\boldsymbol{\sigma} = \frac{1}{J_F}\boldsymbol{P}\boldsymbol{F}^T$ of the first Piola–Kirchhoff stress tensor to the Cauchy stress tensor, see (1.69), Eq. (1.24) can be rewritten

$$\boldsymbol{P} \cdot \text{Grad}\,\boldsymbol{\eta} = J_F\,\boldsymbol{\sigma}\,\boldsymbol{F}^{-T} \cdot \text{Grad}\,\boldsymbol{\eta} = J_F\,\boldsymbol{\sigma} \cdot \text{Grad}\,\boldsymbol{\eta}\,\boldsymbol{F}^{-1} = J_F\,\boldsymbol{\sigma} \cdot \text{grad}\,\boldsymbol{\eta}.$$

Furthermore from (1.10) $dv = J_F\, dV$ follows which is equivalent to $\rho = \rho_0\, J_F$. With these relations the weak form (1.87) can be written in terms of the current configuration

$$\int_{\varphi(B)} \boldsymbol{\sigma} \cdot \text{grad}\,\boldsymbol{\eta}\, dv - \int_{\varphi(B)} \rho\,(\bar{\boldsymbol{b}} - \dot{\boldsymbol{v}}) \cdot \boldsymbol{\eta}\, dv - \int_{\varphi(\partial B_\sigma)} \hat{\boldsymbol{t}} \cdot \boldsymbol{\eta}\, da = 0. \tag{1.93}$$

In this relations equation (1.68) has been used to transform the traction vector $\bar{\boldsymbol{t}}$ to $\varphi(B)$. The symmetry of the Cauchy stress tensor facilitates the replacement of the spatial gradient of the test function $\boldsymbol{\eta}$ by its symmetric part. Hence with the definition

[5] The following formulation can also be applied to inelastic constitutive equations when the elastic stresses are computed from the strain energy function under the constraints of the inelastic evolution equations.

$$\nabla^S \boldsymbol{\eta} = \frac{1}{2}\,(\text{grad}\,\boldsymbol{\eta} + \text{grad}^T\,\boldsymbol{\eta}) \tag{1.94}$$

the weak form follows with respect to the spatial configuration

$$\boxed{\int_{\varphi(B)} \boldsymbol{\sigma}\cdot\nabla^S\boldsymbol{\eta}\,dv - \int_{\varphi(B)} \rho\,(\bar{\boldsymbol{b}} - \dot{\boldsymbol{v}})\cdot\boldsymbol{\eta}\,dv - \int_{\varphi(\partial B_\sigma)} \hat{\boldsymbol{t}}\cdot\boldsymbol{\eta}\,da = 0} \tag{1.95}$$

This relation is, in a formal sense, equivalent to the principle of virtual work of the geometrically linear theory. But here the integral, the stress and virtual strain measures have to be evaluated with respect to the current configuration. Due to this the nonlinearities do appear, however hidden.

In the further variational formulations presented in this section the inertia terms $\rho\,\dot{\boldsymbol{v}}$ are neglected in order to concentrate on static equilibrium equations.

For the automated finite element formulation it is useful and more simple to start from a different derivation of the weak form. As in the weak forms related to the initial configuration (1.90) it is possible to use instead of

$$\boldsymbol{\sigma}\cdot\nabla^S\boldsymbol{\eta} = \boldsymbol{\sigma}\cdot\delta_x\boldsymbol{e} \tag{1.96}$$

where δ_x denotes now the Lie derivative, see (1.41), of the Almansi strain tensor, defined in (1.17).

The first variant of the Lie derivative, see (1.41), is especially attractive when using the automated finite element method. In that case the weak form (1.95) can be written as

$$\boxed{\int_{\varphi(B)} \boldsymbol{\sigma}\cdot\delta_x\boldsymbol{e}\,dv - \int_{\varphi(B)} \rho\,(\bar{\boldsymbol{b}} - \dot{\boldsymbol{v}})\cdot\boldsymbol{\eta}\,dv - \int_{\varphi(\partial B_\sigma)} \hat{\boldsymbol{t}}\cdot\boldsymbol{\eta}\,da = 0} \tag{1.97}$$

with

$$\delta_x\boldsymbol{e} = \left[\frac{d}{d\epsilon}\varphi^*_{\epsilon,0}\,\boldsymbol{e}\right]\bigg|_{\epsilon=0}.$$

1.3.3 Variational Functionals

In this section two variational functionals will be discussed which can alternatively be applied within the discretization process of the finite element method. For a more detailed background, see Washizu (1975).

Principle of stationary elastic potential. In case of a hyper elastic material there exist a strain energy function W, which describes the elastic energy stored in the solid. Based on this strain energy the classical principle of the minimum of potential energy can be formulated in the geometrically linear theory. In finite deformation theory it has in general to be considered that deformations can occur which are non unique. Hence only a stationary value of the potential can be reached. Under the assumption that the applied loads are conservative, which means path independent (non-conservative loads are described in Sect. 6.3.2), the functional

$$\Pi\,(\varphi) = \int\limits_{B} [\,W(\boldsymbol{C}(\varphi)) - \rho_0\,\bar{\boldsymbol{b}}\cdot\varphi\,]\,dV - \int\limits_{\partial B_\sigma} \bar{\boldsymbol{t}}\cdot\varphi\,dA \Longrightarrow STAT \tag{1.98}$$

can be stated for the static problem. The tensor $\boldsymbol{C}$ denotes the right Cauchy–Green strain tensor which depends upon the deformation. The right Cauchy–Green strain tensor is often used to formulate strain energy functions since it leads to the most general formulations.

Out of all possible deformations φ, the ones which make Π stationary fulfil the equilibrium equation. The stationary value of (1.98) can be computed by the variation of Π with respect to the deformation. For this purpose the directional derivative

$$\delta\Pi = D\,\Pi\,(\varphi)\cdot\boldsymbol{\eta} = \frac{d}{d\alpha}\,\Pi\,(\varphi + \alpha\,\boldsymbol{\eta})\Big|_{\alpha=0} \tag{1.99}$$

is applied which is also called first variation of Π. The application of this mathematical operation yields

$$D\,\Pi(\varphi)\cdot\boldsymbol{\eta} = \int\limits_{B} [\,\frac{\partial W}{\partial\varphi}\cdot\boldsymbol{\eta} - \rho_0\,\bar{\boldsymbol{b}}\cdot\boldsymbol{\eta}\,]\,dV - \int\limits_{\partial B_\sigma} \bar{\boldsymbol{t}}\cdot\boldsymbol{\eta}\,dA = 0. \tag{1.100}$$

Note that weak form (1.92) is obtained directly by the variation of (1.99). Hence Eq. (1.100) is equivalent to the weak form (1.90) for a hyperelastic material.

The construction of such principle has several advantages. First of all it can be the basis for mathematical investigations regarding existence and uniqueness of solutions (the latter is however only valid for the linear theory). Secondly, the principle leads to the most efficient finite element formulation when the automated finite element procedure is applied. Finally, it leads to the development of efficient algorithms for the solution of the resulting nonlinear equations on the basis of optimization strategies.

Hu–Washizu principle. Another variational principle is the Hu–Washizu principle, see Washizu (1975). It has gained significance early on for the construction of finite elements. This principle can be derived by writing the weak formulation with additional constraint equations which contain kinematics and constitutive equations. Due to this deformations, strains and stresses occur as independent variables. Contrary, once the principle is constructed its variation with respect to all variables yields the static equilibrium equations, the kinematical relations and the constitutive equation. The formulation of the nonlinear version of the Hu–Washizu principle can be obtained using by any set of work conjugated variables. Here it will be stated in terms of the deformation gradient $\boldsymbol{F}$, the first Piola–Kirchhoff stress tensor $\boldsymbol{P}$ and the deformation φ

$$\Pi(\varphi, \boldsymbol{F}, \boldsymbol{P}) = \int_B [\, W(\boldsymbol{F}) + \boldsymbol{P} \cdot (\, \text{Grad}\, \varphi - \boldsymbol{F}\,)\,]\, dV - \int_B \varphi \cdot \rho_0 \, \bar{\boldsymbol{b}}\, dV - \int_{\partial B_\sigma} \varphi \cdot \hat{\boldsymbol{t}}\, dA \tag{1.101}$$

The variation, according to the definition of the directional derivative, see e.g. (1.89), yields now three independent equations

$$\begin{aligned} D\Pi(\varphi, \boldsymbol{F}, \boldsymbol{P}) \cdot \eta &= \int_B (\, \boldsymbol{P} \cdot \text{Grad}\, \eta - \eta \cdot \rho_0 \, \bar{\boldsymbol{b}}\,)\, dV - \int_{\partial B_\sigma} \eta \cdot \hat{\boldsymbol{t}}\, dA = 0, \\ D\Pi(\varphi, \boldsymbol{F}, \boldsymbol{P}) \cdot \delta \boldsymbol{P} &= \int_B \delta \boldsymbol{P} \cdot (\, \text{Grad}\, \varphi - \boldsymbol{F}\,)\, dV = 0, \\ D\Pi(\varphi, \boldsymbol{F}, \boldsymbol{P}) \cdot \delta \boldsymbol{F} &= \int_B \delta \boldsymbol{F} \cdot (\, \frac{\partial W}{\partial \boldsymbol{F}} - \boldsymbol{P}\,)\, dV = 0. \end{aligned} \tag{1.102}$$

Observe that they represent the weak form (1.87), the kinematical relation (1.5) and a hyperelastic constitutive equation for $\boldsymbol{P}$.

A special form of the Hu–Washizu variational principle can be applied for the construction of finite elements which have to represent nearly incompressible material behaviour. Since incompressibility is associated with a constraint for the volumetric deformation ($J_F \equiv 1$), the split (1.19) can be used to distinguish volumetric and isochoric parts of the deformation. Based on this idea Simo et al. (1985) formulated a three-field functional which is only defined for the volumetric part of the deformation. Hence the independent variable are now the deformation φ, the pressure p and a strain variable θ which is equivalent to J_F. The last variable has to fulfill the constraint condition $\theta = J_F$. With the multiplicative split of the deformation gradient (1.19)

$$\bar{\boldsymbol{F}} = \theta^{\frac{1}{3}}\, \hat{\boldsymbol{F}} \tag{1.103}$$

the split into volumetric and deviatoric parts is achieved. Note that $\hat{\boldsymbol{F}} = J_F^{-\frac{1}{3}}\, \text{Grad}\, \varphi$ can be specified in relation (1.103). Furthermore $\bar{\boldsymbol{C}} = \theta^{\frac{2}{3}}\, J_F^{-\frac{2}{3}}\, \boldsymbol{C} = \theta^{\frac{2}{3}}\, \hat{\boldsymbol{C}}$ holds with (1.20). In the variational principle also the strain energy function $W(\boldsymbol{C})$, see (5.14), has to be defined on the basis of the new variables: $W(\boldsymbol{C}) = W(\theta^{\frac{2}{3}}\, \hat{\boldsymbol{C}})$. Using the additive split $W = W(\theta) + W(\hat{C})$, see (5.14), the following three-field variational functional (selective Hu–Washizu functional) can be constructed

$$\Pi(\varphi, p, \theta) = \int_B [\, W(\hat{C}) + W(\theta) + p(\, J_F - \theta\,)\,]\, dV - \int_B \varphi \cdot \rho_0 \bar{b}\, dV - \int_{\partial B_\sigma} \varphi \cdot \hat{t}\, dA. \tag{1.104}$$

By considering relation (5.16) the Euler–Lagrange equations obtained from this variational principle are

$$\begin{aligned} D\Pi(\varphi, p, \theta) \cdot \eta &= \int_B \left\{ \left(\mathbb{P}\left[2\,\frac{\partial W}{\partial \hat{C}} \right] + p\, J_F\, C^{-1} \right) \cdot \frac{1}{2}\,\delta C - \eta \cdot \rho_0\, \bar{b} \right\} dV \\ &\quad - \int_{\partial B_\sigma} \eta \cdot \hat{t}\, dA = 0 \\ D\Pi(\varphi, p, \theta)\,\delta p &= \int_B \delta p\,(\, J_F - \theta\,)\, dV = 0 \\ D\Pi(\varphi, p, \theta)\,\delta\theta &= \int_B \delta\theta\,(\frac{\partial W}{\partial \theta} - p\,)\, dV = 0. \end{aligned} \tag{1.105}$$

By comparison of relation (5.16) with (5.17) it is clear that the expression $S_{ISO} + S_{VOL}$ can be used in Eq. $(1.105)_1$ for the first term in the integral. This explains the split into isochoric and volumetric parts.

The terms in Eq. $(1.105)_1$ are often written with respect to the spatial configuration to simplify the numerical implementation within the finite element method. However when using the automated fem approach this is not necessary as can be seen in Sect. 4.2.

1.3.4 Mathematical Formalism for Weak Forms

The variational formulation of a linear solid mechanics problem is often written in the mathematical literature using the bi-linear form

$$a(u, \eta) = f(\eta). \tag{1.106}$$

Here u is the displacement or deformation and η denotes the variation. The operator a is related to the stress divergence term and f to the external loads. These operators of the abstract notation can be specified e.g. for the case of linear elasticity by

$$a(\boldsymbol{u}, \boldsymbol{\eta}) = \int_{\Omega} \varepsilon(\boldsymbol{\eta}) \cdot \mathbb{C}\,[\,\varepsilon(\boldsymbol{u})\,]\, d\Omega\,,$$

$$f(\boldsymbol{\eta}) = \int_{\Omega} \rho_0 \bar{\boldsymbol{b}} \cdot \boldsymbol{\eta}\, d\Omega + \int_{\Gamma_\sigma} \hat{\boldsymbol{t}} \cdot \boldsymbol{\eta}\, d\Gamma. \tag{1.107}$$

which is the linear version of (1.90) where $\mathbb{C}$ is the elasticity tensor and $\varepsilon(\boldsymbol{u}) = \frac{1}{2}(\nabla\, \boldsymbol{u} + \nabla^T \boldsymbol{u})$ the linear strain measure.

In this case the linear form of the principle of stationary potential elastic energy, for the nonlinear version see (1.98), assumes a minimum. It can be written as

$$\frac{1}{2}\, a(\boldsymbol{u}, \boldsymbol{u}) - f(\boldsymbol{u}) \Rightarrow MIN. \tag{1.108}$$

with

$$a(\boldsymbol{u}, \boldsymbol{u}) = \int_{\Omega} \varepsilon(\boldsymbol{u}) \cdot \mathbb{C}\,[\,\varepsilon(\boldsymbol{u})\,]\, d\Omega\,,$$

$$f(\boldsymbol{\eta}) = \int_{\Omega} \rho_0 \bar{\boldsymbol{b}} \cdot \boldsymbol{u}\, d\Omega + \int_{\Gamma_\sigma} \hat{\boldsymbol{t}} \cdot \boldsymbol{u}\, d\Gamma. \tag{1.109}$$

In the nonlinear case statement (1.106) looks the same but the definition is different. Then the abstract notation is given with the deformation φ by

$$a(\varphi, \boldsymbol{\eta}) = f(\boldsymbol{\eta}). \tag{1.110}$$

with

$$a(\varphi, \boldsymbol{\eta}) = \int_{\Omega} \boldsymbol{P}(\varphi) \cdot \text{Grad}\, \boldsymbol{\eta}\, d\Omega\,,$$

$$f(\boldsymbol{\eta}) = \int_{\Omega} \rho_0 \bar{\boldsymbol{b}} \cdot \boldsymbol{\eta}\, d\Omega + \int_{\Gamma_\sigma} \hat{\boldsymbol{t}} \cdot \boldsymbol{\eta}\, d\Gamma \tag{1.111}$$

which is equivalent to (1.86). However it is not a bi-linear form.

Note that in the nonlinear case the statement of the stationarity of the potential energy (1.98) is not possible when using the notation of bi-linear forms.

References

Chadwick, P. 1999. *Continuum mechanics, concise theory and problems*. Mineola: Dover Publications.

Ciarlet, P.G. 1988. *Mathematical elasticity I: three-dimensional elasticity*. Amsterdam: North-Holland.

Eringen, A. 1967. *Mechanics of continua*. New York: Wiley.
Flory, P. 1961. Thermodynamic relations for high elastic materials. *Transactions of the Faraday Society* 57: 829–838.
Hencky, H. 1933. The elastic behaviour of vulcanized rubber. *Journal of Applied Mechanics* 1: 45–53.
Holzapfel, G.A. 2000. *Nonlinear solid mechanics*. Chichester: Wiley.
Hudobivnik, B., and J. Korelc. 2016. Closed-form representation of matrix functions in the formulation of nonlinear material models. *Finite Elements in Analysis and Design* 111: 19–32.
Johnson, C. 1987. *Numerical solution of partial differential equations by the finite element method*. Cambridge: Cambridge University Press.
Lubliner, J. 1990. *Plasticity theory*. London: MacMillan.
Malvern, L.E. 1969. *Introduction to the mechanics of a continuous medium*. Englewood Cliffs: Prentice-Hall.
Marsden, J.E., and T.J.R. Hughes. 1983. *Mathematical foundations of elasticity*. Englewood Cliffs: Prentice-Hall.
Ogden, R.W. 1984. *Non-linear elastic deformations*. Chichester: Ellis Horwood and John Wiley.
Simo, J.C., R.L. Taylor, and K.S. Pister. 1985. Variational and projection methods for the volume constraint in finite deformation elasto-plasticity. *Computer Methods in Applied Mechanics and Engineering* 51: 177–208.
Truesdell, C., and W. Noll. 1965. In *The nonlinear field theories of mechanics*, ed. Flügge, S., Handbuch der Physik III/3 Berlin: Springer.
Truesdell, C., and R. Toupin. 1960. *The classical field theories*., Handbuch der Physik III/1 Berlin: Springer.
Washizu, K. 1975. *Variational methods in elasticity and plasticity*, 2nd ed. Oxford: Pergamon Press.

Chapter 2
Automation of Research in Computational Modeling

It is obvious that if we could find characters or signs suited for expressing all our thoughts as clearly and as exactly as arithmetic expresses numbers or geometry expresses lines, we could do in all matters insofar as they are subject to reasoning all that we can do in arithmetic and geometry.

Gottfried Wilhelm Leibniz
Preface to the General Science, 1677

The advances in reliability, generality and interdisciplinary nature of new computational methods derived in recent years are primary result of a holistic approach to computational modeling where advanced software tools and techniques are combined with advanced numerical methods. The holistic approach is playing nowadays a central role in the process that leads to the ultimate goal, i.e. a complete automation of computational modeling. The problem of an automation of the derivation of computational models has been explored by researches from the fields of mathematics, computer science and computational mechanics, resulting in a variety of approaches (e.g. object-oriented approaches, domain specific languages and a hybrid symbolic-numeric methods) and available software tools (e.g. symbolic and algebraic systems, automatic differentiation tools, problem solving environments and numerical libraries). Automation can address all steps of the finite element solution procedure from the strong form of boundary-value problem to the visualization of results, or it can be applied only to the automation of the selected steps of the whole procedure. Consequently, the automation procedures nowadays influence directly how the mechanical problem and corresponding numerical model are postulated and solved.

Alternative approaches to automation of computational modeling are discussed in the first part of the chapter, while an emphasis is given to automatic differentiation. The hybrid symbolic-numeric approach to automation of the finite element method and the system for automatic code generation *AceGen* are described in detail in the second part of the chapter.

J. Korelc and P. Wriggers, *Automation of Finite Element Methods*,
DOI 10.1007/978-3-319-39005-5_2

2.1 Introduction

Finite element technology. As demonstrated throughout this book, there are almost countless ways of how a particular problem can be solved by the finite element method. The complete finite element simulation can be, from the aspect of automation, decomposed into the following steps:

1. formulation of the strong form of an initial boundary-value problem;
2. transformation of the strong form into a weak form or a variational functional;
3. definition of the discretization of the domain and approximation of the field variables and their virtual counterparts (test functions);
4. derivation and solution of additional algebraic equations or differential equations defined at the element level (e.g. constraints or plastic evolution equations);
5. derivation of algebraic equations that describe the contribution of one element to the global residual vector and to the global tangential stiffness matrix;
6. coding of the derived equations in a required computer language;
7. generation of a finite element mesh and its boundary conditions;
8. solution of the global problem;
9. visualization and analysis of the results.

Alternatively, one can also start from the free Helmholtz energy of the problem and derive element equations directly as a gradient of the free energy. This approach is especially appealing for the automation due to the numerical efficiency of the solution when the gradient is obtained by the backward mode of automatic differentiation (see Sect. 2.4). The presented list follows from the engineering approach related to finite element technology where requirements for highly efficient elements (e.g. mixed, under-integrated, etc.) and the variety of elements (etc. solid, structural, coupled) leed to a rather strict separation of the individual finite element level and the global level of the problem.

With respect to the software organization the procedure of solving a problem by the finite element method can be divided into two parts:

1. part independent of the actual physical problem or the *finite element environment*,
2. and the part dependent on the physical problem or the *finite elements*.[1]

Finite elements are provided in open source finite element environments through user defined finite element subroutines or *user subroutines*. A particular finite element subroutine is meant to calculate the contribution of the particular finite element to the global solution of the problem. In the present book only the automation of the generation of user subroutines is presented.

Automation. If the automation of all steps of a finite element process from the strong form of an initial boundary-value problem to the visualization and analysis of results is chosen, then only a very specific subset of possible formulations can be covered.

---

[1] Here the particular finite element is meant to solve a particular physical problem. Large commercial finite element environments can include several hundreds of different finite elements for different physical applications.

Usually, only the standard spatial discretization (see Chap. 4) is considered as presented in Logg (2007). On the other hand, the standard discretization is of little use for problems involving coarse mesh accuracy, locking phenomena and distorted element shapes for which highly problem specific formulations have to be implemented as described in Sects. 6.4 and 6.5. As usual in science, the high uniqueness of a specific formulation renders the whole concept of automation questionable. Making templates or deriving objects for something that is used only once simply does not pay off. This may be the main reason why a complete automation of the finite element method is still not used within the commercial finite element environments. More often, the level of automation used involves only steps that are from the numerical aspect deterministic (e.g. various correctness preserving symbolic manipulations, differentiation and automatic code generation) while the true decisions are left to the researcher.

The automation of computational modeling can be approached having in mind two fundamentally different goals. The automation procedure can aim at the

1. automation of the solution of physical or mathematical problems,
2. or at the automation of scientific research in developing the optimal computational approach for the solution of physical or mathematical problems.

Usually, the difference between the standard description and the algorithmic implementation prevents full automation of scientific research in computational methods. When the automation of the solution is sought then the description and the algorithmic implementation are done within the same framework and the problem is less limiting. However it is questionable whether a complete automation is achievable at the present scientific level. New methods, different ways of discretization, optimal variational principles and optimal algorithmic implementation are still very much under research. While the automation of the solution results in software solutions that are in many ways under control of the software developer the automation of the research into solution requires much higher level of interaction between the user of the software solution and the actual process of automation. In particular, debugging possibilities have to be included in an automatic code generator that can provide a backward link between the automatically generated code and the basic equations of the mathematical model.

2.1.1 Abstract Symbolic Description of a Computational Model

A general requirement for a successful use of automation system is that the basic equations defining the problem have to be written down in an input form that is capable of manipulating equations symbolically. The advantage of automation in computational mechanics becomes apparent only when the description of the problem, which means the way how the basic equations are written down, is appropriate for the symbolic description. When this condition is not met then the automation of the symbolic description of a computational model may require more effort than the classical, well establish approaches. Unfortunately, some of the traditional formulations used in computational mechanics are not appropriate for the symbolic description. The symbolic formulation of computational mechanics problems differs often from the classical formulations described in detail in other chapters of this book and thus brings up the need for rethinking and reformulating known and traditional ways. Despite that, there exist strong arguments why at the end symbolic formulations are indeed beneficial:

1. A symbolic formulation is more compressed and thus gives fewer possibilities for an error.
2. Algebraic operations, such as differentiation, are done automatically.
3. Automatically generated codes are highly efficient and portable.
4. The multi-language and multi-environment capabilities of symbolic systems enable generation of numerical codes for various numerical environments from the same symbolic description.
5. An available collection of prepared symbolic inputs for a broad range of finite elements can be easily adjusted for the user specific problem leading to the on-demand numerical code generation.
6. Multi-field and multi-physic problems can be easily implemented. For example, the symbolic inputs for mechanical analysis and thermal analysis can be combined into a new symbolic input that creates a finite element for a fully coupled and quadratically convergent thermo-mechanical analysis.

An example, underlining the arguments above, is the standard formulation of the tangential stiffness matrix, leading to the matrix form $\mathbf{B}^T\mathbf{DB}$. It can be easily implemented using the symbolic tools. Having in mind that element tangential stiffness matrix is either the Jacobian of the resulting system of discrete algebraic equations or the Hessian of the variational functional, it means that automatic differentiation is sufficient for obtaining the tangent matrix. The work of implementing a $\mathbf{B}^T\mathbf{DB}$ matrix formulation and the efficiency of the resulting code is inferior to the approach when the tangent matrix is derived by backward automatic differentiation. The latter approach requires, regardless on the complexity of the topology and the material model, a single line of symbolic input. The standard $\mathbf{B}^T\mathbf{DB}$ formulation requires much more input for the same result.

Modern finite element simulations are often coupled with optimization procedures that require additionally to the solution of the primal problem also a solution of the sensitivity problem. The aim of sensitivity analysis is to calculate derivatives of arbitrary response functionals with respect to chosen parameters (see e.g. Kleiber et al. 1997; Keulen et al. 2005; Choi and Kim 2005a or 2005b). Sensitivity analysis has become an indispensable part of modern computational algorithms. The use of an analytically exact sensitivity analysis can significantly improve optimization procedures, see Choi and Kim (2005b), Kristanic and Korelc (2008), multi-scale algorithms, see Solinc and Korelc (2015), and the implementation of nonlinear material models, see Korelc and Stupkiewicz (2014), Hudobivnik and Korelc (2016). Thus, any proposed method of automation should address automation of primal as well as sensitivity analysis. However, to the authors knowledge, no large commercial code currently provides a general sensitivity analysis tool. Automation of sensitivity analysis is presented in Chap. 8.

The automation of the finite element methods should not be restricted only to the repetition of the same procedures that are normally done manually on a sheet of paper. However, the ability to describe the problems in terms that are more general does not necessarily mean that the derivation of the specific computational model would be any easier. The true advantages of automation become apparent only if the description of the problem, the notation and the mathematical apparatus are changed as well. It is demonstrated in the book that this can be achieved using the automatic differentiation technique as presented in Sect. 2.4.2. Thus, the basis for the automation of computational modeling is an automatic differentiation based form or ADB form of the basic equations used to describe the problem (see Chap. 3). In the actual implementation of the described methodology a general-purpose automatic code generator *AceGen* Korelc (2011) is used to derive and code characteristic finite element quantities (e.g. residual vector and stiffness matrix) at the level of an individual finite element. This code produces user defined elements for different finite element environments. However there exists also the general-purpose finite element environment *AceFEM* for the solution of the global problem that is taylored for the use together with *AceGen*.

2.2 Advanced Software Tools and Techniques

The use of advanced software technologies plays a central role in the process that leads to the ultimate goal: a complete automation of computational modeling. The problem of automation of computational methods has been explored by researches from the fields of mathematics, computer science and computational mechanics, resulting in a variety of approaches (e.g. the hybrid object-oriented approach by Eyheramendy and Zimmermann 2000 and the hybrid symbolic-numeric approach by Korelc 2002) and available software tools (e.g. computer algebra systems, AD tools by Griewank 2000, problem solving environments and numerical libraries).

The techniques presented in the next sections are the result of rapid development in computer science in the last decades. They are particularly relevant for the description of a nonlinear finite element model on a high abstract level, while preserving the numerical efficiency.

2.2.1 Symbolic and Algebraic Computational Systems

Symbolic and algebraic computational (SAC) systems are tools for the manipulation of mathematical expressions in symbolic form. Widely used SAC systems such as *Mathematica* (www.wolfram.com) or *Maple* (www.maplesoft.com) have become an integrated computing environment that covers all aspects of a computational process, including numerical analysis and graphical presentation of the results. The general SAC systems are also one of the most complex software systems ever developed and the SAC system *Mathematica* is often described as the "world's single largest consumer of algorithms". In the case of complex mechanical models, the direct use of SAC systems is not possible due to several reasons. For the numerical implementation, SAC systems cannot keep up with the run-time efficiency of programming languages such as FORTRAN and C and by no means with highly problem-oriented and efficient numerical environments used for finite element analysis. However, SAC systems can be used for the automatic derivation of appropriate formulas and generation of numerical codes (Korelc 2002; Amberg et al. 1999). The FE method is within the general SAC systems usually implemented as an additional package or toolbox such as *AceGen* Korelc (2011) for *Mathematica.*

The major limitation of the symbolic systems, when applied to complex engineering problems, as pointed out before by many authors (see e.g. Wang 1986; Fritzson and Fritzson 1984; Korelc 1997b and Korelc 2002), is an uncontrollable growth of expressions and consequently redundant operations and inefficient codes. This is especially problematic when a SAC system is used to derive formulas needed in numerical procedures such as the finite element method where the numerical efficiency of the derived formulas and the generated code are of utmost priority.

General SAC systems are very powerful when dealing with one isolated formula, but become very awkward when the code to be generated contains control structures such as If and Do constructs.

2.2.2 Automatic Differentiation Tools

Differentiation is a symbolic operation that plays a crucial role in the development of new numerical procedures. Automatic differentiation or AD is a method to evaluate the derivative of a function specified by a computer program and represents an alternative to more common and widely used numerical differentiation and symbolic differentiation. The AD technique is explained more in detail in Sect. 2.4 due to the central role of AD in automation of the finite element method.

2.2.3 Problem Solving Environments

Problem-solving environments (PSE) are automatic code generators with libraries containing routines for various numerical solution methods. These routines form templates for the generated program codes. The system libraries determine a variety of numerical solution methods available in such systems. They are meant to solve problems, in particular ordinary differential equations or partial differential equations, in an already established way. Several problem-solving environments for a high level abstract description of PDE's based on finite difference method have been derived, such as *SciNapse* (Akers et al. 1998) and *Ctadel* (van Engelen et al. 1995). A comprehensive overview can be found in Gallopoulos et al. (1994).

Numerical libraries. Additionally to the general problem-solving environments, there are also tools that support only numerical operations such as: compiled numerical libraries (e.g. *NAG*, www.nag.co.uk), numerical matrix languages (e.g. *MATLAB*, www.mathworks.com), high-level object oriented languages with object libraries, etc.

Specialized systems for FEM. General finite element environments, such as such as commercial codes like *ABAQUS* (www.hks.com) and *ANSYS* (www.ansys.com) or research codes like *FEAP* (www.ce.berkeley.edu/feap), can also be viewed as a specialized PSE. The general finite element environments can handle, regardless of the type of finite elements, all the phases of a typical FE simulation: pre-processing of the input data, manipulation and organization of the data related to nodes, elements, material characteristics, displacements and stresses, construction of the global matrices by invoking different element subroutines, solving the system of equations, post-processing and analysis of the results.

2.2.4 Hybrid Approaches

The level of automation of FE method can be greatly increased by combining several approaches and tools. Some possible combinations are discussed below.

Hybrid object-oriented approach. The object-oriented approach has brought a new perspective for the development of complex software and in the past decade, numerous object-oriented FE environments have been developed. While the object-oriented approach deals primary with the high level of data abstraction and organization, its principles can be extended also to the complete automation of the finite element method. An overview of the object-oriented hybrid symbolic-numerical approach can be found in Eyheramendy and Zimmermann (2000), Beall and Shephard (1999) and Logg (2007). Modern hybrid object-oriented (HOO) systems, such as *FEniCS* (Logg 2007; Logg and Wells 2012) or *FreeFem++* (Pironneau et al. 2008), provide tools for automation of all FE simulation steps, from the strong form of a PDE to the presentation of the results. A typical HOO system introduces its own domain-specific language and uses built-in C++ libraries for symbolic manipulation. The HOO systems are in general restricted to a particular type of formulations where the general knowledge of the appropriate procedure, that leads from strong form to the element equations, has already been established. This also reduces the expression growth problem since the symbolic code derivation is used only for sub-problems.

Hybrid symbolic-numerical approach. The disadvantage of the hybrid object-oriented approach is the loss of generality and flexibility compared to a general computed algebra systems. Only a small fraction of symbolic manipulation capabilities of the general SAC systems is presented in specialized finite element C++ libraries for symbolic manipulation. The real power of the symbolic approach for testing and applying new, unconventional ideas is provided by general-purpose SAC systems. However, their use is limited for problems that lead to large systems like finite element simulations. Furthermore, the use of large commercial finite element environments for analyzing a variety of problems is standard practice of engineers. The hybrid symbolic-numerical (HSN) approach is a way to combine both. While the hybrid object-oriented systems tends to offer complete FE solution, the idea behind the hybrid symbolic-numerical (HSN) approach is to use a general SAC system for the derivation of the characteristic element quantities and the automatic code generation of user subroutines at the level of one finite element. The automatically generated code is then incorporated into the chosen finite element environment (one or possibly more) and used within the global numerical solution procedures. The HSN approach is explained in more detail in Sect. 2.5.1. An advantage of using a general SAC system is also that it provides well known and defined description language for the derivation of FE equations, generation of FE code and possibly also for the complete FE analysis, as opposed to the hybrid object-oriented systems which introduce their own domain-specific language.

2.3 Automatic Generation of Numerical Codes

Most of the existing numerical methods for solving partial differential equations can be subdivided into two classes: finite difference and finite element methods. In the last years, various approaches to the automation of the two methods were extensively studied. In many ways, the present stage of the automation of the finite difference method is more elaborated and more general than the automation of the finite element method. Various transformations, differentiation, matrix operations, and a large number of degrees of freedom involved in the derivation of characteristic finite element quantities often lead to exponential growth of the expressions in space and time, see e.g. Fritzson and Fritzson (1984). This makes automation of the FE method more complex than automation of the finite difference method. Different approaches have been proposed for the solution of the problem related to expression growth. In the first approach additional low-level knowledge about the problem is introduced and the symbolic code generation is used only for sub-problems, where expression growth does not appear or is not severe, see e.g. Amberg et al. (1999). A pure symbolic approach is often combined with the object-oriented approach and specialized routines for symbolic manipulations of the general CA systems are used instead. An extensive overview of object-oriented hybrid symbolic/numerical approaches can be found in Eyheramendy and Zimmermann (2000) and in Beall and Shephard (1999). The disadvantage of the approach is the loss of generality and flexibility of general CA systems. These specialized systems are often restricted to a particular type of problems where general knowledge of the solution procedure has already been established. The real power of the symbolic approach for testing and applying new, unconventional ideas lies in general purpose systems. Not many attempts have been undertaken to produce a general FE code generator where the key issue of the FE code generation, i.e. expression swell, is treated within the automatic procedure. Techniques such as the use of the symmetric properties of the formulae, the automatic introduction of intermediate variables and the pattern search were applied in Finger Wang (1986), Wang (1991) and Tan et al. (1991).

When the symbolic approach is used in a standard way to describe complex-engineering problems then the common experience of the users of symbolic and algebraic systems is an uncontrollable swell of expression, as pointed out before, leading to inefficient or even unusable codes. The general computer algebra systems come with the built in code optimization capabilities, see e.g. *Maple*, or additional packages for code optimization such as *AceGen* for *Mathematica*. The classical way of optimizing expressions in computer algebra systems is searching for common sub-expressions after all the formulae have been derived and before the generation of the numerical code. This has been proven to be insufficient when applied to general non-linear mechanical problems and only relatively simple finite elements can be derived within such approach.

An alternative approach for automatic code generation is employed in *AceGen* and called Simultaneous Stochastic Simplification of numerical code, see Korelc (1997b). This approach avoids the problem of expression swell by using symbolic and algebraic capabilities of the general computer algebra system *Mathematica* and combining them with the following techniques: automatic differentiation, simultaneous optimization of expressions with automatic selection and introduction of appropriate intermediate variables. Formulae are optimized, simplified and replaced by auxiliary variables simultaneously with the derivation of the problem. A stochastic evaluation of the formulae is applied for determining the equivalence of algebraic expressions, see e.g. Gonnet (1986). The simultaneous approach is appropriate also for problems where intermediate expressions can be subjected to uncontrolled swell.

Using the general finite element environment for analyzing a variety of problems and for incorporating new elements is now already an everyday practice. Although large FE environments often offer a possibility to incorporate user defined elements and material modes, it is time consuming to develop and test these user defined new pieces of software. Practice shows that at the derivation stage of a new numerical model, different languages and different platforms are the best means for the assessment of specific performances and failures of the numerical model. The basic tests, which are performed on a single finite element or on a small patch of elements, can be done most efficiently by using general computer algebra system. Many design flaws, such as element instabilities or poor convergence properties, can be easily identified, if the element quantities are investigated on a symbolic level. Unfortunately a stand-alone CA system becomes very inefficient once there is a larger number of nonlinear finite elements to process or if iterative numerical procedures have to be executed. In order to assess element performances under real conditions, the easiest way is to run the necessary test simulations on sequential machines with good debugging capabilities and with the open source FE environment designed for research purposes, e.g. *FEAP* (www.ce.berkeley.edu/rlt/feap/), *AceFEM* (www.fgg.uni-lj.si/symech/) or *Diffpack* (www.diffpack.com). At the end, for real industrial simulations involving complex geometries a large commercial FE environment has to be used. In order to meet all these demands in an optimal way, an approach is needed that would offer multi-language and multi-environment generation of numerical codes. The automatically generated code is then incorporated into the FE environment that is most suitable for the specific step of the research process. Using the classical code development procedures, re-coding of the element in different languages would be time consuming and is rarely done. With the general SAC systems, re-coding comes practically for free, since the code can be automatically generated for several languages and for several platforms using the same basic symbolic description.

2.4 Automatic Differentiation

The classical way of performing differentiation is to symbolically differentiate a function either manually or by a computer algebra system and to evaluate it at a chosen point. In the case of complex computational models, the symbolic derivatives are difficult to obtain, which is why the numerical differentiation by the finite difference method is often used instead. Automatic differentiation (see e.g. Griewank and Walther 2008; Bartholomew-Biggs et al. 2000; Bischof et al. 2002) is a method to compute the derivative of a function specified by a computer program. It represents an alternative solution to the numerical differentiation as well as to the symbolic differentiation. With the AD technique, one can avoid the problem of expression growth that is associated with the symbolic differentiation performed by a SAC system.

2.4.1 Principles of Automatic Differentiation

Automatic differentiation techniques are based on the fact that every computer program executes a sequence of elementary operations with known derivatives, thus allowing evaluation of exact derivatives via the chain rule for an arbitrary complex formulation. If one has a computer code which allows to evaluate a function f and needs to compute the gradient ∇f of f with respect to arbitrary variables, then the automatic differentiation tools, see e.g. Griewank (2000), Bartholomew-Biggs et al. (2000), Bischof et al. (2002) can be applied to generate the appropriate program code.

There are two approaches for the automatic differentiation of a computer program, often characterized as the forward and the backward mode of automatic differentiation. The procedure is illustrated by a simple example of a function f defined by

$$f = b\,c \quad \text{with} \quad b = \sum_{i=1}^{n} a_i^2 \quad \text{and} \quad c = Sin(b) \tag{2.1}$$

where $a_1, a_2, ..., a_n$ are n independent variables. The forward mode accumulates the derivatives of intermediate variables with respect to the independent variables as follows

$$\begin{aligned}
\nabla b &= \left\{\tfrac{db}{da_i}\right\} = \{2\,x_i\} && i = 1, 2, ..., n \\
\nabla c &= \left\{\tfrac{dc}{dx_i}\right\} = \{Cos(b)\,\nabla b_i\} && i = 1, 2, ..., n \\
\nabla f &= \left\{\tfrac{df}{da_i}\right\} = \{\,\nabla b_i\,c + b\,\nabla c_i\,\} && i = 1, 2, ..., n
\end{aligned} \tag{2.2}$$

Contrary to the forward mode, the backward mode propagates adjoints $\bar{x} = \frac{\partial f}{\partial x}$, which are the derivatives of the final values, with respect to intermediate variables:

$$\begin{aligned}
\bar{f} &= \tfrac{df}{df} = 1 && 1 \\
\bar{c} &= \tfrac{df}{dc} = \tfrac{\partial f}{\partial c}\bar{f} = b\,\bar{f} && 1 \\
\bar{b} &= \tfrac{df}{db} = \tfrac{\partial f}{\partial b}\bar{f} + \tfrac{\partial c}{\partial b}\bar{c} = c\bar{f} + Cos(b)\,\bar{c} && 1 \\
\nabla f &= \{\bar{a}_i\} = \left\{\tfrac{\partial b}{\partial a_i}\,\bar{b}\right\} = \left\{2\,a_i\,\bar{b}\right\} && i = 1, 2, ..., n.
\end{aligned} \tag{2.3}$$

The numerical efficiency of the differentiation of N scalar-valued functions $\mathbf{F} = \{f_i : i \in [1, N]\}$ with respect to M independent variables $\mathbf{a} = \{a_j : j \in [1, M]\}$ can be measured by the numerical work ratio defined as

$$wratio(\mathbf{F}(\mathbf{a})) = \frac{cost(\mathbf{F}(\mathbf{a}), \frac{\partial \mathbf{F}}{\partial \mathbf{a}})}{cost(\mathbf{F}(\mathbf{a}))} \tag{2.4}$$

The ratio is in general proportional to the number of independent variables ($wratio(\mathbf{F}(\mathbf{a})) \propto M$) in the case of forward mode and proportional to the number of functions ($wratio(\mathbf{F}(\mathbf{a})) \propto N$) in the case of backward mode. The upper bound for the ratio in the case of backward mode differentiation of one function with respect to arbitrary number of independent variables is 5 and is usually around 1.5 if care is taken in handling quantities that are common to the function and gradient, see e.g. Griewank (2000). Although numerically superior when the number of functions is small, the backward mode requires potential storage of a large amount of intermediate data during the evaluation of the function that can be as high as the number of numerical operations performed. Additionally, a complete reversal of the program flow is required. This is because the intermediate variables are used in reverse order when related to their computation. For the efficient automation of the FE method, it is desirable that both approaches are available and that the software tool used for automation can automatically select the most efficient approach according to the estimated work ratio.

It should be pointed out that the symbolic differentiation is one of the algebraic operations prone to severe expression growth and it can result even for relatively simple nonlinear elements in hundreds of pages of code. Thus, the use of hybrid system that combines the symbolic tool with the automatic differentiation technique is essential for the high abstract symbolic formulation of FE models. To increase the numerical efficiency of the generated code and to limit the physical size of the generated code, it is essential to minimize the number of calls to the automatic differentiation procedure.

There exist many strategies how the AD procedure can be implemented, see e.g. Bischof et al. (2002). The simplest approach is to use operator overloading and during the evaluation of function f create a trace of all numerical operations and their arguments, later used to evaluate gradients in forward or backward mode. The operator overloading strategy is computationally too inefficient to be used within the finite element procedures. More efficient is a source-to-source transformation strategy that transforms the source code for computing a function into the source code for computing the derivatives of the function. The AD tools based on a source-to-source transformation have been developed for most of the programming languages e.g. *ADIFOR* (www-unix.mcs.anl.gov/autodiff/ADIFOR/) for FORTRAN, *ADOL-C* (www.math.tu-dresden.de/~adol-c/) for C, *MAD* (www.amorg.co.uk/AD/MAD/) for *Matlab* and *AceGen* (www.fgg.uni-lj.si/symech) for *Mathematica.*

2.4.2 Generalized Notation of Automatic Differentiation

Recent advances in the development of automatic differentiation technique, especially the backward mode implementation of the code-to-code approach to automatic differentiation, have rejected the traditional assumption that automatic differentiation is impracticable and that the numerical codes based on automatic differentiation are intrinsically too slow for large-scale numerical computations. However, as powerful as automatic differentiation technique is, the results of the automatic differentiation procedure might not automatically correspond to the specific mathematical formalism used to describe the mechanical problem. One of the primal goals of the present book is to introduce an appropriate notation that builds a bridge between the classical mathematical notation of computational models and the actual computer implementation. It is essential for the generalized notation that it can be directly translated into the program code and that the derived program code is numerically efficient. The idea presented in the book is to achieve the unification by means of extended automatic differentiation technique combined with the automatic code generation. The introduced notation does not only simplify the derivation of the corresponding equations, but also reflects much more closely the actual algorithmic implementation. In this way, the mathematical formulation and computer implementation become indistinguishable.

The introduced generalized notation is comprised of three components that:

- define the mathematical operation,
- indicate the algorithm used to obtain the quantity and
- specify the flow of information within the algorithm.

Let **a** be a set of mutually independent variables and f an arbitrary function of **a**. The "computational derivative" is then defined by the following formalism

$$\frac{\hat{\delta} f(\mathbf{a})}{\hat{\delta}\mathbf{a}} \tag{2.5}$$

The above formalism has to be viewed in an algorithmic way. The operator $\frac{\hat{\delta} f(\mathbf{a})}{\hat{\delta}\mathbf{a}}$ represents a differentiation of a function f with respect to variables **a** performed by the automatic differentiation algorithm. Thus, in the context of the book the operator has a dual purpose. It indicates the mathematical operation of differentiation as well as it indicates the algorithm used to obtain the required quantity. It also denotes that, in the process of derivation, the AD procedure has been called. The mathematical formalisms that are part of the traditional FE formulation such as partial derivatives $\frac{\partial(\bullet)}{\partial(\bullet)}$, total derivatives $\frac{D(\bullet)}{D(\bullet)}$, directional derivatives, consistent derivatives etc., can all be represented by the AD procedure. However, the result of an AD procedure might not automatically correspond to any of the above mathematical formalisms. Thus, there exists a need for an extended functionality of the traditional AD procedure as well as a need for an extended notation that would correspond to an extended functionality. This extended functionally can be provided by introducing additional information used within the process of automatic differentiation. It thus defines exceptions within the AD procedure.

Let **b** be a set of mutually independent intermediate variables that are part of the evaluation of a function f. $\mathbf{f}_b$ is a set of arbitrary functions of **a** such that $\mathbf{b} := \mathbf{f}_b(\mathbf{a})$, and **M** is an arbitrary matrix. AD exceptions are then introduced by the following formalism

$$\left.\frac{\hat{\delta}\,(\bullet)}{\hat{\delta}\,(\bullet)}\right|_{\frac{D\mathbf{b}}{D\mathbf{a}}=\mathbf{M}} \tag{2.6}$$

where $\frac{\hat{\delta}(\bullet)}{\hat{\delta}(\bullet)}$ stands for the various forms of AD operators that will be introduced later. Notation (2.6) indicates that during the AD procedure, the total derivatives of an arbitrary set of mutually independent intermediate variables **b** with respect to independent variables **a** are set to be equal to the matrix **M**. The AD exceptions can be viewed as a bridge inside the chain-rule that goes directly from **b** to **a**.

In a special case, **M** represents the actual total derivatives of **b** with respect to **a**. Thus $\mathbf{M} = \frac{D\mathbf{b}}{D\mathbf{a}}$, however **M** can not be derived from the actual program code considered by the AD procedure. The matrix **M** has to be in this special case calculated by an alternative computational procedure or imported as an input data of the problem. The situation is indicated by the following formalism

$$\left.\frac{\hat{\delta}\,(\bullet)}{\hat{\delta}\,(\bullet)}\right|_{\frac{D\mathbf{b}}{D\mathbf{a}}=\widehat{D_{\mathbf{a}}\mathbf{b}}} \tag{2.7}$$

The formalism $\frac{D\mathbf{b}}{D\mathbf{a}} = \widehat{D_{\mathbf{a}}\mathbf{b}}$ in (2.7) denotes that the actual total derivatives of **b** with respect to **a** were obtained by an alternative procedure, stored in a matrix $\widehat{D_{\mathbf{a}}\mathbf{b}}$ and used during the AD procedure.

When no AD exceptions are defined for any of the intermediate variables that appear as part of evaluation of function f, then the computational derivative (2.5) yields the same result as the partial derivative, thus

$$\frac{\hat{\delta} f(\mathbf{a})}{\hat{\delta}\mathbf{a}} = \frac{\partial f(\mathbf{a})}{\partial \mathbf{a}}. \tag{2.8}$$

The use of the proposed AD formalism instead of the standard partial derivative formalism also indicates that the AD procedure is being used to obtain the result. Thus, the $\frac{\hat{\delta}(\bullet)}{\hat{\delta}(\bullet)}$ formalism is used instead of the $\frac{\partial(\bullet)}{\partial(\bullet)}$ formalism wherever the automation of the specific computational procedure is proposed.

If the computer program that defines the function is relatively simple, the AD can be applied at the global level of the program. More often, the computer program is complex (e.g. finite element environment) with some parts available only as compiled libraries and thus unavailable for automatic differentiation. In that case the AD can still be applied at the level of particular subroutines (e.g. finite element subroutines). When the AD procedure is used within the subroutine then the AD procedure recognizes only dependencies that appear explicitly within the derived subroutine. Thus, all the input parameters of the subroutine are automatically considered as independent variables with all derivatives set to zero. Any dependency of the input parameters of the subroutines has to be explicitly introduced as an AD exception.

For further derivations it is important to note that in the backward mode of automatic differentiation the expression $\frac{\hat{\delta} f_1(\mathbf{a})}{\hat{\delta}\mathbf{a}} + \frac{\hat{\delta} f_2(\mathbf{a})}{\hat{\delta}\mathbf{a}}$ can result in a code that is twice as large as and twice times slower than the code produced by the equivalent expression $\frac{\hat{\delta}(f_1(\mathbf{a})+f_2(\mathbf{a}))}{\hat{\delta}\mathbf{a}}$. In general, the proposed automation procedure would be optimal if the number of AD calls is kept to a minimum.

The additional information introduced by the AD exception (2.6) is associated with the intermediate variables **b**. The point in the algorithm where this additional information is introduced is important as well. Two typical situations are considered. The simplest case is when additional information is introduced together with a particular call of AD procedure. This situation is referred to as "local definition of AD exception". In the case of complex computational models that involve a large number of differentiations and several types of variables for which AD exceptions have to be specified, local definition of AD exception becomes cumbersome. The information about all AD exceptions relevant for the intermediate variables **b** can also be introduced at the point in algorithm where **b** is introduced. The situation is referred to as "global definition of AD exceptions".

2.4.3 Local Definition of AD Exceptions

The local definition of an AD exception is provided by

$$\nabla f_A = \left.\frac{\hat{\delta} f(\mathbf{a}, \mathbf{b}(\mathbf{a}))}{\hat{\delta} \mathbf{a}}\right|_{\frac{D\mathbf{b}}{D\mathbf{a}}=\mathbf{M}} . \tag{2.9}$$

Notation (2.9) indicates that during the AD procedure, the total derivatives of intermediate variables $\mathbf{b}$ with respect to independent variables $\mathbf{a}$ are set to be equal to matrix $\mathbf{M}$. The use of AD exception (2.9) can be beneficial also when $\mathbf{M} = \frac{D\mathbf{b}}{D\mathbf{a}}$ and explicit algorithmic dependency of $\mathbf{b}$ with respect to $\mathbf{a}$ exists. Thus, in principle the required total derivatives can be obtained automatically by the AD procedure using the chain-rule. This is often the case when there exists a profound mathematical relationship that enables the evaluation of total derivatives $\frac{D\mathbf{b}}{D\mathbf{a}}$ in a more efficient way (e.g. when the evaluation of $\mathbf{b}$ involves iterative loops, inverse matrices, etc.).

Several special cases and generalizations of (2.9) can also be introduced. The first special case is when the intermediate variables $\mathbf{b}$ algorithmically depend on $\mathbf{a}$ but have to be considered constant during the AD procedure. In this case, the direct use of AD would not give the correct result without intervention of the user. The correct results can be obtained by setting the matrix $\mathbf{M}$ in (2.9) to zero as follows

$$\nabla f_B = \left.\frac{\hat{\delta} f(\mathbf{a}, \mathbf{b}(\mathbf{a}))}{\hat{\delta} \mathbf{a}}\right|_{\frac{D\mathbf{b}}{D\mathbf{a}}=\mathbf{0}} = \left.\frac{\hat{\delta} f(\mathbf{a}, \mathbf{b}(\mathbf{a}))}{\hat{\delta} \mathbf{a}}\right|_{\mathbf{b}=const.} . \tag{2.10}$$

Situation (2.10) frequently appears in the formulation of mechanical problems where, instead of the total variation, some arbitrary variation of a given quantity has to be evaluated.

The second special case of (2.9) appears when intermediate variables $\mathbf{b}$ algorithmically do not depend on $\mathbf{a}$, however from the mathematical formulation of the problem it is apparent that some additional (usually implicit) dependencies have to be considered for the differentiation leading to

$$\nabla f_C = \left.\frac{\hat{\delta} f(\mathbf{a}, \mathbf{b})}{\hat{\delta} \mathbf{a}}\right|_{\frac{D\mathbf{b}}{D\mathbf{a}}=\mathbf{M}} . \tag{2.11}$$

Exception (2.9) can also be generalized. Let $\mathbf{c}$ be an additional set of intermediate variables that appear during the evaluation of function f such that $\mathbf{b} := \mathbf{f}_b(\mathbf{c})$ and let $\mathbf{f}_c$ be a set of arbitrary functions of $\mathbf{a}$ such that $\mathbf{c} := \mathbf{f}_c(\mathbf{a})$. A bridge in the chain-rule can now be formed that goes directly from $\mathbf{b}$ to $\mathbf{c}$. The chain-rule from $\mathbf{c}$ to $\mathbf{a}$ is then propagated automatically by the AD procedure to complete the evaluation of computational derivatives. The situation is described by the following formalism

$$\nabla f_D = \left. \frac{\hat{\delta} f(\mathbf{a}, \mathbf{b}(\mathbf{c}(\mathbf{a})))}{\hat{\delta}\mathbf{a}} \right|_{\frac{D\mathbf{b}}{D\mathbf{c}}=\mathbf{M}} . \tag{2.12}$$

It is important for the uniqueness of AD procedure that $\frac{\partial \mathbf{b}}{\partial \mathbf{a}}$ is zero in this case, thus **b** depends explicitly only on **c** and it does not depend directly on **a**.

2.4.4 Global Definition of AD Exceptions

AD exceptions (2.9)–(2.12) are imposed during the execution of the particular call of the AD procedure to which they are attached. The global definition of AD exceptions relevant for the intermediate variables **b** appears at the point in the algorithm where **b** is introduced and is indicated by the following formalism

$$\begin{aligned} \mathbf{b} &= \mathbf{f}_b(\mathbf{a})|_{\frac{D\mathbf{b}}{D\mathbf{a}}=\mathbf{M}(\mathbf{a})} \\ \nabla f_A &= \frac{\hat{\delta} f(\mathbf{a}, \mathbf{b}(\mathbf{a}))}{\hat{\delta}\mathbf{a}} \end{aligned} . \tag{2.13}$$

The global definition of the AD exception (2.13) is equivalent to the local definition of the AD exception (2.9) except that the AD exception (2.13) is defined globally, thus valid for every subsequent call of the AD procedure while (2.9) applies only for the specific call of the AD procedure. The formalism (2.13) does not change the value of the variables **b**. However, the definition (2.13) forces the AD procedure to ignore all dependencies of **b** as defined by function $\mathbf{f}_b$ and specifies that during all subsequent calls of the AD procedure the total derivatives of variables **b** with respect to **a** are set to be equal to matrix **M**.

As for the local definition of AD exceptions, there also exist special cases and generalizations for the global definition of AD exceptions. In Table 2.1 the various types of local definitions of AD exceptions introduced in previous sections are summarized and accompanied by the corresponding global definitions. The letters A to D that identify the specific types that are introduced for later use.

Obviously, multiple definitions of AD exceptions for the same variables can clash. For example, it is possible to define first a global AD exception (2.13) and later, at the actual call of the AD procedure, additionally a local AD exception (2.9). When local and global exceptions clash, the local definition of an AD exception takes precedence over the global definition of an AD exception for the same variable.

2.4.5 Differentiation with Respect to Variables with an Index

If $f(\mathbf{a})$ is an explicit function of all components of **a** then the physical size of the generated code for a derivative $\frac{\hat{\delta} f(\mathbf{a})}{\hat{\delta}\mathbf{a}}$ depends on the number of independent

Table 2.1 Typical automatic differentiation exceptions

Type	Local definition of AD exception	Global definition of AD exception
A	$\nabla f_A = \left.\frac{\hat{\delta} f(\mathbf{a},\mathbf{b}(\mathbf{a}))}{\hat{\delta}\mathbf{a}}\right\|_{\frac{D\mathbf{b}}{D\mathbf{a}}=\mathbf{M}}$	$\mathbf{b} = \mathbf{f}_b(\mathbf{a})\|_{\frac{D\mathbf{b}}{D\mathbf{a}}=\mathbf{M}(\mathbf{a})}$ $\nabla f_A = \frac{\hat{\delta} f(\mathbf{a},\mathbf{b}(\mathbf{a}))}{\hat{\delta}\mathbf{a}}$
B	$\nabla f_B = \left.\frac{\hat{\delta} f(\mathbf{a},\mathbf{b}(\mathbf{a}))}{\hat{\delta}\mathbf{a}}\right\|_{\mathbf{b}=const.}$	$\mathbf{b} = \mathbf{f}_b(\mathbf{a})\|_{\frac{D\mathbf{b}}{D\mathbf{a}}=\mathbf{0}}$ $\nabla f_B = \frac{\hat{\delta} f(\mathbf{a},\mathbf{b}(\mathbf{a}))}{\hat{\delta}\mathbf{a}}$
C	$\nabla f_C = \left.\frac{\hat{\delta} f(\mathbf{a},\mathbf{b})}{\hat{\delta}\mathbf{a}}\right\|_{\frac{D\mathbf{b}}{D\mathbf{a}}=\mathbf{M}}$	$\mathbf{b} = \mathbf{f}_b\|_{\frac{D\mathbf{b}}{D\mathbf{a}}=\mathbf{M}}$ $\nabla f_C = \frac{\hat{\delta} f(\mathbf{a},\mathbf{b})}{\hat{\delta}\mathbf{a}}$
D	$\nabla f_D = \left.\frac{\hat{\delta} f(\mathbf{a},\mathbf{b}(\mathbf{c}(\mathbf{a})))}{\hat{\delta}\mathbf{a}}\right\|_{\frac{D\mathbf{b}}{D\mathbf{c}}=\mathbf{M}}$	$\mathbf{c} = \mathbf{f}_c(\mathbf{a})$ $\mathbf{b} = \mathbf{f}_b(\mathbf{c})\|_{\frac{D\mathbf{b}}{D\mathbf{c}}=\mathbf{M}}$ $\nabla f_D = \frac{\hat{\delta} f(\mathbf{a},\mathbf{b}(\mathbf{c}(\mathbf{a})))}{\hat{\delta}\mathbf{a}}$

variables. This can again lead to swell of expressions and is not acceptable for larger numbers of independent variables. Thus, explicit code for all terms of the gradient can be generated only if the number of independent variables is small. The size of the generated code can be reduced by an introduction of *representative formulas*. A representative formula is one general formula, that can be used for the evaluation of several other formulas. It is a formula with an index. If the index represents a loop counter then the representative formula can be evaluated within the loop several times in succession. The concept is of course well known and often used in mechanics to reduce the complexity of derivations. It is here extended to the automatic differentiation procedure. The generation of a representative formula is indicated by the following formalism

$$\frac{\hat{\delta} f(\mathbf{a})}{\hat{\delta} a_m} \tag{2.14}$$

The operator $\frac{\hat{\delta} f(\mathbf{a})}{\hat{\delta} a_m}$ represents differentiation of the function $f(\mathbf{a})$ with respect to the mth element of the vector of the independent variables $\mathbf{a}$ performed by an automatic differentiation algorithm. The result is an algorithm that, when evaluated in a loop with the loop counter m, yields the elements of the gradient vector $\nabla_{\mathbf{a}} f$.

2.5 Automatic Code Generation with AceGen

2.5.1 *Hybrid Symbolic-Numerical System AceGen*

The idea implemented in *AceGen* is not to try to combine different tools, but to combine different techniques inside one system in order to avoid the above mentioned problems. Thus, the main objective is to combine techniques in such a way that will lead to an optimal environment for the design and coding of numerical subroutines. Among the presented systems the most versatile are indeed the SAC systems. They normally contain, besides algebraic manipulation, graphics and numerical capabilities, also powerful programming languages. It is therefore quite easy to simulate other techniques inside the SAC system. The approach of automatic code generation used in *AceGen* is called *Simultaneous Stochastic Simplification of numerical code* (Korelc 1997a). This approach combines the general computer algebra system *Mathematica* with an automatic differentiation technique and an automatic theorem proving by examples. To alleviate the problem of the growth of expressions and redundant calculations, simultaneous simplification of symbolic expressions is used. Stochastic evaluation of the formulae is applied for determining the equivalence of algebraic expressions, instead of the conventional pattern matching technique. *AceGen* was designed to approach especially to hard problems, where the general strategy for an efficient formulation of numerical procedures, such as analytical sensitivity analysis of complex multi-field problems, has not yet been established.

The structure of *AceGen* is presented in Fig. 2.1. General characteristics of the *AceGen* code generator are:

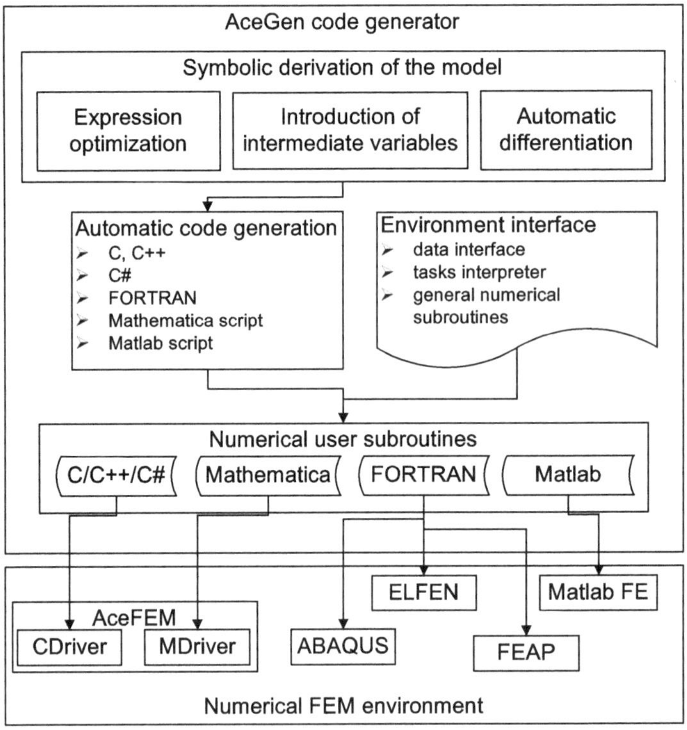

Fig. 2.1 Hybrid symbolic-numeric approach to automation of the finite element method

1. simultaneous optimization of expressions immediately after they have been derived,
2. automatic differentiation technique,
3. automatic selection of the appropriate intermediate variables,
4. generation of the whole program structure,
5. appropriate for large problems where also intermediate expressions can be subjected to the uncontrolled swell,
6. global expression optimization with stochastic evaluation of expressions,
7. differentiation with respect to indexed variables,
8. automatic interface to other numerical environments (by using the Splice command of *Mathematica*),
9. multi-language code generation (FORTRAN/FORTRAN90, C/C++/C#, *Mathematica* language, *Matlab* language),
10. advanced methods for exploring and debugging of generated formulas,
11. system is written in the symbolic language of *Mathematica.*

2.5.2 Typical AceGen Automatic Code Generation Procedure

A simple example is considered here to illustrate the standard *AceGen* procedure. It is a subroutine that returns the determinant of the Jacobi matrix of a nonlinear transformation from the reference to initial configuration for quadrilateral element topology (for details see Sect. 4.2.1). The syntax of the *AceGen* script language is the same as the syntax of the *Mathematica* script language with some additional functions. A short description of the *AceGen* syntax is given in Appendix A while the detailed description can be found in Korelc (2011). The input for *AceGen* that produces the required subroutine is as follows

```
<<AceGen`;
SMSInitialize["DetJSub","Language"->"C"];
SMSModule["DetJ",Real[X$$[2,4],k$$,e$$,Jg$$]];
{ξ,η}⊢SMSReal[{k$$,e$$}];
{𝕏c,𝕐c}⊢Table[SMSReal[X$$[i,j]],{i,2},{j,4}];
Nh⊨{(1-ξ) (1-η),(1+ξ) (1-η),(1+ξ) (1+η),(1-ξ) (1+η)}/4;
Jg⊨SMSD[{Nh.𝕏c,Nh.𝕐c},{ξ,η}];
SMSExport[Det[Jg],J$$];
SMSWrite["LocalAuxiliaryVariables"→True];
```
(2.15)

This input can be divided into six characteristic steps:

Step 1: Initialization

```
<<AceGen`;
SMSInitialize["DetJSub","Language"->"C"];
```

At the beginning of the session the `SMSInitialize` function initializes the system and specifies the name of the generated source code file. The additional option `"Language"->"C"` specifies that the subroutine is generated in C language.

Step 2: Definition of input and output parameters

```
SMSModule["DetJ",Real[X$$[2,4],k$$,e$$,Jg$$]];
```

The `SMSModule` function defines the name (`DetJ`) and input/output parameters of the subroutine. Input parameter `X$$` is a 2×4 matrix of nodal coordinates, `k$$` and `e$$` are ξ and η coordinates of the material point in reference frame and `Jg$$` is an output parameter of the subroutine. Double `$` character always indicates that the symbol is an input or output parameter of the generated subroutine.

Step 3: Definition of numeric-symbolic interface variables

```
{ξ,η}⊢SMSReal[{k$$,e$$}];
{𝕏c,𝕐c}⊢Table[SMSReal[X$$[i,j]],{i,2},{j,4}];
```

The `SMSReal` function assigns random signature to input parameters `k$$` and `e$$`. The $\vdash$ operator then creates new auxiliary variables ξ and η. Similarly, vectors of auxiliary variables `𝕏c` and `𝕐c` hold X and Y coordinates of element nodes.

Step 4: Derivation of the problem

```
ℕh⊨{(1-ξ) (1-η),(1+ξ) (1-η),(1+ξ) (1+η),(1-ξ) (1+η)}/4;
Jg⊨SMSD[{ℕh.𝕏c,ℕh.𝕐c},{ξ,η}];
```

During the description of the problem a special operator $\models$ is used that performs global optimization of expressions and then creates new auxiliary variables if *AceGen* finds out that the introduction of a new auxiliary variable is necessary. The `SMSD` function performs the automatic differentiation of one or several expressions with respect to one or a vector of auxiliary variables by simultaneously enhancing the already derived code.

Step 5: Exporting results to output parameters

```
SMSExport[Det[Jg],Jg$$];
```

The results of the derivation are assigned to the output parameter `Jg$$` of the subroutine by `SMSExport` function.

Step 6: Code generation

```
SMSWrite["LocalAuxiliaryVariables"→True];
```

At the end of the session the `SMSWrite` function writes the contents of the vector of the generated formulas to file `DetJSub.c` in a prescribed language format.

Due to the advantage of the simultaneous optimization procedure we can execute each step separately and examine intermediate results. This is also the basic way how to trace errors that might occur during the *AceGen* session. The generated source code in C language is as follows.

```
/******************* S U B R O U T I N E *********************/
void DetJ(double X[2][4],double (*k),double (*e),double (*Jg))
{
double v[36];
v[31]=(-1e0+(*e))/4e0;
v[30]=(1e0+(*e))/4e0;
v[29]=(-1e0-(*k))/4e0;
v[28]=(-1e0+(*k))/4e0;
(*J)=(v[31]*(X[0][0]-X[0][1])+v[30]*(X[0][2]-X[0][3]))*(v[29]*(X[1][1]-
X[1][2])+v[28]*(X[1][0]
 -X[1][3]))-(v[29]*(X[0][1]-X[0][2])+v[28]*(X[0][0]-X[0][3]))*(v[31]*(X[1][0]-
X[1][1])+v[30]*
 (X[1][2]-X[1][3]));
};
```

With the change of language option to `"Language"->"Fortran"` the following source code in FORTRAN language is generated.

```
!******************* S U B R O U T I N E **********************
      SUBROUTINE DetJ(X,k,e,Jg)
      IMPLICIT NONE
      include 'sms.h'
      DOUBLE PRECISION v(36),X(2,4),k,e,Jg
      v(31)=((-1d0)+e)/4d0
      v(30)=(1d0+e)/4d0
      v(29)=((-1d0)-k)/4d0
      v(28)=((-1d0)+k)/4d0
      J=(v(31)*(X(1,1)-X(1,2))+v(30)*(X(1,3)-X(1,4)))*(v(29)*(X(2,2)
     &-X(2,3))+v(28)*(X(2,1)-X(2,4)))-(v(29)*(X(1,2)-X(1,3))+v(28)*(X
     &(1,1)-X(1,4)))*(v(31)*(X(2,1)-X(2,2))+v(30)*(X(2,3)-X(2,4)))
      END
```

2.5.3 Simultaneous Simplification Procedure

During the description of the problem the special operator $\models$ was used to perform the simultaneous optimization of expressions and the creation of new intermediate variables.

A typical *AceGen* function takes the expression provided by the user, either interactively or from a file, and returns an optimized version of the expression. Optimized version of the expression can result in a newly created auxiliary symbol (v_i), or in an original expression in parts replaced by previously created auxiliary symbols. In the first case, *AceGen* stores the new expression in an internal database. The database contains a global vector of all expressions, information about dependencies of the symbols, labels and names of the symbols, partial derivatives, etc. The database is a global object that maintains information during the *AceGen* session.

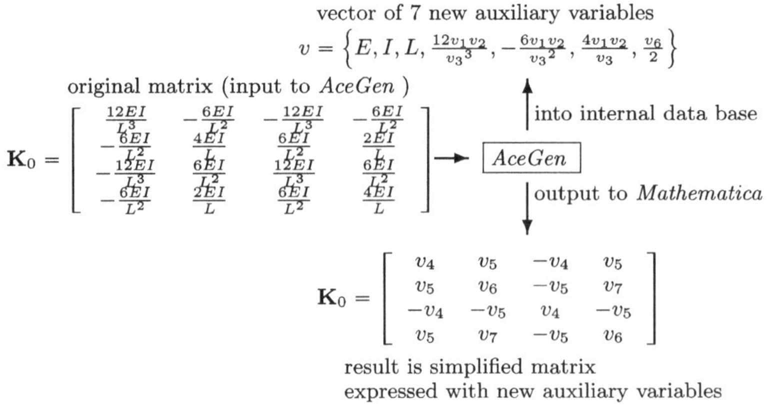

Fig. 2.2 Simultaneous simplification procedure

The classical way of optimizing expressions in a computer algebra systems is searching for common sub-expressions at the end of the derivation, before the generation of the numerical code. In the numerical code common sub-expressions appear as auxiliary variables. An alternative approach is implemented in *AceGen* where formulas are optimized, simplified and replaced by the auxiliary variables simultaneously with the derivation of the problem. The optimized version is then used in further operations (Fig. 2.2). When the optimization is performed simultaneously, the explicit form of an expression is obviously lost, since some parts are replaced by intermediate variables.

In real problems it is almost impossible to recognize the identity of two expressions (for example the symmetry of the tangent stiffness matrix in nonlinear mechanical problems) automatically using only the pattern matching mechanisms. Normally our goal is to recognize the identity automatically without introducing additional knowledge into the derivation such as tensor algebra, matrix transformations, etc. Commands in *Mathematica* such as `Simplify`, `Together`, and `Expand`, are useless in the case of large expressions. Additionally, these commands are efficient only when the whole expression is considered. When optimization is performed simultaneously, the explicit form of the expression is lost. The only possible way at this stage of computer technology seems to be an algorithm which finds equivalence of expressions numerically. This relatively old idea (see for example Martin 1971 or Gonnet 1986) is rarely used, although it is essential for dealing with especially hard problems. However, to show identity numerically is not a mathematically rigorous proof for the identity of two expressions. Thus the correctness of the simplification can be determined only within a certain degree of probability. With regard to our experience the proof of numerical identity of expressions is sufficient in mechanical analysis when dealing with more or less "smooth" functions.

2.5.4 *Efficiency and Limitations of Automation of Computational Modeling*

An important question arises: how to understand and analyze automatically generated formulas? The automatically generated codes should not act like a "black box". For example, after using the automatic differentiation tools we have no insight in the actual structure of the derivatives. One might argue that this is not needed, however if the finite element codes are to be derived with an efficiency comparable with the manually written codes then an efficient debugging procedure becomes a necessity. The standard way in which *Mathematica* works is to evaluate a single expression completely, which then can be analyzed and printed in a required format. If the problem is complex with potentially hundreds of expressions and it includes loops and branching, then more sophisticated procedures for the analysis and interactive run time debugging are required. Additionally, when optimization of the derived numerical code is performed simultaneously with the derivation of formulas, the explicit form of the expressions is lost.

The automatically generated finite element subroutines are difficult to analyze and to understand directly. Thus, it is in general difficult to debug the generated subroutine after it is included into the commercial finite element environment. For an efficient interactive debugging it is important where the debugging actually takes place. The *Mathematica* based high-level symbolic interface standard for inter-program communication called *MathLink* allows to run external programs under whatever debugger is provided for a particular compiler. However, this approach is not applicable for automatically generated codes, since the structure of the automatically generated codes is rather "unreadable", hence difficult to be debugged directly. The approach where the generated code is not debugged directly but where the programmer can look at the analytical expressions and their current numerical values when using the symbolic environment is essential for the debugging of automatically generated codes.

2.6 Automatic Differentiation and Finite Element Method

The AD tools were primarily developed for the evaluation of the gradient of an objective function used within gradient-based optimization procedures. In general, the objective function can be defined by a program composed of many subroutines including a complete FE environment. Thus, one can apply the AD tools directly to the complete FE environment to obtain the required derivatives when the evaluation of the objective function involves FE simulation. The AD tools have been successfully applied to FE environments with several 100,000 lines of code; see e.g. Bischof et al. (2003).

A large finite element environment usually employs a variety of finite elements, solution procedures. Commonly commercial numerical libraries are used within such system for which the source codes are not readily available. For these finite element environments it is difficult to directly apply the AD tools to get e.g. the global stiffness matrix of a large-scale problem. However, the AD technology can still be used for the evaluation of specific quantities that appear as a part of the FE simulation. For example, one can use AD at the individual element level to evaluate element specific quantities such as:

- strain and stress tensors,
- nonlinear coordinate transformations,
- residual vectors,
- consistent tangent stiffness matrices and
- sensitivity pseudo-load vectors.

A direct use of automatic differentiation tools for the development of nonlinear finite elements turns out to be complex and not straightforward. Furthermore the numerical efficiency of the resulting codes is poor. One solution, followed mostly in hybrid object-oriented systems, is to use problem specific solutions to evaluate local tangent matrix in a optimal way, see. e.g. Kirby et al. (2005). Another solution, pursued in hybrid symbolic-numeric systems, is to combine a general computer algebra system and the AD technology, see e.g. Korelc (2002).

An implementation of the AD procedure has to fulfill specific requirements in order to obtain element source code that is as efficient as manually written codes. Some basic requirements are:

- The AD procedure can be initiated at any time and at any point of the derivation of the formulas and as many times as required (e.g. in the example at the end the AD is used 13 times during the generation of an element subroutine). The recursive use of standard AD tools on the same code, if allowed at all, leads to numerically inefficient source code. This requirement limits the use of standard AD tools. An alternative approach is implemented in Korelc (2002) where the source-to-source transformation strategy is replaced by a method that consistently enhances the existing code rather than producing a new one.
- The storage of the intermediate variables is not a limitation when the backward differentiation method is employed at single element level. Finite element formulations at the single element level involve a relatively small set of independent and intermediate variables.
- For reasons of efficiency, the results of all previous uses of AD have to be accounted for when automatic differentiation is applied several times inside the same subroutine.
- The user has to be able to use all capabilities of the symbolic system on the final and the intermediate results of the AD procedure.
- The AD procedure must offer a mechanism for the descriptions of various mathematical formalisms applied within a finite element formulation.

In implicit methods, the bulk of computational time is spent on performing differentiations, thus the efficiency of the automation of differentiation is most important. Two effects that have to be considered:

- the number of recursive uses of differentiation within the derivation of the same subroutine and
- the approach employed to perform differentiation.

As presented in Sect. 2.4 there are two approaches for the implementation of the automatic differentiation of a computer program, the forward and the backward mode of automatic differentiation. The numerical efficiency of the differentiation of N scalar-valued functions with respect to M independent variables can be measured by the numerical work ratio (2.4). The ratio is in general proportional to the number of independent variables in the case of forward mode and proportional to the number of functions in the case of backward mode (see also Sect. 2.4). For the efficient automation of the FE method, it is desirable that both approaches are available and that the software tool used for automation can automatically select the most efficient approach according to the estimated work ratio, though the backward mode would be preferable in most of the cases.

Contrary to the classical implementation of automatic differentiation where the AD works as a code-to-code translator, the implementation in *AceGen* enhances the currently derived code. However, the result of repeated use of AD within the same subroutine is that the simultaneous code optimization procedures become less and less efficient. To summarize, the automatically generated code would be numerically efficient if the number of functions to differentiate and the number of calls to the AD procedure is kept at minimum. One consequence of the rule is that in general formulations where the element residual vector is derived as the gradient of a scalar function, e.g. variational potential, leads to a more efficient numerical code than formulations based on the weak form of the equilibrium equations where the variation of the kinematic quantities e.g. strains tensor or tangential gap vector requires differentiation of several scalar functions. Thus, the possibility of transforming the weak form into the "pseudo-potential" scalar function is always worth to explore. The pseudo-potential has to be formed in a way that the automatic differentiation procedure of the pseudo-potential accompanied with the proper automatic differentiation exceptions leads to the correct equations of the problem.

The book demonstrates by means of several examples direct consequences of using automatic differentiation exceptions. It will show how the mechanical problem and its corresponding numerical model are formulated and solved. The ADB form of the problem description is rather straightforward for e.g. total-Lagrangian displacement-based finite element formulation of hyperelastic problems. However, it can be nontrivial for e.g. spatial formulation of finite strain elasto-plastic finite elements.

2.6.1 ADB Form of General Potential Form

Assume that the solution of a problem is defined as the minimum of the potential $\Pi(\mathbf{p}) = \int_{\Omega} W(\mathbf{p})\, d\Omega$ where $\mathbf{p} = \{p_1\,,\, p_2\,,\, \ldots\,,\, p_{n_{tp}}\}^T$ is a set of unknown parameters of the problem. The variation of $\Pi(\mathbf{p})$ is computed as

$$\delta \Pi(\mathbf{p}) = \frac{\partial \Pi(\mathbf{p})}{\partial \mathbf{p}} \delta \mathbf{p} = \int_{\Omega} \frac{\partial W(\mathbf{p})}{\partial \mathbf{p}} \, d\Omega \, \delta \mathbf{p} = 0 \tag{2.16}$$

where $\delta \mathbf{p} = \{\delta p_1, \delta p_2, \ldots, \delta p_{n_{tp}}\}^T$ is a variation of unknown parameters. Equation (2.16) leads to a set of nonlinear algebraic equations of the form

$$\mathbf{R} = \int_{\Omega} \frac{\partial W(\mathbf{p})}{\partial \mathbf{p}} \, d\Omega = \mathbf{0}. \tag{2.17}$$

The ADB form of the general problem with potential is obtained from (2.17) where the partial derivative is directly replaced by the computational derivative

$$\mathbf{R} = \int_{\Omega} \frac{\hat{\delta} W(\mathbf{p})}{\hat{\delta} \mathbf{p}} \, d\Omega = \mathbf{0}. \tag{2.18}$$

Furthermore, the individual element contribution $\mathbf{R}_e$ to the global residual vector $\mathbf{R}$ is in standard finite element formulations obtained by a numerical integration rule—typically Gauss integration

$$\mathbf{R}_e \approx \sum_{g=1}^{n_g} w_{gp} \, \mathbf{R}_g \tag{2.19}$$

where w_{gp} stands for the Gauss point weights, n_g is the number of Gauss points and $\mathbf{R}_g$ is the Gauss point contribution to the residual vector or **Gauss** point residual given by

$$\mathbf{R}_g = \frac{\hat{\delta} W(\mathbf{p})}{\hat{\delta} \mathbf{p}} \tag{2.20}$$

The corresponding ADB form of the Gauss point contribution to the global tangent matrix $\mathbf{K}$ is

$$\mathbf{K}_g = \frac{\hat{\delta} \mathbf{R}_g}{\hat{\delta} \mathbf{p}_e}. \tag{2.21}$$

A schematic *AceGen* input for the general potential form (2.20) and (2.21) is provided by the following *AceGen* input segment

```
𝕡e⊢SMSReal[Table[p$$[i],{i,1,np}]];
W⊨fW[𝕡e];
ℝg⊨SMSD[W,𝕡e];
𝕂g⊨SMSD[ℝg,𝕡e];
```

(2.22)

2.6.2 ADB Form of General Weak Form

Assume the weak form $\int_{\Omega} \mathbf{a} \cdot \delta \mathbf{b}\, d\Omega + \cdots = 0$ where $\mathbf{a}$ and $\mathbf{b}$ are tensors of an arbitrary order and $\delta \mathbf{b}$ is a directional derivative or variation of $\mathbf{b}$. The symbolic formulation of the weak form is not straightforward, since $\delta \mathbf{b}$ is not a real but rather fictitious quantity and AD cannot be applied directly.[2] But AD can be applied directly after the weak form is discretized. The variation of $\mathbf{b}$ is computed as

$$\delta \mathbf{b}(\mathbf{p}) = D\,\mathbf{b}(\mathbf{p})\,\delta \mathbf{p} = \frac{\partial \mathbf{b}(\mathbf{p})}{\partial \mathbf{p}}\,\delta \mathbf{p} \tag{2.23}$$

and the scalar product $\mathbf{a}(\mathbf{p}) \cdot \delta \mathbf{b}(\mathbf{p})$ as

$$\mathbf{a}(\mathbf{p}) \cdot \delta \mathbf{b}(\mathbf{p}) = \mathbf{a}(\mathbf{p}) \cdot \frac{\partial \mathbf{b}(\mathbf{p})}{\partial \mathbf{p}} \delta \mathbf{p} = \left(\mathbf{a}(\mathbf{p}) \cdot \frac{\partial \mathbf{b}(\mathbf{p})}{\partial \mathbf{p}} \right) \delta \mathbf{p}. \tag{2.24}$$

The product $\mathbf{a}(\mathbf{p}) \cdot \frac{\partial \mathbf{b}(\mathbf{p})}{\partial \mathbf{p}}$ can also be written component wise

$$\left(\mathbf{a}(\mathbf{p}) \cdot \frac{\partial \mathbf{b}(\mathbf{p})}{\partial \mathbf{p}} \right)_m = \mathbf{a}(\mathbf{p}) \cdot \frac{\partial \mathbf{b}(\mathbf{p})}{\partial p_m} = tr\left(\mathbf{a}(\mathbf{p})^T \frac{\partial \mathbf{b}(\mathbf{p})}{\partial p_m} \right) \tag{2.25}$$

by using the trace operator, see Appendix B.1.3 and B.1.5. The discretized weak form is then given by

$$\int_{\Omega} \mathbf{a}(\mathbf{p}) \cdot \delta \mathbf{b}(\mathbf{p})\, d\Omega + \cdots = \sum_{m=1}^{n_{tp}} \left(\int_{\Omega} \mathbf{a}(\mathbf{p}) \cdot \frac{\partial \mathbf{b}(\mathbf{p})}{\partial p_m}\, d\Omega \right) \delta p_m + \cdots = 0 \tag{2.26}$$

and leads to a set of n_{tp} algebraic equations of the form

$$\mathbf{R} = \int_{\Omega} \mathbf{a}(\mathbf{p}) \cdot \frac{\partial \mathbf{b}(\mathbf{p})}{\partial \mathbf{p}}\, d\Omega + \cdots = \mathbf{0}. \tag{2.27}$$

The ADB form of the discretized weak form is obtained from (2.27) where the partial derivative is directly replaced by the computational derivative

$$\mathbf{R} = \int_{\Omega} \mathbf{a}(\mathbf{p}) \cdot \frac{\hat{\delta} \mathbf{b}(\mathbf{p})}{\hat{\delta} \mathbf{p}}\, d\Omega + \cdots = \mathbf{0}. \tag{2.28}$$

---

[2] Of course, a general computer algebra system can be used for building the necessary apparatus to deal with the virtual quantities in a traditional way and then automatic code generation can be applied on the results. However, the elegance of using automatic differentiation would then be lost.

The vector $\frac{\hat{\delta}\mathbf{b}(\mathbf{p})}{\hat{\delta}\mathbf{p}}$ in Eq. (2.28) has to be derived explicitly (all n_{tp} components in closed form) for the evaluation of the scalar product. This operation can lead to an expression swell problem. Additionally, small differences in the actual form of the derived expressions can disrupt the code optimization procedure. The form (2.28) is also not optimal for the use of backward mode of AD since the cost of backward AD depends linearly on the number of scalar functions to be differentiated. For the present case, the numerical cost depends linearly on the number of components of **b**. Thus, as mentioned before, the possibility of transforming the weak form into the pseudo-potential scalar function is worth exploring. The pseudo-potential has to be formed in a way that automatic differentiation of the pseudo-potential accompanied by proper automatic differentiation exceptions leads to the correct set of discretized equations of the problem. For the case (2.26) this can be achieved by introducing the AD exception that hides the dependency $\mathbf{a}(\mathbf{p})$ from the AD procedure as follows

$$\mathbf{R} = \int_{\Omega} \mathbf{a}(\mathbf{p}) \cdot \frac{\hat{\delta}\mathbf{b}(\mathbf{p})}{\hat{\delta}\mathbf{p}} d\Omega + \cdots = \int_{\Omega} \left. \frac{\hat{\delta}(\mathbf{a}(\mathbf{p}) \cdot \mathbf{b}(\mathbf{p}))}{\hat{\delta}\mathbf{p}} \right|_{\frac{D\mathbf{a}}{D\mathbf{p}}=\mathbf{0}} d\Omega + \cdots = \mathbf{0}. \tag{2.29}$$

With the introduction of the pseudo-potential

$$W^P = \mathbf{a}(\mathbf{p}) \cdot \mathbf{b}(\mathbf{p}) = tr(\mathbf{a}(\mathbf{p})^T \mathbf{b}(\mathbf{p})) \tag{2.30}$$

the final ADB form of the discretized weak form (2.26) is obtained as

$$\mathbf{R} = \int_{\Omega} \left. \frac{\hat{\delta} W^P(\mathbf{p})}{\hat{\delta}\mathbf{p}} \right|_{\mathbf{a}=\text{const.}} d\Omega + \cdots = \mathbf{0}. \tag{2.31}$$

Assuming that the integral is evaluated by a numerical integration rule as in the previous section, the Gauss point contribution to the residual vector or **Gauss** point residual leads to

$$\mathbf{R}_g = \mathbf{a}(\mathbf{p}_e) \cdot \frac{\hat{\delta}\mathbf{b}(\mathbf{p}_e)}{\hat{\delta}\mathbf{p}_e} \tag{2.32}$$

or when using the pseudo-potential form to

$$\mathbf{R}_g = \left. \frac{\hat{\delta} W^P(\mathbf{p}_e)}{\hat{\delta}\mathbf{p}_e} \right|_{\mathbf{a}=\text{const.}} . \tag{2.33}$$

The corresponding ADB form of Gauss point contribution to the global tangent matrix **K** is for both cases given by

$$\mathbf{K}_g = \frac{\hat{\delta}\mathbf{R}_g}{\hat{\delta}\mathbf{p}_e}. \tag{2.34}$$

The above procedure is valid for tensors of arbitrary rank. A schematic *AceGen* input for the standard weak form (2.32) for the case when **a** and **b** are scalars is

```
𝕡e⊢SMSReal[Table[p$$[i],{i,1,np}]];
a⊨fa[𝕡e];
b⊨fb[𝕡e];
δb⊨SMSD[b,𝕡e];
ℝg⊨a δb;
𝕂g⊨SMSD[ℝg,𝕡e];
```

(2.35)

in the case of rank-one tensors it can be written as

```
𝕡e⊢SMSReal[Table[p$$[i],{i,1,np}]];
𝕒⊨fa[𝕡e];
𝕓⊨fb[𝕡e];
δ𝕓⊨SMSD[𝕓,𝕡e];
ℝg⊨𝕒.δ𝕓;
𝕂g⊨SMSD[ℝg,𝕡e];
```

(2.36)

and in the case of rank-two tensors (2.32) leads to

```
𝕡e⊢SMSReal[Table[p$$[i],{i,1,np}]];
𝕒⊨fa[𝕡e];
𝕓⊨fb[𝕡e];
δ𝕓⊨Transpose[SMSD[𝕓,𝕡e],{2,3,1}];
ℝg⊨Table[Tr[𝕒ᵀ δ𝕓⟦i⟧],{i,1,Length[𝕡e]}];
𝕂g⊨SMSD[ℝg,𝕡e];
```

(2.37)

Note that in the case of rank-two tensors, the presented formulation requires only one execution of the AD procedure.

Similarly, a schematic *AceGen* input for the pseudo-potential ADB form (2.33) of the generalized weak form for the scalar product leads to[3]

```
𝕡e⊢SMSReal[Table[p$$[i],{i,1,np}]];
a⊢SMSFreeze[fa[𝕡e]];
b⊨fb[𝕡e];
WP⊨a b;
ℝg⊨SMSD[WP,𝕡e,"Constant"→a];
𝕂g⊨SMSD[ℝg,𝕡e];
```

(2.38)

[3]By using the SMSFreeze operator a new auxiliary variable is created that can be safely used as independent variable within differentiation, see also Appendix A.2.2.

for the rank-one tensors to

```
𝕡𝕖⊢SMSReal[Table[p$$[i],{i,1,np}]];
𝕒⊢SMSFreeze[fa[𝕡𝕖]];
𝕓⊨fb[𝕡𝕖];
WP⊨𝕒.𝕓;
ℝg⊨SMSD[WP,𝕡𝕖,"Constant"→𝕒];
𝕂g⊨SMSD[ℝg,𝕡𝕖];
```
(2.39)

and in the case of rank-two tensors to

```
𝕡𝕖⊢SMSReal[Table[p$$[i],{i,1,np}]];
𝕒⊢SMSFreeze[fa[𝕡𝕖]];
𝕓⊨fb[𝕡𝕖];
WP⊨Tr[𝕒ᵀ.𝕓];
ℝg⊨SMSD[WP,𝕡𝕖,"Constant"→𝕒];
𝕂g⊨SMSD[ℝg,𝕡𝕖];
```
(2.40)

Note that in the case when **b** and **a** are scalars, the efficiency of the pseudo-potential ADB form (2.33) is the same as efficiency of the standard ADB weak form (2.32).

2.6.3 *Representative Formulas for Residual and Tangent Matrix*

If n_p is the number element global degrees of freedom (DOF) then $\mathbf{R}_g$ has n_p components and $\mathbf{K}_g$ has n_p^2 components. The physical size of the automatically generated code is consequently proportional to the square of the number of element DOF's. This can lead to a swell of expressions and is not acceptable for higher order elements. Thus, an explicit code for all terms of the residual and tangent matrix, as presented in Sect. 2.6.1 for a general potential based formulations and in Sect. 2.6.2 for a general weak form formulations, can be generated only if the number of unknowns is small. The size of the generated code can be reduced by the generation of representative formulas as described in Sect. 2.4.5.

The characteristic mth term of the Gauss point residual is given by

$$R_{gm} = \frac{\hat{\delta} W(\mathbf{p}_e)}{\hat{\delta} p_{em}}. \tag{2.41}$$

for the potential based form (2.20) and for the weak form (2.32) by

$$R_{gm} = \mathbf{a}(\mathbf{p}) \cdot \frac{\hat{\delta}\mathbf{b}(\mathbf{p})}{\hat{\delta} p_{em}} = tr\left(\mathbf{a}(\mathbf{p})^T \frac{\hat{\delta}\mathbf{b}(\mathbf{p})}{\hat{\delta} p_{em}}\right) \tag{2.42}$$

and for the pseudo-potential form (2.33) by

$$R_{gm} = \left.\frac{\hat{\delta} W^P(\mathbf{p}_e)}{\hat{\delta} p_{em}}\right|_{\mathbf{a}=\text{const.}} . \tag{2.43}$$

The corresponding representative formula for the $(m, n)^{th}$ term of the tangent matrix at a Gauss point is for all cases given by

$$K_{gmn} = \frac{\hat{\delta} R_{gm}}{\hat{\delta} p_{en}}. \tag{2.44}$$

A schematic *AceGen* input where one representative formula is generated for an arbitrary element of the residual and one representative formula for an arbitrary element of the tangent matrix for the standard weak form (2.42) is presented in Box 2.1.

```
pe⊢SMSReal[Table[p$$[i],{i,1,np}]];
W⊨fW[pe];
SMSDo[m,1,np]
  Rgm⊨SMSD[W,pe,m];
  SMSExport[wgp Rgm,R$$[m],"AddIn"→True];
  SMSDo[n,1,np]
    Kgmn⊨SMSD[Rgm,pe,n];
    SMSExport[wgp Kgmn,K$$[m,n],"AddIn"→True];
  SMSEndDo[];
SMSEndDo[];
```

Box 2.1. *AceGen* input for generation of representative formulas for a general potential based *ADB* formulation

The *AceGen* command `SMSD[W,pe,m]` in Box 2.1 performs automatic differentiation of the potential W with respect to the mth element of the set of unknowns $\mathbf{p}_e$. The resulting representative formula is evaluated within the `SMSDo[..., {m,1,np}]` loop n_p times. Since the successive evaluations override the results of the previous evaluations, the results have to be simultaneously stored or exported by the `SMSExport` command into externally defined or allocated array (`$$R`). The same is true for the inner loop that evaluates the elements of the tangent matrix.

A similar schematic *AceGen* input is presented for the general weak *ADB* formulation (2.42) in Box 2.2 and for the general pseudo-potential *ADB* formulation (2.43) in Box 2.3.

```
pe⊢SMSReal[Table[p$$[i],{i,1,np}]];
a⊨fa[pe];b⊨fb[pe];
SMSDo[
  δbm⊨SMSD[b,pe,m];
  Switch[order
   ,"scalars",Rgm⊨a δbm ;
   ,"rank-one tensors",Rgm⊨a.δbm;
   ,"rank-two tensors",Rgm⊨Tr[aᵀ.δbm];
  ];
  SMSExport[wgp Rgm,R$$[m],"AddIn"→True];
  SMSDo[
   Kgmn⊨SMSD[Rgm,pe,m];
   SMSExport[wgp Kgmn,K$$[m,n],"AddIn"→True];
   ,{n,1,np}]
  ,{m,1,np}];
```

Box 2.2. *AceGen* input for generation of representative formulas for a general weak form *ADB* formulation

```
pe⊢SMSReal[Table[p$$[i],{i,1,np}]];
a⊢SMSFreeze[fa[pe]];b⊨fb[pe];
Switch[order
  ,"scalars",WP⊨a b;
  ,"rank-one tensors",WP⊨a.b;
  ,"rank-two tensors",WP⊨Tr[aᵀ.b];
 ];
SMSDo[
  Rgm⊨SMSD[WP,pe,m,"Constant"→a];
  SMSExport[wgp Rgm,R$$[m],"AddIn"→True];
  SMSDo[
   Kgmn⊨SMSD[Rgm,pe,n];
   SMSExport[wgp Kgmn,K$$[m,n],"AddIn"→True];
   ,{n,1,np}]
  ,{m,1,np}];
```

Box 2.3. *AceGen* input for generation of representative formulas for a general pseudo-potential *ADB* formulation

Representative formulas for multi-field problems. The set of unknowns $\mathbf{p}_e$ of a finite element is commonly used to discretize more than one scalar field (e.g. three displacements u, v and w). In that case the generation of one very general representative formula for an arbitrary element of the residual and tangent matrix can lead to redundant numerical operations and consequently slower codes. The solution of the problem is to split a set of independent variables $\mathbf{p}_e$ into a union of disjoint subsets in a way that each subset discretizes only one scalar field. The representative formulas are then generated for an arbitrary element of a particular subset.

The kth subset of unknowns, denoted by $\mathbf{p}_{ek}$, is defined by

$$\mathbf{p}_{ek} \subset \mathbf{p}_e, \quad \bigcup_{k=1}^{n_s} \mathbf{p}_{ek} = \mathbf{p}_e, \quad \mathbf{p}_{ek} \bigcap \mathbf{p}_{el} = \mathbf{0} : \forall k \neq l \tag{2.45}$$

where n_s is the number of subsets. Let n_{pk} be the length of the kth subset of unknowns $\mathbf{p}_{ek}$, p_{ekm_k} the m_kth element of $\mathbf{p}_{ek}$ and $\bar{R}_{gm_k}$ a representative formula for an arbitrary element of a subset of the residual $\mathbf{R}_g$ that belongs to the subset of unknowns $\mathbf{p}_{ek}$. A set of n_s representative formulas for the evaluation of the residual is then given by

$$\bar{\mathbf{R}}_g = \left\{ \bar{R}_{gm_k} = \frac{\hat{\delta} W(\mathbf{p}_e)}{\hat{\delta}(p_{ekm_k})} : k = 1, \ldots, n_s \right\}^T . \tag{2.46}$$

In the same way a set of $n_s \times n_s$ representative formulas is obtained for the evaluation of the tangent matrix

$$\bar{\mathbf{K}}_g = \left[\bar{K}_{gm_k n_l} = \frac{\hat{\delta} \bar{R}_{gm_k}}{\hat{\delta}(p_{eln_l})} : k = 1, \ldots, n_s, \ l = 1, \ldots, n_s \right] . \tag{2.47}$$

Consider an example where a set of unknowns is split into two subsets with the lengths n_{p1} and n_{p2} as follows

$$\begin{aligned} \mathbf{p}_{e1} &= \left\{ p_{e1i}, \ i = 1, \ldots, n_{p1} \right\}^T . \\ \mathbf{p}_{e2} &= \left\{ p_{e2i}, \ i = 2, \ldots, n_{p2} \right\}^T . \end{aligned} \tag{2.48}$$

The set $\bar{\mathbf{R}}_g$ contains in this case two representative formulas and a matrix of representative formulas for the evaluation of the tangent matrix $\bar{\mathbf{K}}_g$ has dimension 2×2. A schematic *AceGen* input for the evaluation of the residual and tangent matrix for the general two subset case is given in Box 2.4.

```
pe1⊢SMSReal[Table[p$$[i],{i,np1}]];
pe2⊢SMSReal[Table[p$$[np1+i],{i,np2}]];
W⊨fW[pe1,pe2];
SMSDo[m1,1,np1];
  Rgm1⊨SMSD[W,pe1,m1];
  SMSExport[wgp Rgm1,p$$[m1],"AddIn"→True];
  SMSDo[n1,1,np1];
    Kgm1n1⊨SMSD[Rgm1,pe1,n1];
    SMSExport[wgp Kgm1n1,s$$[m1,n1],"AddIn"→True];
  SMSEndDo[];
  SMSDo[n2,1,np2];
    Kgm1n2⊨SMSD[wgp Rgm1,pe2,n2];
    SMSExport[Kgm1n2,s$$[m1,np1+n2],"AddIn"→True];
  SMSEndDo[];
SMSEndDo[];
SMSDo[m2,1,np2];
  Rgm2⊨SMSD[W,pe2,m2];
  SMSExport[wgp Rgm2,p$$[np1+m2],"AddIn"→True];
  SMSDo[n1,1,np1];
    Kgm2n1⊨SMSD[Rgm2,pe1,n1];
    SMSExport[wgp Kgm2n1,s$$[np1+m2,n1],"AddIn"→True];
  SMSEndDo[];
  SMSDo[n2,1,np2];
    Kgm2n2⊨SMSD[Rgm2,pe2,n2];
    SMSExport[wgp Kgm2n2,s$$[np1+m2,np1+n2],"AddIn"→True];
  SMSEndDo[];
SMSEndDo[];
```

Box 2.4. *AceGen* input for generation of representative formulas for a general two-subsets case

The algorithm presented in in Box 2.4 contains 6 loops and is rather complex. The algorithm can be significantly simplified in a special case when a set of unknowns is divided into subsets of equal lengths, thus $n_{p1} = n_{p2} = \frac{n_p}{2}$. A schematic *AceGen* input for the evaluation of the residual and tangent matrix for the a two-subsets case with the same length is given in Box 2.5. An example for the first case are continuum elements based on mixed variational principles were unknowns can be divided into two subsets: a subset of displacement degrees of freedom and a subset of mixed modes. An example for the second case are isoparametric continuum elements where subsets can be formed based on nodal degrees of freedom. An extensive study how different implementations of the same element effect efficiency of the generated code is presented in Chap. 4.

```
nps=np/2;
pe1⊢SMSReal[Table[p$$[i],{i,nps}]];
pe2⊢SMSReal[Table[p$$[nps+i],{i,nps}]];
W⊨fW[pe1,pe2];
SMSDo[m,1,np/2];
  Rgm⊨{SMSD[W,pe1,m],SMSD[W,pe2,m]};
  SMSExport[wgp Rgm,{p$$[m],p$$[nps+m]},"AddIn"→True];
  SMSDo[n,1,np/2];
    Kgmn⊨{SMSD[Rgm,pe1,n],SMSD[Rgm,pe1,n]}ᵀ;
    SMSExport[wgp Kgmn,{{s$$[m,n],s$$[m,nps+n]},
      {s$$[nps+m,n],s$$[nps+m,nps+n]}},"AddIn"→True];
  SMSEndDo[];
SMSEndDo[];
```

Box 2.5. *AceGen* input for generation of representative formulas for a two-subsets case with the equal lengths

2.7 Automatic Generation of FE User Subroutines

The procedure described above and the code generated is not intended to be included into a specific FE environment. FE environments that enable the use of user defined finite elements usually require a strict form of a list of input and output parameters. *AceGen* can automatically generate the list of input and output parameters.

The standard procedure to generate finite element source code using *AceGen* is comprised of four major phases:

1. *AceGen* initialization,
2. template initialization,
3. definition of standard user subroutines,
   The definition of each user subroutine consists of several steps:
   (a) declaration of a standard user subroutine,
   (b) definition of numeric-symbolic interface variables,
   (c) derivation of the problem,
   (d) exporting results to output parameters.
4. code generation.

The specific *AceGen* input can be divided into six characteristic steps. As an example, the *AceGen* input that generates three-dimensional, 8 noded, isoparametric, hyperelastic solid element is presented here. Only the structure of the *AceGen* input is explained, while the theoretical background will be given later in Chap. 4. Here only the most common and basic features and data structures are depicted. More extensive information about the interactions between the *AceGen* code generator and the target FE environment is given in Appendix A.2.4.

Step 1: *AceGen* initialization

```
<<AceGen`;
SMSInitialize["H1element","Environment"->"AceFEM"];
```

At the beginning of the session the `SMSInitialize` function initializes the system and specifies the name of the generated source code file. The additional option `"Environment"->"AceFEM"` specifies the finite element environment for which the element source code will be generated. The *AceFEM* finite element environment is chosen in this case. Additionally to the *AceFEM* environment one can also specify e.g. the research code *FEAP* or the commercial code *ABAQUS*.

Step 2: Template initialization

```
SMSTemplate[
  "SMSTopology"→"H1"
  ,"SMSDomainDataNames"→
   {"E -elastic modulus","ν -poisson ratio",…}
  ,"SMSDefaultData"->{21000,0.3,…}
  ,"SMSSymmetricTangent"→True
 ];
nen=SMSNoNodes;ndim=SMSNoDimensions;
np=SMSNoDOFGlobal;
```

The `SMSTemplate` function initializes constants that are needed to create a proper interface between the generated user subroutines and the finite element environment. Typically, a minimum of four constants has to be specified:

`"SMSTopology"` constant specifies a keyword that defines the element topology. For example the "H1" keyword defines a three-dimensional, 8 noded, hexahedral element with 3 degrees of freedom per node. Setting the topology constant also sets other constants that directly depend on topology, e.g. the number of nodes (`"SMSNoNodes"`), the spatial dimension (`"SMSNoDimensions"`), etc. Their values are available after the `SMSTemplate` command and they can be used to make the *AceGen* input more general and problem independent.

`"SMSDomainDataNames"` constant specifies a list of keywords that characterize the material constants. Material constants appear as an input data for finite element simulation.

`"SMSDefaultData"` constant specifies default (most common) values of the material constants.

`"SMSSymmetricTangent"` constant specifies whether the tangent matrix is symmetric or not. Only the upper triangular matrix has to be explicitly formed when the tangent matrix is symmetric.[4]

[4]This option also effects the selection of the algorithm applied to solve the resulting system of linear equations when *AceFEM* is used.

Step 3.a: Declaration of standard user subroutine

```
SMSStandardModule["Tangent and residual"];
```

While the `SMSModule` command used in Sect. 2.5.2 starts the definition of a user subroutine with a user supplied set of input/output parameters, the `SMSStandardModule` command starts the definition of the subroutine with a predefined set of input/output parameters that is adjusted for the needs of the specified finite element environment. A standard set of user subroutines supported by *AceGen* is:

"Tangent and residual" standard user subroutine returns the tangent matrix (stored in variable `s$$` with dimensions $(n_p + n_{eh}) \times (n_p + n_{eh})$) and the residual or internal load vector (stored in variable `p$$` with dimension $n_p + n_{eh}$) for the current state of element related data.

"Postprocessing" user subroutine returns two arrays with an arbitrary number of post-processing quantities as follows: `gpost$$` is an array of post-processing quantities at an integration point with dimensions *number of integration points*×*number of integration point quantities*; `npost$$` is an array of the nodal point quantities with dimensions *number of nodes* × *number of nodal point quantities*.

"Sensitivity pseudo-load" user subroutine returns a matrix of pseudo-load vectors (stored in the variable `s$$` with dimensions $n_\phi \times (n_p + n_{eh})$) that is used in sensitivity analysis to calculate the sensitivity of the global unknowns with respect to an arbitrary parameter.

"Dependent sensitivity" is a standard user subroutine that is employed to calculate the sensitivity of locally coupled unknowns at element level. Details about the sensitivity analysis are presented in Chap. 8.

Step 3.b: Definition of numeric-symbolic interface variables

```
{Em,ν}⊢SMSReal[Table[es$$["Data",i],{i,2}]];
{λ,μ}⊨SMSHookeToLame[Em,ν];
𝕏IO⊢Table[SMSReal[nd$$[i,"X",j]],{i,nen},{j,ndim}];
𝕦IO⊢SMSReal[Table[nd$$[i,"at",j],{i,nen},{j,ndim}]];
𝕡e=Flatten[𝕦IO];
SMSDo[Ig,1,SMSInteger[es$$["id","NoIntPoints"]]];
Ξ={ξ,η,ζ}⊢Table[SMSReal[es$$["IntPoints",i,Ig]],{i,3}];
wgp⊢SMSReal[es$$["IntPoints",4,Ig]];
```

The input parameters of the standard user subroutines allow access to all data stored in the environment data base that relates to a specific element. The most frequently used input data is:

`nd$$[i,"X", j]` or X_{Ij} is jth component of the initial coordinates of the Ith node,

`nd$$[i,"at",j]` or p_{Ij} is the jth unknown at the Ith element node,

`es$$["Data",i]` is the ith material constant (as previously defined by `"SMSDomainDataNames"` constant),
`es$$["IntPoints",i,j]` is the ith coordinate of the jth Gauss point,
`es$$["IntPoints",4,j]` is the Gauss point weight at the jth Gauss point,
`es$$["id","NoIntPoints"]` is the number of Gauss points.

Step 3.c: Derivation of the problem

```
Ξn={{-1,-1,-1},{1,-1,-1},{1,1,-1},{-1,1,-1},{-1,-1,1},
   {1,-1,1},{1,1,1},{-1,1,1}};
Nh⊧Table[1/8 (1+ξ Ξn[[i,1]]) (1+η Ξn[[i,2]]) (1+ζ Ξn[[i,3]]),
   {i,1,nen}];
X⊢SMSFreeze[Nh.XIO];u⊧Nh.uIO;Je⊧SMSD[X,Ξ];Jed⊧Det[Je];
H⊧SMSD[u,X,"Dependency"→{Ξ,X,SMSInverse[Je]}];
F⊧IdentityMatrix[3]+H;Ce⊧Fᵀ.F;JF⊧Det[F];
W⊧1/2 λ (JF-1)^2+μ (1/2 (Tr[Ce]-3)-Log[JF]);
Rg⊧Jed SMSD[W,pe];
Kg⊧SMSD[Rg,pe];
```

In this part, the shape functions are defined, the steps (computation of displacement gradient, deformation gradient and right Cauchy–Green strain tensor) are followed to formulate the potential form W. Furthermore the residual vector and the tangent matrix are computed. Details and theoretical background will be described later in Chap. 4.

Step 3.d: Exporting results to output parameters

```
SMSExport[wgp Rg,p$$,"AddIn"→True];
SMSExport[wgp Table[Kg[[i,j]],{i,1,np},{j,i,np}],
  Table[s$$[i,j],{i,1,np},{j,i,np}],"AddIn"→True];
SMSEndDo[];
```

The results of the derivation are assigned to the output parameters `s$$` and `p$$` of the *"Tangent and residual"* standard user subroutine by the `SMSExport` function. Here the symmetry of the tangent matrix is accounted for by exporting only the upper triangle matrix.

Steps 3.a–3.d: Definition of second user subroutines Steps 3.a–3.d can be repeated several times for the definition of all required user subroutines.

Step 4: Code generation

```
SMSWrite[];
```

At the end of the session the `SMSWrite` function writes the generated formulas together with the code that is related to the interface of the chosen finite element environment. This output is written to a file in the programming language of the target finite element environment.

References

Akers, R., P. Baffes, E. Kant, C. Randall, and R.Y. Steinberg. 1998. Automatic synthesis of numerical codes for solving partial differential equations. *Mathematics and Computers in Simulation* 45: 3–22.

Amberg, G., R. Tonhardt, and C. Winkler. 1999. Finite element simulations using symbolic computing. *Mathematics and Computers in Simulation* 49: 257–274.

Bartholomew-Biggs, M., S. Brown, B. Christianson, and L. Dixon. 2000. Automatic differentiation of algorithms. *Journal of Computational and Applied Mathematics* 124: 171–190.

Beall, M., and M. Shephard. 1999. Object-oriented framework for reliable numerical simulations. *Engineering with Computers* 15: 61–72.

Bischof, C., P. Hovland, and B. Norris. 2002. Implementation of automatic differentiation tools. In *Proceedings of the ACM SIGPLAN Workshop on Partial Evaluation and Semantics-Based Program Manipulation*, ed. C. Norris, and J.J.B. Fenwick. New York: ACM Press.

Bischof, C., H.M. Buecker, B. Lang, A. Rasch, and J.W. Risch. 2003. Extending the functionality of the general-purpose finite element package sepran by automatic differentiation. *International Journal for Numerical Methods in Engineering* 58: 2225–2238.

Choi, K.K., and N.H. Kim. 2005a. *Structural sensitivity analysis and optimization 1, Linear systems*. New York: Springer Science+Business Media.

Choi, K.K., and N.H. Kim. 2005b. *Structural sensitivity analysis and optimization 2, Nonlinear systems and applications*. New York: Springer Science+Business Media.

Eyheramendy, D., and T. Zimmermann. 2000. Object-oriented symbolic derivation and automatic programming of finite elements in mechanics. *Engineering with Computers* 15(1): 12–36.

Fritzson, P., and D. Fritzson. 1984. The need for high-level programming support in scientific computing applied to mechanical analysis. *Computers and Structures* 45: 387–395.

Gallopoulos, E., E. Houstis, and J. Rice. 1994. Problem-solving environments for computational science. *IEEE Computational Science in Engineering* 1: 11–23.

Gonnet, G. 1986. New results for random determination of equivalence of expression. In *Proceedings of 1986 ACM Symposium on Symbolic and Algebraic Computation*, ed. B.W. Char, 127–131. Waterloo, ON: ACM.

Griewank, A. 2000. *Evaluating derivatives: principles and techniques of algorithmic differentiation*. Philadelphia: SIAM.

Griewank, A., and A. Walther. 2008. *Evaluating derivatives: principles and techniques of algorithmic differentiation*, 2nd ed. Philadelphia: SIAM.

Hudobivnik, B., and J. Korelc. 2016. Closed-form representation of matrix functions in the formulation of nonlinear material models. *Finite Elements in Analysis and Design* 111: 19–32.

Keulen, F., R. Haftka, and N. Kim. 2005. Review of options for structural design sensitivity analysis. part 1: Linear systems. *Computer Methods in Applied Mechanics and Engineering* 194: 3213–3243.

Kirby, R.C., M. Knepley, A. Logg, and L.R. Scott. 2005. Optimizing the evaluation of finite element matrices. *SIAM Journal on Scientific Computing* 27: 741–758.

Kleiber, M., H. Antúnez, T. Hien, and P. Kowalczyk. 1997. *Parameter sensitivity in nonlinear mechanics*. Chichester: Wiley.

Korelc J. 1997a. Application of computer algebra systems in structural analysis. In *Proceedings of 7tf Conference Computer in Structural Engineering*, pages 21–28. University of Ljubljana.

Korelc, J. 1997b. Automatic generation of finite-element code by simultaneous optimization of expressions. *Theoretical Computer Science* 187: 231–248.

Korelc, J. 2002. Multi-language and multi-environment generation of nonlinear finite element codes. *Engineering with Computers* 18: 312–327.

Korelc, J. 2011. *AceGen and AceFEM user manual*. Technical report, University of Ljubljana www.fgg.uni-lj.si/symech/.

Korelc, J., and S. Stupkiewicz. 2014. Closed-form matrix exponential and its application in finite-strain plasticity. *International Journal for Numerical Methods in Engineering* 98: 960–987.

Kristanic, N., and J. Korelc. 2008. Optimization method for the determination of the most unfavorable imperfection of structures. *Computational Mechanics* 42: 859–872.

Logg, A. 2007. Automating the finite element method. *Archives of Computational Methods in Engineering* 14: 93–138.

Logg, A.M.K.A., and G. Wells. 2012. *Automated solution of differential equations by the finite element method*. Berlin: Springer.

Martin, W. A. 1971. Determining the equivalence of algebraic expressions by hash coding, *Proceedings of the 2nd Symposium on Symbolic and Algebraic Manipulation*, ed. S.R. Petrick, 305–310. Los Angeles, CA: ACM.

Pironneau O., F. Hecht, and A. Hyaric. 2008. Freefem++. Technical report www.freefem.org.

Solinc, U., and J. Korelc. 2015. A simple way to improved formulation of fe2 analysis. *Computational Mechanics* 56: 905–915.

Tan, H., T. Chang, and D. Zheng. 1991. On symbolic manipulation and code generation of a hybrid 3-dimensional solid element. *Engineering with Computers* 7: 47–59.

van Engelen, R.A., L. Wolters, and G. Cats. 1995. *Ctadel: A Generator of Efficient Code for PDE-based Scientific Applications*. Technical report, Leiden Institute of Advanced Computer Science. http://www.liacs.nl/TechRep/1995/tr95-26.ps.gz.

Wang, P.S. 1986. Finger: A symbolic system for automatic generation of numerical programs in finite element analysis. *Journal of Symbolic Computation* 2: 305–316.

Wang P.S. 1991. Symbolic computation and parallel software. Technical report, Department of Mathematics and Computer Science, Kent State University, USA.

Chapter 3
Automation of Primal Analysis

In order to formulate nonlinear finite elements symbolically in a general but simple way, a clear mathematical formulation is needed at the highest abstract level possible. The book presents a hybrid symbolic-numeric approach to automation of primal as well as sensitivity analysis. Appropriate problem descriptions for the fully implicit primal analysis of non-linear, coupled, time-dependent problems are presented here accompanied by the automation of the sensitivity analysis in Chap. 8.

3.1 Classification of Nonlinear Computational Problems

The mathematical modeling of technical applications in solid mechanics leads in general to nonlinear partial differential equations that characterize the associated initial and boundary value problems. When dealing with nonlinear partial differential equation which describe the deformation process of solids then the change of state variables and deformations in time has to be considered. These problems are known as initial boundary value problems which additionally depend upon the time. Engineering applications, related to initial value problems, include vibration analysis of structures or impact problems like car crash simulations. In such cases the inertia term in the linear momentum equation, cannot be neglected. Another class of problems is related to inelastic constitutive behavior, like elasto-plasticity, visco-plasticity or visco-elasticity. The inelastic response is governed by evolution equations and thus results in general in a time (real or pseudo time) dependent process.

Spatial discretization of a nonlinear, time dependent partial differential equation by finite elements leads to a system of ordinary differential equations in time. If the time dependency can be neglected (time-independent problems) then the system of nonlinear ordinary differential equations reduces to a nonlinear algebraic equations system stemming from the finite element discretization. For the time-dependent problems a system of nonlinear algebraic equations is obtained after an approximation of the time derivatives within a given time step.

J. Korelc and P. Wriggers, *Automation of Finite Element Methods*,
DOI 10.1007/978-3-319-39005-5_3

The numerical efficiency of a solution procedure can be greatly improved if some of the resulting algebraic equations are local at a specific finite element. Those equations can be partially resolved at the individual finite element level before the global equations are assembled and solved leading to a locally coupled system of nonlinear algebraic equations. Consequently, the number of global unknowns of the problem is reduced. However the complexity of the solution algorithm is increased. The resulting coupled system of equations can be solved by the nested iterative-sub iterative procedure described in Sect. 3.3 and presented in Box 3.4. Here, the term "coupled" refers only to the problems where coupled systems of equations are solved through the nested iterative-sub iterative procedure. For problems where the coupled systems of equations are solved within a single iteration procedure, the same automation procedure as for uncoupled problems applies. Another approach to coupled problems is the "staggered solution strategy", see e.g. Zienkiewicz and Taylor (2000a), where a discrete model is formulated for each field and solved separately. In a staggered solution strategy the effect of coupling is handled by data transfer between the separate models. A fully consistent linearization is not required, thus the automation approach for the uncoupled problems can be applied at each sub problem independently.

For the purpose of automation, it is convenient to classify computational models in a way that the same automation algorithm can be applied for all problems of the same class. Based on previous considerations the nonlinear computational problems can be classified with respect to the formation, assembly and solution of the discretized equations. The classical solution procedures for primal and sensitivity analysis of the typical computational models that arise in the description of mechanical problems are discussed in detail in Michaleris et al. (1994). These authors subdivide the computational models in steady-state, steady-state coupled, transient and transient coupled classes. Since the terms steady-state and transient relate to the characteristics of the physical problem rather than to the characteristics of a computational scheme the classes are named here differently. The problems are classified by the way how the discretized equations are formed. Depending on the response of the system one distinguishes uncoupled and coupled problems and with respect to time (real or pseudo time) time-independent and time-dependent problems. Such classification yields the following categories:

1. time-independent problems,
2. time-dependent problems,
3. time-independent coupled problems,
   (a) time-independent locally coupled problems,
   (b) time-independent Gauss point coupled problems,
4. time-dependent coupled problems.
   (a) time-dependent locally coupled problems,
   (b) time-dependent Gauss point coupled problems,

The definitions and the solution methods of all the classes are discussed for the primal analysis in the present chapter and for the sensitivity analysis in the Chap. 8.

The advantage of the adopted classification is that the same automation algorithm can be applied to all problems of the same class. For each class, the traditional derivation of element residual vector and tangent matrix, needed within the numerical solution procedure, is provided. The traditional notation is then transformed into automatic differentiation based or ADB formulation.

In all cases the formulation leads to the system of nonlinear algebraic equations

$$\mathbf{R} = \mathbf{0} \tag{3.1}$$

stemming from the finite element discretization. The global residual vector $\mathbf{R}$ of the problem is formed in finite element methods by the standard assembly procedure

$$\mathbf{R} = \mathop{\mathsf{A}}_{e=1}^{n_e} \mathbf{R}_e \tag{3.2}$$

where $\mathbf{R}_e$ represents the contribution of the eth element to the global residual or **element residual** and n_e is the total number of finite elements used to discretize the domain of the problem. The formulation (3.2) originates from the localized nature of finite element interpolations and additive nature of the free energy function or virtual work principle. Furthermore, the element residual vector is obtained in standard finite element formulations by a numerical integration rule

$$\mathbf{R}_e = \sum_{g=1}^{n_g} w_{gp}\, \mathbf{R}_g \tag{3.3}$$

where w_{gp} stands for the Gauss point weights, n_g is the number of Gauss points and $\mathbf{R}_g$ is the Gauss point contribution to the element residual vector or Gauss **point residual**. The formulation (3.3) is a special case of a more general formulation (3.2) and (3.2) again a special case of (3.1). From the mathematical point of view it would be sufficient to automatize only the most general Eq. (3.1). However, from the numerical point of view as well as from the aspect of automation it is convenient to automatize the lowest level not yet satisfactory efficiently solved by the functionality of the finite element environment. In general any task that is independent of a specific problem can be implemented once and used for arbitrary problems. Thus, the automation of such task does not increase the overall level of automation.

Formulation (3.1) represents the problem that results in a completely arbitrary system of nonlinear algebraic equations. Linearization of (3.1) would in that case require the use of general automatic differentiation tools at the global level of the numerical environment. The benefits of the specific data and algorithmic structure of the finite element method are in that case not accounted for and the result is not numerically efficient. Formulation (3.3) naturally stems from the finite element formulation of problems in solid mechanics and structures. Thus only (3.2) and (3.3) formulations are considered here.

3.1.1 Time-Independent Problems

The response of a time-independent problem is defined by a system of nonlinear algebraic equations stemming from the finite element discretization

$$\mathbf{R}(\mathbf{p}) = \mathbf{0}. \tag{3.4}$$

Here the unknown variables $\mathbf{p} \in \mathbb{R}^{n_{tp}}$ have to be determined where n_{tp} is the total number of global unknowns of the problem. Let $\mathbf{p}_e \subseteq \mathbf{p}$ be a subset of the global vector of unknowns $\mathbf{p}$ on which the eth element depends explicitly. Both, the element $\mathbf{R}_e$ and the GAUSS point $\mathbf{R}_g$ contributions to the global residual, are an explicit functions of $\mathbf{p}_e$

$$\begin{aligned} \mathbf{R}_e &= \mathbf{R}_e\left(\mathbf{p}_e\right) \\ \mathbf{R}_g &= \mathbf{R}_g\left(\mathbf{p}_e\right). \end{aligned} \tag{3.5}$$

3.1.2 Time-Dependent Problems

In the case of time-dependent problems, the response of the system is a function of time. The standard way of solving time-dependent problems is using a time-following procedure where the time domain $0 \leq t \leq \bar{t}$ is discretized into n_{step} time steps with associated solution vectors $\mathbf{p}_0, \mathbf{p}_1, \ldots, \mathbf{p}_{n-1}, \mathbf{p}_n, \mathbf{p}_{n+1}, \ldots, \mathbf{p}_{n_{\text{step}}}$. In the following it is assumed that $\mathbf{p}_n = \mathbf{p}(t_n)$ is the solution vector at the end of the last calculated time step, $\mathbf{p}_{n+1} = \mathbf{p}(t_{n+1})$ the current vector of unknowns and $\mathbf{p}_{n_{\text{step}}} = \mathbf{p}(\bar{t})$ the vector of unknowns at the terminal time. When the time-dependent problem is solved incrementally, the only unknowns at the chosen time increment belong to the current vector of unknowns $\mathbf{p}_{n+1}$. The same is valid for time-independent problems. In order to simplify the notation and to enable a unified description of the solution procedures for time-dependent and time-independent problems the index $n+1$ that indicates the current time will be omitted for the current time quantities, thus $\mathbf{p} = \mathbf{p}_{n+1}$. The notation used here to describe the time-dependent behavior is depicted in Fig. 3.1 which shows the function of component $p_i(t)$ of the solution vector $\mathbf{p}(t)$ and its evaluation at different times.

In general, the current residual vector $\mathbf{R} = \mathbf{R}_{n+1} = \mathbf{R}(t_{n+1})$ can implicitly depend on all solution vectors. However, for the process of automation, only those solution vectors are relevant that explicitly appear as a part of the formulation. All solution algorithms for time-dependent problems rely on an approximation of the time derivatives which have to be chosen within a given time step. As an example the velocity $\boldsymbol{v}(t)$ can be approximated by a difference quotient using the displacements $\boldsymbol{u}(t) \subseteq \mathbf{p}(t)$ at different times

$$\boldsymbol{v}(t) = \frac{d}{dt}\boldsymbol{u}(t) \approx \frac{1}{\Delta t}\left[\alpha \boldsymbol{u}(t_{n-1}) + \beta \boldsymbol{u}(t_n) + \gamma \boldsymbol{u}(t_{n+1})\right]. \tag{3.6}$$

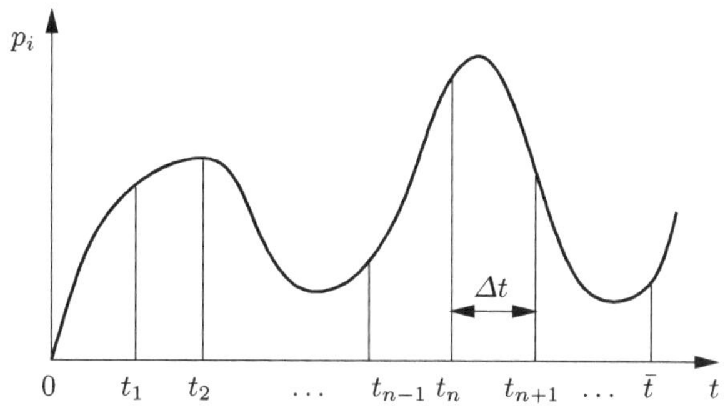

Fig. 3.1 Time-dependent behaviour of a component p_i

This leads to the following system of algebraic equations

$$\mathbf{R}(\mathbf{p}, \mathbf{p}_n, \mathbf{p}_{n-1}) = \mathbf{0}. \tag{3.7}$$

In the simplest case, when the time derivatives are approximated by a first-order finite difference scheme, the time-dependent problem explicitly depends only on two solution vectors as follows

$$\mathbf{R}(\mathbf{p}, \mathbf{p}_n) = \mathbf{0}. \tag{3.8}$$

The time-dependent problem can in general explicitly depend on arbitrary number of solution vectors (e.g. some visco-elastic material models depend additionally on $\mathbf{p}_0$). However, since the solution vectors $\mathbf{p}(t) : t \leq t_n$ are known, the additional dependency does not change the automation of the primal problem, but it does require additional terms to be added for each additional time step to the sensitivity pseudo-load terms introduced in Chap. 8.

3.1.3 Time-Independent Coupled Problems

The residuals of the time-independent coupled system are given by

$$\begin{aligned} \mathbf{R}(\mathbf{p}, \mathbf{h}) &= \mathbf{0}, \\ \mathbf{Q}(\mathbf{p}, \mathbf{h}) &= \mathbf{0} \end{aligned} \tag{3.9}$$

where residuals $\mathbf{R}$ and $\mathbf{Q}$ have to be simultaneously solved for the unknown vectors $\mathbf{p} \in \mathbb{R}^{n_{tp}}$ and $\mathbf{h} \in \mathbb{R}^{n_{th}}$ and n_{th} is the total number of coupled unknowns of the problem.

Time-independent locally coupled problems. When $\mathbf{R}$ and $\mathbf{Q}$ are coupled at the global level then the automation procedures are the same as for the uncoupled problems. Important for the automation is the situation where there is one residual ($\mathbf{R}$)

defined and assembled at the global level of the problem and $\mathbf{Q}$ is composed of n_e local residuals defined on the individual finite element level. The situation is named a "time-independent locally coupled" problem. Let e be the element number, $\mathbf{p}_e \subseteq \mathbf{p}$ a subset of the global vector of unknowns $\mathbf{p}$ on which the eth element explicitly depends and $\mathbf{Q}_e$ the local residual related to the eth element. Consequently, $\mathbf{Q}_e$ depends explicitly only on the corresponding vector of unknowns $\mathbf{h}_e \subseteq \mathbf{h}$ and on the subset $\mathbf{p}_e$ of the global vector of unknowns. The time-independent locally coupled problem is defined by

$$\begin{aligned} \mathbf{R}(\mathbf{p}, \mathbf{h}) &= \mathbf{0}, \\ \mathbf{Q}_e(\mathbf{p}_e, \mathbf{h}_e) &= \mathbf{0} : e = 1, \ldots, n_e. \end{aligned} \tag{3.10}$$

The vector of coupled unknowns $\mathbf{h}$ and residuals $\mathbf{Q}$ are then composed of n_e local vectors of unknowns $\mathbf{h}_e$ and local residuals $\mathbf{Q}_e$.

$$\mathbf{h} = \bigcup_{e=1}^{n_e} \mathbf{h}_e, \ \mathbf{Q} = \bigcup_{e=1}^{n_e} \mathbf{Q}_e \tag{3.11}$$

A typical example is the implementation of finite element formulations based on mixed variational principles where, additionally to the global degrees of freedom, degrees of freedom and corresponding equilibrium equations are defined locally at each finite element. For time-independent locally coupled problem the individual finite element residual $\mathbf{R}_e$ depends only on a vector of unknowns of the following local problem

$$\mathbf{R}_e = \mathbf{R}_e\left(\mathbf{p}_e, \mathbf{h}_e\right). \tag{3.12}$$

Time-independent Gauss point coupled problems. These type of problems occur when the element residual is result of a numerical integration (3.3) and each Gauss point residual $\mathbf{R}_g$ is coupled with one local problem. Let $\mathbf{Q}_g$ be the local residual at the gth Gauss point and $\mathbf{h}_g$ the corresponding vector of local unknowns. Also assume that $\mathbf{Q}_g$ does not depend directly on the element unknowns $\mathbf{p}_e$ but indirectly through an additional set of intermediate variables $\mathbf{r}_g$ defined at each local problem $\mathbf{Q}_g$. The time-independent Gauss point coupled problem is then defined by

$$\begin{aligned} \mathbf{R}(\mathbf{p}, \mathbf{h}) &= \mathbf{0}, \\ \mathbf{Q}_g(\mathbf{r}_g(\mathbf{p}_e), \mathbf{h}_g) &= \mathbf{0} : g = 1, \ldots, n_{tg} \end{aligned} \tag{3.13}$$

where n_{tg} as the total number of Gauss points of the problem. It is assumed that the local residual $\mathbf{Q}_g$ appears as a part of the eth element. Consequently, $\mathbf{Q}_g$ explicitly depends only on the corresponding vector of unknowns $\mathbf{h}_g$ and on a set of intermediate variables $\mathbf{r}_g$ that depends explicitly on the subset $\mathbf{p}_e$ of the global vector of unknowns. The reason for the introduction of an additional set of intermediate variables is numerical efficiency of the implementation of implicit solution procedures as presented in next section.

The residual $\mathbf{R}_g$ at the Gauss points contributes to the individual finite element residual $\mathbf{R}_e$. In the case of time-independent Gauss point coupled problems $\mathbf{R}_g$ depends on the unknowns $\mathbf{h}_g$ of the associated local problem and the element residual $\mathbf{R}_e$ depends on a sub-set of unknowns $\bigcup_{g \in G_e} \mathbf{h}_g$ where $G_e \subseteq \{1, \ldots, n_{tg}\}$ is a set of indices of Gauss points of the eth element.

$$\mathbf{R}_g = \mathbf{R}_g(\mathbf{p}_e, \mathbf{h}_g(\mathbf{r}_g)) \tag{3.14}$$

$$\mathbf{R}_e = \mathbf{R}_e \left(\mathbf{p}_e, \bigcup_{g \in G_e} \mathbf{h}_g \right) \tag{3.15}$$

The vectors of the coupled unknowns $\mathbf{h}$ and the residuals $\mathbf{Q}$ are then composed of n_{tg} local vectors of unknowns $\mathbf{h}_g$ and local residuals $\mathbf{Q}_g$.

$$\mathbf{h} = \bigcup_{g=1}^{n_{tg}} \mathbf{h}_g, \ \ \mathbf{Q} = \bigcup_{g=1}^{n_{tg}} \mathbf{Q}_g \tag{3.16}$$

A typical example of time-independent Gauss point coupled problems is the implementation of the plane stress condition in the case of complex hyper-elastic material models for two-dimensional solid and shell elements. The enforcement of the plane stress condition at the integration point level leads to an additional nonlinear algebraic equation defined at each integration point.

3.1.4 Time-Dependent Coupled Problems

Time-dependent locally coupled problems. The combination of time-dependent (3.8) and locally coupled (3.10) formulation leads to a time-dependent locally coupled problem. The residuals $\mathbf{R}$ and $\mathbf{Q}_e$ of the time-dependent locally coupled problem are given by

$$\begin{aligned} \mathbf{R}(\mathbf{p}, \mathbf{h}, \mathbf{p}_n, \mathbf{h}_n) &= \mathbf{0}, \\ \mathbf{Q}_e(\mathbf{p}_e, \mathbf{h}_e, \mathbf{p}_{en}, \mathbf{h}_{en}) &= \mathbf{0} : \ e = 1, \ldots, n_e, \end{aligned} \tag{3.17}$$

and the element residual is for the time-dependent locally coupled problems given by

$$\mathbf{R}_e = \mathbf{R}_e \left(\mathbf{p}_e, \mathbf{h}_e, \mathbf{p}_{en}, \mathbf{h}_{en} \right). \tag{3.18}$$

Time-dependent Gauss point coupled problems. Similarly, the combination of a time-dependent (3.8) and a Gauss point coupled (3.13) formulation leads to the definition of the time-dependent Gauss point coupled problem. The residuals $\mathbf{R}$ and $\mathbf{Q}_g$ of the time-dependent Gauss point coupled problem are given by

$$\begin{aligned} \mathbf{R}(\mathbf{p}, \mathbf{h}, \mathbf{p}_n, \mathbf{h}_n) &= \mathbf{0}, \\ \mathbf{Q}_g(\mathbf{r}_g(\mathbf{p}_e), \mathbf{h}_g, \mathbf{p}_{en}, \mathbf{h}_{gn}) &= \mathbf{0} : \ g = 1, \ldots, n_{tg}, \end{aligned} \tag{3.19}$$

and the Gauss point contribution to the element residual by

$$\mathbf{R}_g = \mathbf{R}_g(\mathbf{p}_e, \mathbf{h}_g(\mathbf{r}_g), \mathbf{p}_{en}, \mathbf{h}_{gn}) \tag{3.20}$$

and the element residual by

$$\mathbf{R}_e = \mathbf{R}_e\left(\mathbf{p}_e, \bigcup_{g \in G_e} \mathbf{h}_g, \mathbf{p}_{en}, \bigcup_{g \in G_e} \mathbf{h}_{gn}\right) \tag{3.21}$$

A typical example of a time-dependent Gauss point coupled problem is related to inelastic constitutive behavior. The inelastic response is governed by evolution equations that after time discretization lead to an additional system of algebraic equations. These are defined locally and are independent at each integration point.

3.2 Solution of Nonlinear Systems of Equations

In the case of time-independent problems a system of nonlinear algebraic equations $\mathbf{R}(\mathbf{p}) = \mathbf{0}$ has to be solved for the unknown vector $\mathbf{p}$. Two different aspects have to be considered when solving the nonlinear system of equations equation. These are

1. the general solvability of the nonlinear equation systems and
2. the formulation of adequate numerical methods and algorithms.

The first aspect involves examination of

- existence of solutions in a defined region,
- number of solutions in this region and
- the influence of the change of function $\mathbf{R}$ with respect to the solution.

Clarification of these questions needs methods of nonlinear functional analysis which have to be applied to the nonlinear partial differential equations resulting from the modeling procedure. This area cannot be treated in depth in the framework of this book. However for applications that fall in the range of the nonlinear theory of elasticity results can be found in Marsden and Hughes (1983), Ciarlet (1988), Johnson (1987), Brenner and Scott (2002) and Braess (2007). Further books which contain a general treatment of the above mentioned questions are provided by Vainberg (1964) or Ortega and Rheinboldt (1970).

We will assume in the following that solutions of the equation system $\mathbf{R}(\mathbf{p}) = \mathbf{0}$ exist in the considered regions and will devote ourself to the numerical methods for the determination of solutions of $\mathbf{R}(\mathbf{p}) = \mathbf{0}$. The approximation of solutions of the nonlinear equation system $\mathbf{R}(\mathbf{p}) = \mathbf{0}$ can be obtained by different methods. Among these are

1. methods for the construction of sets that contain solutions,
2. procedures to find all solutions and
3. methods which just approximate one solution.

The first two methods need in general deeper knowledge of the underlying mathematical structure of the associated partial differential equations. Due to that fact only procedures will be considered which yield one approximate solution at a time, but which can be applied to find successively other solutions of $\mathbf{R}(\mathbf{p}) = \mathbf{0}$. Since a direct solution of $\mathbf{R}(\mathbf{p}) = \mathbf{0}$ is in general not possible, iterative solution procedures have to be constructed. These methods permit different ways to solve the problem which will be described in detail in the next sections. In general different procedures can be distinguished:

- methods which base on linearization,
- minimization schemes or
- reduction methods, which lead to simpler nonlinear equation systems.

When choosing a solution method the following fundamental questions have to be clarified which constitute the success of an iterative method and its associated algorithm:

1. Does the iterative method converge to the solution?
2. How fast is the convergence? Does the rate of convergence depend upon the problem size?
3. How efficient is the algorithm?
   (a) How many numerical operations are needed within one iteration step?
   (b) How many iterations are necessary to converge within a given accuracy?
   (c) How much memory of the CPU does the iterative method need?

The first question concerns the global convergence characteristics of the iterative method and is essential for the user who needs a robust and reliable iterative method for his problem. Also the other questions are of importance. The efficiency depends on several factors which are determined e.g. by the linear solver within an iterative method, the finite element formulation itself and the convergence properties of the iterative solution method. In case of large problem sizes a nonlinear finite element equation system with a great number of unknowns is obtained which requires a lot of memory and many numerical operations in the solution process. When the number of operations increases quadratically with respect to the number of unknowns the method is said to be of order $O(N^2)$. Such method cannot be applied to large problems. Here methods that need only $O(N)$ operations are advantageous and thus present a vivid research area, for an overview see e.g. Rheinboldt (1984), Hackbusch (1994), Elman et al. (2005), Douglas et al. (2003) and Saad (2003).

In solid mechanics and nonlinear structural mechanics the range of problems is wide and areas are quite different (geometrical nonlinearity, physical nonlinearity, stability etc.). Thus there exists up to now no iterative method which can be applied to all different problem areas in an efficient and robust way. Additionally, for highly nonlinear problems the solution of time-independent problems cannot in general be achieved in one step. More efficient procedures can be derived when the resulting system of equations can be naturally parameterized in a way that for some given value of parameter the solution is trivial. The system of equations $\mathbf{R}(\mathbf{p}) = \mathbf{0}$ will be parameterized for the following considerations in the form

$$\mathbf{R}(\mathbf{p}, \lambda) = \mathbf{0}. \tag{3.22}$$

With the introduction of parameter λ the final solution is achieved in n_{step} incremental steps with associated solution vectors $\mathbf{p}_0, \ldots, \mathbf{p}_{n-1}, \mathbf{p}_n, \mathbf{p}_{n+1}, \ldots, \mathbf{p}_{n_{\text{step}}}$. As an example, problems in solid mechanics and nonlinear structural mechanics subjected to quasi-static proportional load are frequently parameterized by introducing the loading parameter λ as follows

$$\mathbf{R}(\mathbf{p}, \lambda) = \mathbf{R}^{\text{int}}(\mathbf{p}) - \lambda\, \mathbf{R}^{\text{ref}} = 0 \tag{3.23}$$

where $\mathbf{R}^{\text{int}}$ denotes the contribution of internal forces to the nodal force vector and $\mathbf{R}^{\text{ref}}$ is the reference load vector associated with the pattern of the applied nodal forces. The loading parameter is introduced to be able to change the load level with an iterative method. The value of parameter λ is usually determined by the problem at hand, e.g. as total given load. In that case the total load, related load parameter $\bar{\lambda}$, will be split as follows: $\bar{\lambda} = \sum_{i=1}^{n_{\text{step}}} \Delta\lambda_i$. However in special iterative methods, like the arc-length method, it makes sense to consider λ as an unknown variable.

The choice of the best method for a special application is related to the aspects discussed above. For nonlinear problems in structural mechanics there exist a large number of algorithms and solution procedures. Most common ones, that are applied within finite element methods, are

- Newton–Raphson methods,
- quasi-Newton methods,
- dynamical relaxation and
- continuation or arc-length methods.

With the now existing computer power and the possibilities related to automation of finite element development it is almost natural to only use the Newton method as a standard tool for solving nonlinear finite element problems. Another essential method is the arc-length method since it allows computing complex solution paths, see e.g. Riks (1972), Crisfield (1981) and Schweizerhof and Wriggers (1986).

Linear equation solvers provide an essential ingredient for the efficient solution of nonlinear finite element equation systems. This stems from the fact that linearization is used to construct iterative solution schemes leading to very large finite element linear equation systems. The iterative procedures then arrives at the global solution via the solution of several linear sub-problems. While standard elimination methods are often successfully applied to solve the linear equation system of small and middle size finite element discretizations, see e.g. Taylor (2000), one relies for large systems on sparse solvers, see e.g. Duff et al. (1989), Duff (2004) and Schenk and Gärtner (2004). For large systems also iterative equation solvers can be applied successfully. Here the method of pre-conditioned conjugate gradients and multi-grid methods are often employed for symmetrical matrix systems, for mathematical details, see e.g. Ciarlet (1989), Hackbusch (1994), Schwetlick and Kretschmar (1991), Douglas et al. (2003) and Saad (2003). Applications within the method of finite elements in solid mechanics can be found e.g. in Braess (2007), Kickinger (1996), Korneev et al. (2003). The method of dynamical relaxation makes a "detour" via dynamics to construct a memory saving iterative solver based on an explicit integration method.

The efficiency of the different methods for the solution of nonlinear equation systems depends also upon the application and its size in terms of number of unknowns. As an example, the method of Newton–Raphson can be very efficient in combination with direct elimination methods for problems with low number of unknowns. For problems with large number of unknowns quasi-Newton methods or the dynamical relaxation can be more efficient since, even with more iterations, they need less computation time. However also a combination of the Newton–Raphson method together with iterative linear solvers can be very efficient and faster than quasi-Newton procedures, see e.g. Hackbusch (1994), Meyer (1990), Boersma and Wriggers (1997) and Jung and Langer (2001).

3.2.1 Newton–Raphson Method

The most frequently applied scheme for the iterative solution of systems of nonlinear algebraic equation is the Newton–Raphson algorithm. It is based on a Taylor series expansion of (3.4) at an already known state $\mathbf{p}^{(i)}$

$$\mathbf{R}(\mathbf{p}^{(i)} + \Delta\mathbf{p}^{(i)}) = \mathbf{R}(\mathbf{p}^{(i)}) + D\mathbf{R}(\mathbf{p}^{(i)}) \cdot \Delta\mathbf{p}^{(i)} + \mathbf{r}(\mathbf{p}^{(i)}) \tag{3.24}$$

where the upper index (i) denotes the quantities at the ith iteration and

$$\Delta\mathbf{p}^{(i)} = \mathbf{p}^{(i+1)} - \mathbf{p}^{(i)}. \tag{3.25}$$

In (3.24) $D\mathbf{R}(\mathbf{p}^{(i)}) \cdot \Delta\mathbf{p}^{(i)}$ characterizes the directional derivative of $\mathbf{R}(\mathbf{p})$ at $\mathbf{p}^{(i)}$, also referred to as linearization. The linearization of the vector $\mathbf{R}$ yields a matrix, which is also known as Hesse-, Jacobi- or tangent matrix. This matrix will be abbreviated in the following by $\mathbf{K} = \mathbf{K}^{(i)} = D\mathbf{R}(\mathbf{p}^{(i)})$. In further derivations all quantities without an upper index are evaluated at the already known state $\mathbf{p}^{(i)}$. $\mathbf{K}$ is obtained by a standard finite element assembly procedure of the **element tangent** operators

$$\mathbf{K} = \mathop{\mathsf{A}}_{e=1}^{n_e} \mathbf{K}_e, \tag{3.26}$$

where $\mathbf{K}_e$ represents the contribution of the eth element to the global tangent matrix $\mathbf{K}$ obtained by linearization of the element residual vector $\mathbf{R}_e(\mathbf{p}_e^{(i)})$

$$\mathbf{K}_e = \frac{\partial \mathbf{R}_e}{\partial \mathbf{p}_e}. \tag{3.27}$$

The tangent matrix is obtained by numerical integration over the element domain like the residual

$$\mathbf{K}_e = \sum_{g=1}^{n_g} w_{gp} \mathbf{K}_g \tag{3.28}$$

where $\mathbf{K}_g$ is the Gauss **point tangent matrix** defined by

$$\mathbf{K}_g = \frac{\partial \mathbf{R}_g}{\partial \mathbf{p}_e}. \tag{3.29}$$

The vector $\mathbf{r}$ in (3.24) is the residuum of the Taylor series. By neglecting the residuum the linear equation system

$$\mathbf{R}(\mathbf{p}^{(i)} + \Delta\mathbf{p}^{(i)}) \approx \mathbf{K}(\mathbf{p}^{(i)})\Delta\mathbf{p}^{(i)} + \mathbf{R}(\mathbf{p}^{(i)}) = \mathbf{0} \tag{3.30}$$

is obtained from (3.24) and a new value of the unknowns can be obtained from (3.25)

$$\mathbf{p}^{(i+1)} = \mathbf{p}^{(i)} + \Delta\mathbf{p}^{(i)}. \tag{3.31}$$

Equations (3.30) and (3.31) are the basis of the iterative algorithm presented in Box 3.1 for the solution of Eq. (3.4).

The rate of convergence of the Newton–Raphson scheme is characterized by the inequality $\left\|\mathbf{p}^{(i+1)} - \mathbf{p}\right\| \leq C\left\|\mathbf{p}^{(i)} - \mathbf{p}\right\|^2$ where $\mathbf{p}$ is solution of (3.4), see e.g. Isaacson and Keller (1966), pp. 115 or Schwetlick and Kretschmar (1991), pp. 195. The quadratic convergence of the Newton–Raphson scheme, which is apparent from the above inequality, has a local character since it is only valid near the solution point. This convergence behavior is advantageous since most of the time only a few iterations are needed to obtain the solution of (3.4). A drawback of the Newton–Raphson scheme stems from the fact that, in every iteration step, the tangent matrix

Input: $\mathbf{p}_0$ `// starting value for` $\mathbf{p}$ `(e.g.` $\mathbf{0}$`)`
$\mathbf{p} \leftarrow \mathbf{p}_0$
repeat
  foreach *element* **do**
    $\mathbf{R}_e \leftarrow \mathbf{R}_e(\mathbf{p}_e)$
    $\mathbf{K}_e \leftarrow \frac{\partial \mathbf{R}_e}{\partial \mathbf{p}_e}$ .... $\overset{\texttt{automation}}{\longrightarrow}$ .... $\mathbf{K}_e \leftarrow \frac{\hat{\delta} \mathbf{R}_e}{\hat{\delta} \mathbf{p}_e}$
    add $\mathbf{R}_e$ to $\mathbf{R}$ and $\mathbf{K}_e$ to $\mathbf{K}$
  end foreach
  solve $\mathbf{K}\Delta\mathbf{p} + \mathbf{R} = \mathbf{0}$ for unknown $\Delta\mathbf{p}$
  $\mathbf{p} \leftarrow \mathbf{p} + \Delta\mathbf{p}$
until *error criterion for* $\|\mathbf{R}\|$ *and* $\|\mathbf{p}\|$ *is fulfilled*
Result: $\mathbf{p}$ `// converged solution`

Box 3.1. Newton–Raphson iterative scheme for solution of time-independent problems

$\mathbf{K}$ has to be computed and a linear equation system has to be solved, which can be quite time consuming and hence expensive. Thus it is essential to construct tangent matrices at element level which can be evaluated efficiently. To shorten notation the term Newton scheme will be used instead of the historically more correct Newton–Raphson scheme.

Often it is not possible to converge to the final solution within one step within a nonlinear solution. To circumvent this difficulty the solution is split into n_{step} incremental steps. Hence the nonlinear problem is solved n_{step}-times which results in a considerable additional expenditure with respect to the original problem. A possibility to improve the convergence radii of the Newton–Raphson scheme and thus decrease the number of incremental steps required to reach the final solution is to construct a global method by damping the Newton–Raphson iteration. This leads to line-search algorithms, see e.g. Crisfield (1991), Luenberger (1984) and Bazaraa et al. (1993). However often in time or history dependent nonlinear problems (like elasto-plastic or frictional contact problems) it is anyway necessary to apply the load in several steps to capture the correct physical behavior.

Furthermore path-following methods like the arc-length method can be introduced to overcome complex shapes of the load deflections curves. Here different methods were developed starting with the work by Riks (1972) and later Keller (1977). The following papers discuss variants of the general method Ramm (1981), Crisfield (1981), Schweizerhof and Wriggers (1986) and Wagner (1991) and thus introduce different approaches. Overviews related to the arc-length procedures and their application to engineering problems can be found in Riks (1984), Wagner and Wriggers (1988) and Crisfield and Shi (1991) and in the books Crisfield (1991) and Wriggers (2008). Note also that many problems that exhibit decreases in the load deflection curves are no longer quasi-static but dynamic. Hence the best way and closest to reality is to treat such by using a dynamic approach or by switching from static analysis to dynamic analysis when such decrease in the load deflection curve is observed.

3.2.2 Automation of Solution of Time-Independent and Time-Dependent Problems

The automation of the parameterized time-independent and time-dependent systems involves only the derivation of the element tangent matrix.[1] The element tangent matrix can be automatically derived by directly applying the automatic differentiation procedure to (3.27) leading to

$$\mathbf{K}_e = \frac{\hat{\delta}\mathbf{R}_e}{\hat{\delta}\mathbf{p}_e} \tag{3.32}$$

and the Gauss point tangent matrix by using the computational derivative procedure in (3.29) as follows

$$\mathbf{K}_g = \frac{\hat{\delta}\mathbf{R}_g}{\hat{\delta}\mathbf{p}_e}. \tag{3.33}$$

In the case of the parametrized time-independent problem (3.22) and the time-dependent problem (3.8) the solution is obtained in n_{step} steps with the associated solution vectors $\mathbf{p}_1, \mathbf{p}_2, \ldots, \mathbf{p}_{n_{\text{step}}}$. Let us introduce the notation $\mathbf{p} = \mathbf{p}_{n+1} = \mathbf{p}(\lambda_{n+1}) = \mathbf{p}(\lambda)$ for the current vector of unknowns for parametrized time-independent problems and $\mathbf{p} = \mathbf{p}_{n+1} = \mathbf{p}(t_{n+1}) = \mathbf{p}(t)$ for the current vector of unknowns for the time-dependent problems. The implementation of the Newton–Raphson iterative solution procedure and its automation for parametrized time-independent problems is presented in Box 3.2 and summarized in Table 3.1.

The parameter of the problem λ is increased in Algorithm 3.2 in $\frac{\bar{\lambda}}{\Delta\lambda}$ constant steps. Only one solution vector ($\mathbf{p}$) has to be stored in the computer memory at any given time for the implementation of Algorithm 3.2. However, the strategy is not very efficient when the level of non-linearity of the problem is changing along the path and often leads to unnecessary small parameter steps. A more efficient procedure is presented in Box 3.3. The algorithm in Box 3.3 deals with the primal analysis of time-dependent non-linear systems using an adaptive time stepping scheme. Two solution vectors, $\mathbf{p}_n$ at time t_n and $\mathbf{p}$ at current time t_{n+1} are stored in the memory at any given time. If the error criterion for $\|\mathbf{R}\|$ and $\|\mathbf{p}\|$ is not fulfilled within the maximum number of Newton–Raphson iterations then the time increment Δt is decreased and the particular step is repeated. When the number of Newton–Raphson iterations is small the time increment Δt can be increased again.

[1] One possibility is to apply difference quotients which approximate the tangent matrix $\boldsymbol{K}$, see e.g. Wriggers (2008).

```
Input: p0 // starting value for p (e.g. 0)
Input: Δλ // parameter increment
λ ← 0; p ← p0
repeat
    λ ← λ + Δλ
    begin iterative solution of one parameter step
        repeat
            if maximum number of iterations exceeded then
                break enclosing loop // simulation has failed
            end if
            foreach element do
                Re ← Re (pe, λ)
                Ke ← ∂Re/∂pe  ......... automation ⟶ .........  Ke ← δ̂Re/δ̂pe
                add Re to R and Ke to K
            end foreach
            solve KΔp + R = 0 for unknown Δp
            p ← p + Δp
        until error criterion for ||R|| and ||Δp|| is fulfilled
until terminal value of parameter (λ̄) is reached
Result: p // converged solution in last step
```

Box 3.2. Primal analysis of parametrized time-independent non-linear systems with constant stepping

Table 3.1 Primal analysis of time-independent and time-dependent non-linear problems

Primal time-independent problem	$\mathbf{R}(\mathbf{p}) = \mathbf{0}$
Primal time-dependent problem	$\mathbf{R}(\mathbf{p}, \mathbf{p}_n) = \mathbf{0}$
Solution of primal problem	$\mathbf{K} = \dfrac{\partial \mathbf{R}}{\partial \mathbf{p}}$ $\mathbf{K}\Delta\mathbf{p} + \mathbf{R} = \mathbf{0}$ $\mathbf{p} \leftarrow \mathbf{p} + \Delta\mathbf{p}$
Automation of element tangent matrix	$\mathbf{K}_e = \dfrac{\hat{\delta}\mathbf{R}_e}{\hat{\delta}\mathbf{p}_e}$
Automation of Gauss point tangent matrix	$\mathbf{K}_g = \dfrac{\hat{\delta}\mathbf{R}_g}{\hat{\delta}\mathbf{p}_e}$

3.3 Solution of Coupled Nonlinear Systems of Equations

There exist several possibilities to solve a coupled nonlinear system of equations. When $\mathbf{R}$ and $\mathbf{Q}$ are coupled at the global level, like in (3.9), then the automation procedures are the same as for the uncoupled problems. This is called **monolithic approach** where a global residual vector and its tangent matrix are formed that encompass all equations $\mathbf{R} \cup \mathbf{Q}$ and all unknowns $\mathbf{p} \cup \mathbf{h}$ of the problem. Standard Newton iterative scheme presented in Box 3.1 can then be applied.

Input: $\mathbf{p}_0$ `// starting value for` $\mathbf{p}$ `(e.g.` $\mathbf{0}$`)`
Input: Δt_0 `// starting value for time increment`
$t_n \leftarrow 0$; $\Delta t \leftarrow \Delta t_0$;$\mathbf{p}_n \leftarrow \mathbf{p}_0$
repeat
  $t \leftarrow t_n + \Delta t$
  begin iterative solution of one time step
    $\mathbf{p} \leftarrow \mathbf{p}_n$ `// last converged solution is starting value for current step (other strategies are possible)`
    repeat
      foreach *element* **do**
        $\mathbf{R}_e \leftarrow \mathbf{R}_e\,(\mathbf{p}_e, \mathbf{p}_{en}, \Delta t)$
        $\mathbf{K}_e \leftarrow \frac{\partial \mathbf{R}_e}{\partial \mathbf{p}_e}$ .... `automation` $\longrightarrow$ .... $\mathbf{K}_e \leftarrow \frac{\hat{\delta} \mathbf{R}_e}{\hat{\delta} \mathbf{p}_e}$
        add $\mathbf{R}_e$ to $\mathbf{R}$ and $\mathbf{K}_e$ to $\mathbf{K}$
      end foreach
      solve $\mathbf{K}_e \Delta \mathbf{p} + \mathbf{R} = \mathbf{0}$ for unknown $\Delta \mathbf{p}$
      $\mathbf{p} \leftarrow \mathbf{p} + \Delta \mathbf{p}$
    until *error criterion for* $\|\mathbf{R}\|$ *and* $\|\mathbf{p}\|$ *is fulfilled* **and** *maximum number of iterations is not exceeded*
  if *error criterion is fulfilled* **then**
    $\mathbf{p}_n \leftarrow \mathbf{p}$; $t_n \leftarrow t$ `// update solution for next time step`
    adjust time step Δt
  else
    cut time step Δt
  end if
until *terminal time* $\bar{t}$ *is reached*
Result: $\mathbf{p}$ `// converged solution in last step`

Box 3.3. Primal analysis of time-dependent non-linear systems with adaptive time stepping

In the case of locally coupled problems (3.10), a special variant of the **monolithic approach** can be applied where the number of global unknowns is reduced by the **static condensation** of the unknowns that are local for the specific element as presented in Sect. 3.3.3. The static condensation procedure is in general independent of a specific problem. It can be implemented once and then used for arbitrary problems. Thus, the automation of the static condensation procedure does not increase the complexity of automation.

Using automation partitioned Newton schemes can be consistently linearized. Within a **partitioned scheme** a nested iterative procedure (Newton algorithm) is applied at global and local level to solve the problem. Here the local problem is e.g. solved at Gauss point level as part of the coupled problem. This algorithm is called nested iterative-sub iterative Newton scheme. In order to achieve quadratic convergence of the overall partitioned scheme, the tangent operators have to be derived consistently with the actual solution algorithm. The resulting operator is called consistent or algorithmic tangent matrix.

3.3.1 Solution of Locally Coupled Problems

First, the coupled system of nonlinear algebraic equations (3.10) is decomposed into a dependent problem and an independent problem by considering the current solution vector of the eth local problem $\mathbf{h}_e$ as a function of the solution vector of the global problem $\mathbf{p}$. Equations (3.10) and (3.12) are then rewritten as

$$\mathbf{R}(\mathbf{p},\mathbf{h}(\mathbf{p})) = \mathop{\mathsf{A}}_{e=1}^{n_e} \mathbf{R}_e = \mathbf{0}, \tag{3.34}$$

$$\mathbf{Q}_e(\mathbf{p}_e,\mathbf{h}_e(\mathbf{p}_e)) = \mathbf{0} : \ e = 1,\ldots,n_e, \tag{3.35}$$

$$\mathbf{R}_e = \mathbf{R}_e\left(\mathbf{p}_e,\mathbf{h}_e(\mathbf{p}_e)\right). \tag{3.36}$$

where $\mathbf{R}$ denotes the independent residual and $\mathbf{Q}_e$ the eth dependent residual. Similarly, the time-dependent locally coupled system of nonlinear algebraic equations (3.17) can be decomposed into a dependent problem and an independent problem

$$\mathbf{R}(\mathbf{p},\mathbf{h}(\mathbf{p}),\mathbf{p}_n,\mathbf{h}_n) = \mathop{\mathsf{A}}_{e=1}^{n_e} \mathbf{R}_e = \mathbf{0}, \tag{3.37}$$

$$\mathbf{Q}_e(\mathbf{p}_e,\mathbf{h}_e(\mathbf{p}_e),\mathbf{p}_{en},\mathbf{h}_{en}) = \mathbf{0} : \ e = 1,2,\ldots,n_e, \tag{3.38}$$

$$\mathbf{R}_e = \mathbf{R}_e\left(\mathbf{p}_e,\mathbf{h}_e(\mathbf{p}_e),\mathbf{p}_{en},\mathbf{h}_{en}\right). \tag{3.39}$$

The partitioned Newton–Raphson scheme is the same for the time-independent as for the time-dependent systems, thus only the time-independent case will be considered in further derivations. Let the upper index (j) denote the quantities at the jth iteration of the inner Newton–Raphson loop, and let the upper index (i) denote quantities at the ith iteration of the outer Newton–Raphson loop. Furthermore the same quantity without an upper index stands for a quantity at the converged state of either inner or outer loop. First, the dependent problem (3.35) is solved for the unknown current solution vector $\mathbf{h}_e$ and fixed solution vector $\mathbf{p}_e$ in the inner loop. The linearization of the dependent residual (3.35) yields

$$\mathbf{A}_e(\mathbf{h}_e^{(j)})\Delta\mathbf{h}_e^{(j)} + \mathbf{Q}_e(\mathbf{h}_e^{(j)}) = \mathbf{0} \tag{3.40}$$

where $\mathbf{A}_e$ stands for the dependent tangent operator defined by

$$\mathbf{A}_e = \frac{\partial \mathbf{Q}_e(\mathbf{h}_e^{(j)})}{\partial \mathbf{h}_e} \tag{3.41}$$

The linear system (3.40) is solved for the unknown increment $\Delta \mathbf{h}_e^{(j)}$ and used to update the current dependent solution vector $\mathbf{h}_e$

$$\mathbf{h}_e^{(j+1)} = \mathbf{h}_e^{(j)} + \Delta \mathbf{h}_e^{(j)}. \tag{3.42}$$

After convergence of the inner loop is achieved, the linearization of the independent residual (3.34) yields

$$\mathbf{K}(\mathbf{p}^{(i)}, \mathbf{h})\Delta \mathbf{p}^{(i)} + \mathbf{R}(\mathbf{p}^{(i)}, \mathbf{h}) = \mathbf{0} \tag{3.43}$$

where $\mathbf{K}$ stands for the independent tangent operator obtained by a standard finite element assembly procedure (3.44) of the element tangent operators $\mathbf{K}_e$

$$\mathbf{K} = \mathop{\mathsf{A}}_{e=1}^{n_e} \mathbf{K}_e. \tag{3.44}$$

The independent solution vector is updated after the linear system (3.43) has been solved for the unknown increment $\Delta \mathbf{p}$ of the independent solution vector.

$$\mathbf{p}^{(i+1)} = \mathbf{p}^{(i)} + \Delta \mathbf{p}^{(i)} \tag{3.45}$$

In order to achieve quadratic convergence, the tangent operator has to be evaluated consistently with the presented numerical procedure. Due to the additive nature of assembly operator in (3.34) it is sufficient to linearize individual element contribution to the global residual. The linearization of the element residual (3.36) yields the element tangent operator

$$\mathbf{K}_e = \frac{\partial \mathbf{R}_e}{\partial \mathbf{p}_e} + \frac{\partial \mathbf{R}_e}{\partial \mathbf{h}_e}\frac{D\mathbf{h}_e}{D\mathbf{p}_e} \tag{3.46}$$

The unknown derivative $\frac{D\mathbf{h}_e}{D\mathbf{p}_e}$ in (3.46) is obtained by differentiating (3.35) with respect to $\mathbf{p}_e$

$$\mathbf{A}_e \frac{D\mathbf{h}_e}{D\mathbf{p}_e} = -\frac{\partial \mathbf{Q}_e}{\partial \mathbf{p}_e}. \tag{3.47}$$

Inserting (3.47) into (3.46) yields the final form of the independent element tangent operator $\mathbf{K}_e$

$$\mathbf{K}_e = \frac{\partial \mathbf{R}_e}{\partial \mathbf{p}_e} - \frac{\partial \mathbf{R}_e}{\partial \mathbf{h}_e}\mathbf{A}_e^{-1}\frac{\partial \mathbf{Q}_e}{\partial \mathbf{p}_e}. \tag{3.48}$$

The presented solution procedure for the nested iterative-sub iterative Newton scheme is summarized in Box 3.4. The algorithm in Box 3.4 only presents a solution procedure for one time or incremental step. The solution for this step can then be used within a general path-following procedure with constant steps (Box 3.2) or adaptive (Box 3.3) stepping.

3.3.2 *Automation of the Solution of Locally Coupled Problems*

The automation of the derivation of residuals $\mathbf{R}_e$ and $\mathbf{Q}_e$ in (3.10) depends on the actual problem solved. For example, in Sect. 2.6 the general *ADB* form of $\mathbf{R}_e$ is presented for the problems with potential and for the problems defined by the generalized weak form equation. The automation of the corresponding tangent operators is straightforward after the automation of $\mathbf{R}_e$ and $\mathbf{Q}_e$ is known. The automation of the dependent tangent operator $\mathbf{A}_e$ is achieved by replacing the partial derivative in (3.41) with the computational derivative as follows

$$\mathbf{A}_e = \frac{\hat{\delta}\mathbf{Q}_e}{\hat{\delta}\mathbf{h}_e}. \tag{3.49}$$

The independent element tangent operator (3.48) can be derived by directly applying the automatic differentiation procedure on the sub-iterative loop in Box 3.4 leading to

Input: $\mathbf{p}_n$ `// starting value for` $\mathbf{p}$
Input: $\mathbf{h}_n$ `// starting value for` $\mathbf{h}$
begin iterative loop
  $\mathbf{p} \leftarrow \mathbf{p}_n$
  repeat
    foreach *element* **do**
      begin sub iterative loop
        $\mathbf{h}_e \leftarrow \mathbf{h}_{en}$
        repeat
          $\mathbf{Q}_e \leftarrow \mathbf{Q}_e(\mathbf{p}_e, \mathbf{h}_e)$
          $\mathbf{A}_e \leftarrow \frac{\partial \mathbf{Q}_e}{\partial \mathbf{h}_e}$ .... $\xrightarrow{\text{automation}}$ .... $\mathbf{A}_e \leftarrow \frac{\hat{\delta}\mathbf{Q}_e}{\hat{\delta}\mathbf{h}_e}$
          solve $\mathbf{A}_e \Delta\mathbf{h}_e + \mathbf{Q}_e = \mathbf{0}$ for unknown $\Delta\mathbf{h}_e$
          $\mathbf{h}_e \leftarrow \mathbf{h}_e + \Delta\mathbf{h}_e$
        until *error criterion for* $\|\mathbf{Q}_e\|$ *and* $\|\mathbf{h}_e\|$ *is fulfilled*
      $\mathbf{R}_e \leftarrow \mathbf{R}_e(\mathbf{p}_e, \mathbf{h}_e(\mathbf{p}_e))$
      $\mathbf{K}_e \leftarrow \frac{\partial \mathbf{R}_e}{\partial \mathbf{p}_e} - \frac{\partial \mathbf{R}_e}{\partial \mathbf{h}_e}\mathbf{A}_e^{-1}\frac{\partial \mathbf{Q}_e}{\partial \mathbf{p}_e}$ ... $\xrightarrow{\text{automation}}$ ... $\mathbf{K}_e \leftarrow \left.\frac{\hat{\delta}\mathbf{R}_e}{\hat{\delta}\mathbf{p}_e}\right|_{\frac{D\mathbf{h}_e}{D\mathbf{p}_e} = -\mathbf{A}_e^{-1}\frac{\hat{\delta}\mathbf{Q}_e}{\hat{\delta}\mathbf{p}_e}}$
      add $\mathbf{R}_e$ to $\mathbf{R}$ and $\mathbf{K}_e$ to $\mathbf{K}$
    end foreach
    solve $\mathbf{K}\Delta\mathbf{p} + \mathbf{R} = \mathbf{0}$ for unknown $\Delta\mathbf{p}$
    $\mathbf{p} \leftarrow \mathbf{p} + \Delta\mathbf{p}$
  until *error criterion for* $\|\mathbf{R}\|$ *and* $\|\mathbf{p}\|$ *is fulfilled*
Result: $\mathbf{p}$ and $\mathbf{h}$ `// converged solution`

Box 3.4. Newton iterative-sub iterative scheme for primal analysis of time-independent locally coupled problems

$$\mathbf{K}_e = \frac{\hat{\delta}\mathbf{R}_e}{\hat{\delta}\mathbf{p}_e} \tag{3.50}$$

The independent element tangent operator derived in this way is already "consistent" or "algorithmic" and no additional procedures to derive consistent tangent modulus are required. However, this would involve differentiation of all arithmetic operations involved in execution of the sub-iterative scheme. This can be avoided if a local AD exception of type A is used to prescribe the derivatives of the dependent solution vector $\mathbf{h}_e$ with respect to the independent solution vector

$$\mathbf{K}_e = \left.\frac{\hat{\delta}\mathbf{R}_e}{\hat{\delta}\mathbf{p}_e}\right|_{\frac{D\mathbf{h}_e}{D\mathbf{p}_e}=-\mathbf{A}_e^{-1}\frac{\hat{\delta}\mathbf{Q}_e}{\hat{\delta}\mathbf{p}_e}} . \tag{3.51}$$

Alternatively, the global AD exception of type D, imposed on the definition of the current converged dependent solution vector $\mathbf{h}_e$, can be applied

$$\begin{aligned} \mathbf{h}_e &\leftarrow \mathbf{h}_e|_{\frac{D\mathbf{h}_e}{D\mathbf{p}_e}=-\mathbf{A}_e^{-1}\frac{\hat{\delta}\mathbf{Q}_e}{\hat{\delta}\mathbf{p}_e}} \\ \mathbf{K}_e &= \frac{\hat{\delta}\mathbf{R}_e}{\hat{\delta}\mathbf{p}_e} \end{aligned} \tag{3.52}$$

which yields the same result as (3.51).

Equation (3.52) effectively redefines the derivatives of $\mathbf{h}_g$ by attaching an additional AD exception to the definition. The purpose of the use of the global form of the AD exception in (3.52) is also to have the ability to apply several disjunctive definitions for the converged dependent solution vector $\mathbf{h}_e$. Note that Eqs. (3.50)–(3.52) are all numerically equivalent and that several other variants are possible. Solution and automation of time-independent and time-dependent locally coupled problems are summarized in Table 3.2 and presented in Box 3.4.

3.3.3 *Solution and Automation of Locally Coupled Problems Using Static Elimination of Local Unknowns*

Let $\mathbf{g} = \mathbf{p}_u \cup \mathbf{p}_e \cup \mathbf{h}_e$ be a set of global unknowns $\mathbf{p}$ extended by a set of local unknowns $\mathbf{h}_e$ of the eth element where $\mathbf{p}_u = \mathbf{p} \backslash \mathbf{p}_e$ is a set of global unknowns with the nodal unknowns of the eth element omitted and let $\mathbf{G} = \mathbf{R}_u \cup \mathbf{R}_{p_e} \cup \mathbf{Q}_e$ be a corresponding residual. The residual vector $\mathbf{R}_{p_e}$ that corresponds to a set of unknowns $\mathbf{p}_e$ of eth element is further split into a contribution of the eth element $\mathbf{R}_e$ and a contribution $\mathbf{R}_p$ of the rest of elements, thus $\mathbf{R}_{p_e} = \mathbf{R}_e + \mathbf{R}_p$. Linearization of $\mathbf{G}$ then leads to the following extended system of linear equations

Table 3.2 Primal analysis of nonlinear locally coupled problems

	Dependent problems	Independent problem
Primal time-independent problem	$\mathbf{Q}_e(\mathbf{p}_e, \mathbf{h}_e(\mathbf{p}_e)) = \mathbf{0}$	$\mathbf{R}(\mathbf{p}, \mathbf{h}(\mathbf{p})) = \mathbf{0}$
Primal time-dependent problem	$\mathbf{Q}_e(\mathbf{p}_e, \mathbf{h}_e(\mathbf{p}_e), \mathbf{p}_{en}, \mathbf{h}_{en}) = \mathbf{0}$	$\mathbf{R}(\mathbf{p}, \mathbf{h}(\mathbf{p}), \mathbf{p}_n, \mathbf{h}_n) = \mathbf{0}$
Solution of primal problem	$\mathbf{A}_e \Delta\mathbf{h}_e + \mathbf{Q}_e = \mathbf{0}$ $\mathbf{A}_e = \dfrac{\partial \mathbf{Q}_e}{\partial \mathbf{h}_e}$ $\mathbf{h}_e \leftarrow \mathbf{h}_e + \Delta\mathbf{h}_e$	$\mathbf{K}\Delta\mathbf{p} + \mathbf{R} = \mathbf{0}$ $\mathbf{K} = \mathsf{A}_{e=1}^{n_e} \dfrac{\partial \mathbf{R}_e}{\partial \mathbf{p}_e}$ $\mathbf{p} \leftarrow \mathbf{p} + \Delta\mathbf{p}$
Automation of element tangent using local AD exceptions	$\mathbf{A}_e = \frac{\hat{\delta}\mathbf{Q}_e}{\hat{\delta}\mathbf{h}_e}$	$\mathbf{K}_e = \frac{\hat{\delta}\mathbf{R}_e}{\hat{\delta}\mathbf{p}_e}\Big\vert_{\frac{D\mathbf{h}_e}{D\mathbf{p}_e} = -\mathbf{A}^{-1}\frac{\hat{\delta}\mathbf{Q}_e}{\hat{\delta}\mathbf{p}_e}}$
Automation of element tangent using global AD exceptions	$\mathbf{A}_e = \frac{\hat{\delta}\mathbf{Q}_e}{\hat{\delta}\mathbf{h}_e}$	$\mathbf{h}_e \leftarrow \mathbf{h}_e\vert_{\frac{D\mathbf{h}_e}{D\mathbf{p}_e} = -\mathbf{A}^{-1}\frac{\hat{\delta}\mathbf{Q}_e}{\hat{\delta}\mathbf{p}_e}}$ $\mathbf{K}_e = \dfrac{\hat{\delta}\mathbf{R}_e}{\hat{\delta}\mathbf{p}_e}$

$$
\begin{bmatrix} \mathbf{K}_{uu} & \mathbf{K}_{up} & \mathbf{0} \\ \mathbf{K}_{pu} & \mathbf{K}_e + \mathbf{K}_{pp} & \mathbf{K}_{ph} \\ \mathbf{0} & \mathbf{K}_{hp} & \mathbf{A}_e \end{bmatrix} \begin{Bmatrix} \Delta\mathbf{p}_u \\ \Delta\mathbf{p}_e \\ \Delta\mathbf{h}_e \end{Bmatrix} = - \begin{Bmatrix} \mathbf{R}_u \\ \mathbf{R}_e + \mathbf{R}_p \\ \mathbf{Q}_e \end{Bmatrix} \tag{3.53}
$$

where

$$
\begin{aligned}
\mathbf{K}_{uu} &= \frac{\partial \mathbf{R}_u}{\partial \mathbf{p}_u}, \ \mathbf{K}_{up} = \frac{\partial \mathbf{R}_u}{\partial \mathbf{p}_e}, \ \mathbf{K}_{pu} = \frac{\partial \mathbf{R}_{p_e}}{\partial \mathbf{p}_u}, \ \mathbf{K}_e = \frac{\partial \mathbf{R}_e}{\partial \mathbf{p}_e}, \\
\mathbf{K}_{pp} &= \frac{\partial \mathbf{R}_p}{\partial \mathbf{p}_e}, \ \mathbf{K}_{ph} = \frac{\partial \mathbf{R}_e}{\partial \mathbf{h}_e}, \ \mathbf{K}_{hp} = \frac{\partial \mathbf{Q}_e}{\partial \mathbf{p}_e}, \ \mathbf{A}_e = \frac{\partial \mathbf{Q}_e}{\partial \mathbf{h}_e}.
\end{aligned} \tag{3.54}
$$

Numerical efficiency of the finite element solution as well as the condition number of the global tangent matrix can be greatly improved if the local unknowns $\mathbf{h}_e$ are eliminated from (3.53) before the assembly of the global matrix. The fact that $\Delta\mathbf{h}_e$ and $\Delta\mathbf{p}_u$ are uncoupled can be used to efficiently eliminate $\Delta\mathbf{h}_e$ from (3.53). Term $\Delta\mathbf{h}_e$ expressed from the third equation leads to

$$
\Delta\mathbf{h}_e = -\mathbf{A}_e^{-1} \left(\mathbf{Q}_e + \mathbf{K}_{hp}\, \Delta\mathbf{p}_e\right). \tag{3.55}
$$

By inserting (3.55) into second equation in (3.53) a reduced set of global equations is obtained from (3.53) as follows

$$
\begin{bmatrix} \mathbf{K}_{uu} & \mathbf{K}_{up} \\ \mathbf{K}_{pu} & \mathbf{K}_{e,cond} + \mathbf{K}_{pp} \end{bmatrix} \begin{Bmatrix} \Delta\mathbf{p}_u \\ \Delta\mathbf{p}_e \end{Bmatrix} = - \begin{Bmatrix} \mathbf{R}_u \\ \mathbf{R}_{e,cond} + \mathbf{R}_p \end{Bmatrix} \tag{3.56}
$$

where $\mathbf{K}_{e,cond}$ and $\mathbf{R}_{e,cond}$ are the statically condensed element tangent matrix and residual, respectively

$$\mathbf{K}_{e,cond} = \mathbf{K}_e - \mathbf{K}_{ph}\,\mathbf{A}_e^{-1}\,\mathbf{K}_{hp}, \tag{3.57}$$

$$\mathbf{R}_{e,cond} = \mathbf{R}_e - \mathbf{K}_{ph}\,\mathbf{A}_e^{-1}\,\mathbf{Q}_e. \tag{3.58}$$

After Eq. (3.56) is solved for the unknown $\Delta\mathbf{p}_u$ and $\Delta\mathbf{p}_e$, Eq. (3.55) is applied to evaluate the unknown $\Delta\mathbf{h}_e$. With $\mathbf{K}_{e,cond}$ and $\mathbf{R}_{e,cond}$ evaluated at element level the global solution procedure remains the same as for uncoupled problems. Since the static condensation procedure is independent of a specific problem it can be implemented in a general way and used for arbitrary problems. From the point of automation it is convenient to consider the static condensation procedure as a part of the algorithm that assembles global tangent matrix and residual. In that case, the element nodal unknowns $\mathbf{p}_e$ are extended by the local unknowns $\mathbf{h}_e$ and all procedures for the automation of uncoupled problems can be applied directly. Even more convenient is that the solution procedures for locally coupled problems can be directly combined with the solution procedure for Gauss point coupled problems. A typical example are mixed finite elements combined with inelastic materials. A mixed finite element formulation results in a locally coupled problem and at the same time inelastic material results in Gauss point coupled problem.

Let $\breve{\mathbf{p}}_e = \mathbf{p}_e \cup \mathbf{h}_e$ be a set of nodal and locally coupled element degrees of freedom (DOFs). An extended element residual $\breve{\mathbf{R}}_e$ and an extended element tangent matrix $\breve{\mathbf{K}}_e$ is then defined by

$$\breve{\mathbf{R}}_e = \begin{Bmatrix} \mathbf{R}_e \\ \mathbf{Q}_e \end{Bmatrix}, \tag{3.59}$$

$$\breve{\mathbf{K}}_e = \frac{\hat{\delta}\breve{\mathbf{R}}_e}{\hat{\delta}\breve{\mathbf{p}}_e}. \tag{3.60}$$

Equation (3.60) requires only one call to the AD procedures. Consequently, it is in general optimal from the point of numerical efficiency of the resulting code. However, when the algorithms to evaluate $\mathbf{R}_e$ and $\mathbf{Q}_e$ follow significantly different paths, the code optimization procedure might be able to find numerically a more efficient solution if the sub-matrices of $\breve{\mathbf{K}}_e$ are evaluated by separate calls to the AD procedure

$$\breve{\mathbf{K}}_e = \begin{bmatrix} \frac{\hat{\delta}\mathbf{R}_e}{\hat{\delta}\mathbf{p}_e} & \frac{\hat{\delta}\mathbf{R}_e}{\hat{\delta}\mathbf{h}_e} \\ \frac{\hat{\delta}\mathbf{Q}_e}{\hat{\delta}\mathbf{p}_e} & \frac{\hat{\delta}\mathbf{Q}_e}{\hat{\delta}\mathbf{h}_e} \end{bmatrix}. \tag{3.61}$$

The integration point contribution to the element residual and tangent matrix is similarly defined by

$$\breve{\mathbf{R}}_g = \begin{Bmatrix} \mathbf{R}_g \\ \mathbf{Q}_g \end{Bmatrix}, \; \breve{\mathbf{K}}_g = \frac{\hat{\delta} \breve{\mathbf{R}}_g}{\hat{\delta} \breve{\mathbf{p}}_e}, \text{ and } \breve{\mathbf{K}}_g = \begin{bmatrix} \frac{\hat{\delta}\mathbf{R}_g}{\hat{\delta}\mathbf{p}_e} & \frac{\hat{\delta}\mathbf{R}_g}{\hat{\delta}\mathbf{h}_e} \\ \frac{\hat{\delta}\mathbf{Q}_g}{\hat{\delta}\mathbf{p}_e} & \frac{\hat{\delta}\mathbf{Q}_g}{\hat{\delta}\mathbf{h}_e} \end{bmatrix}. \tag{3.62}$$

Note that the element residual and tangent matrix have to be integrated over the element domain before the static condensation of $\breve{\mathbf{R}}_e$ and $\breve{\mathbf{K}}_e$ is performed.

3.3.4 Solution of Gauss Point Coupled Problems

The solution procedure of Gauss point coupled problems is similar to the solution procedure of locally coupled problems. The time-independent Gauss point coupled problem (3.13) is decomposed into a dependent and an independent problem by considering the current solution vector of the gth local problem $\mathbf{h}_g$ as a function of the solution vector of the global problem $\mathbf{p}$

$$\begin{aligned} \mathbf{R}(\mathbf{p}, \mathbf{h}(\mathbf{p})) &= \mathop{\mathsf{A}}_{e=1}^{n_e} \left(\sum_{g \in G_e} w_{gp} \mathbf{R}_g \right) = \mathbf{0}, \\ \mathbf{Q}_g(\mathbf{r}_g(\mathbf{p}_e), \mathbf{h}_g(\mathbf{r}_g(\mathbf{p}_e))) &= \mathbf{0} : \; g = 1, \ldots, n_{tg}, \\ \mathbf{R}_g &= \mathbf{R}_g \left(\mathbf{p}_e, \mathbf{h}_g(\mathbf{r}_g(\mathbf{p}_e)) \right). \end{aligned} \tag{3.63}$$

The time-dependent Gauss point coupled problem is similarly decomposed into

$$\begin{aligned} \mathbf{R}(\mathbf{p}, \mathbf{h}(\mathbf{p}), \mathbf{p}_n, \mathbf{h}_n) &= \mathop{\mathsf{A}}_{e=1}^{n_e} \left(\sum_{g \in G_e} w_{gp} \mathbf{R}_g \right) = \mathbf{0}, \\ \mathbf{Q}_g(\mathbf{r}_g(\mathbf{p}_e), \mathbf{h}_g(\mathbf{r}_g(\mathbf{p}_e)), \mathbf{p}_{en}, \mathbf{h}_{gn}) &= \mathbf{0} : \; g = 1, \ldots, n_{tg}, \\ \mathbf{R}_g &= \mathbf{R}_g \left(\mathbf{p}_e, \mathbf{h}_g(\mathbf{r}_g(\mathbf{p}_e)), \mathbf{p}_{en}, \mathbf{h}_{gn} \right). \end{aligned} \tag{3.64}$$

Due to the additive nature of the assembly operator and the Gauss integration used to form the global residual $\mathbf{R}$ in (3.63a) it is sufficient to linearized an individual Gauss point contribution to the element residual and consequently global residual. Thus, the partitioned Newton scheme derived in the previous section for locally coupled problems can be directly applied for Gauss point coupled problems. The linearisation of the dependent residual (3.63b) leads to the following update procedure for the Gauss point unknowns $\mathbf{h}_g$

$$\begin{aligned} \mathbf{A}_g(\mathbf{h}_g^{(j)}) \Delta \mathbf{h}_g^{(j)} + \mathbf{Q}_g(\mathbf{h}_g^{(j)}) &= \mathbf{0} \\ \mathbf{A}_g &= \frac{\partial \mathbf{Q}_g}{\partial \mathbf{h}_g} \\ \mathbf{h}_g^{(j+1)} &= \mathbf{h}_g^{(j)} + \Delta \mathbf{h}_g^{(j)} \end{aligned} \tag{3.65}$$

Table 3.3 Primal analysis of nonlinear Gauss point coupled problems

	Dependent problems	Independent problem
Primal time-independent problem	$\mathbf{Q}_g(\mathbf{r}_g(\mathbf{p}_e), \mathbf{h}_g) = \mathbf{0}$	$\mathbf{R}(\mathbf{p}, \mathbf{h}(\mathbf{p})) = \mathbf{0}$
Primal time-dependent problem	$\mathbf{Q}_g(\mathbf{r}_g(\mathbf{p}_e), \mathbf{h}_g, \mathbf{p}_{en}, \mathbf{h}_{gn}) = \mathbf{0}$	$\mathbf{R}(\mathbf{p}, \mathbf{h}(\mathbf{p}), \mathbf{p}_n, \mathbf{h}_n) = \mathbf{0}$
Solution of primal problem	$\mathbf{A}_g \Delta \mathbf{h}_g + \mathbf{Q}_g = \mathbf{0}$ $\mathbf{A}_g = \frac{\partial \mathbf{Q}_g}{\partial \mathbf{h}_g}$ $\mathbf{h}_g \leftarrow \mathbf{h}_g + \Delta \mathbf{h}_g$	$\mathbf{K}\Delta\mathbf{p} + \mathbf{R} = \mathbf{0}$ $\mathbf{K} = \mathop{\mathsf{A}}_{e=1}^{n_e} \left(\sum_{g \in G_e} w_{gp} \frac{\partial \mathbf{R}_g}{\partial \mathbf{p}_e} \right)$ $\mathbf{p} \leftarrow \mathbf{p} + \Delta\mathbf{p}$
Automation of element tangent using local AD exceptions	$\mathbf{A}_g = \frac{\hat{\delta} \mathbf{Q}_g}{\hat{\delta} \mathbf{h}_g}$	$\mathbf{K}_g = \left. \frac{\hat{\delta} \mathbf{R}_g}{\hat{\delta} \mathbf{p}_e} \right\rvert_{\frac{D\mathbf{h}_g}{D\mathbf{r}_g} = -\mathbf{A}^{-1} \frac{\hat{\delta}\mathbf{Q}_g}{\hat{\delta}\mathbf{r}_g}}$
Automation of element tangent using global AD exceptions	$\mathbf{A}_g = \frac{\hat{\delta} \mathbf{Q}_g}{\hat{\delta} \mathbf{h}_g}$	$\mathbf{h}_g \leftarrow \mathbf{h}_g \rvert_{\frac{D\mathbf{h}_g}{D\mathbf{r}_g} = -\mathbf{A}^{-1} \frac{\hat{\delta}\mathbf{Q}_g}{\hat{\delta}\mathbf{r}_g}}$ $\mathbf{K}_g = \frac{\hat{\delta} \mathbf{R}_g}{\hat{\delta} \mathbf{p}_e}$

and the linearisation of independent residual (3.63a) leads to an update of the global unknowns $\mathbf{p}$

$$
\begin{aligned}
\mathbf{K}(\mathbf{p}^{(i)}, \mathbf{h})\Delta\mathbf{p}^{(i)} + \mathbf{R}(\mathbf{p}^{(i)}, \mathbf{h}) &= \mathbf{0}, \\
\mathbf{K} &= \mathop{\mathsf{A}}_{e=1}^{n_e} \left(\sum_{g \in G_e} w_{gp} \mathbf{K}_g \right) \\
\mathbf{p}^{(i+1)} &= \mathbf{p}^{(i)} + \Delta\mathbf{p}^{(i)}.
\end{aligned}
\tag{3.66}
$$

The two formulations (locally coupled and Gauss point coupled) differ in the derivation of the consistent tangent matrix. The Gauss point residual explicitly depends only on a set of intermediate variables $\mathbf{r}_g$, thus the linearization of a Gauss point contribution (3.63c) to the independent residual yields

$$
\mathbf{K}_g = \frac{\partial \mathbf{R}_g}{\partial \mathbf{p}_e} + \frac{\partial \mathbf{R}_g}{\partial \mathbf{h}_g} \frac{D\mathbf{h}_g}{D\mathbf{r}_g} \frac{\partial \mathbf{r}_g}{\partial \mathbf{p}_e} = \frac{\partial \mathbf{R}_g}{\partial \mathbf{p}_e} - \frac{\partial \mathbf{R}_g}{\partial \mathbf{h}_g} \mathbf{A}_g^{-1} \frac{\partial \mathbf{Q}_g}{\partial \mathbf{r}_g} \frac{\partial \mathbf{r}_g}{\partial \mathbf{p}_e}. \tag{3.67}
$$

The two formulations are identical for $\mathbf{r}_g = \mathbf{p}_e$. In general Eq. (3.67) provides a numerically more efficient way of evaluating the element tangent matrix than Eq. (3.48) when the cardinality of $\mathbf{r}_g$ is less than cardinality of $\mathbf{p}_e$ ($|\mathbf{r}_g| < |\mathbf{p}_e|$).

3.3.5 Automation of the Solution of Gauss Point Coupled Problems

The automation of the dependent tangent operator $\mathbf{A}_g$ follows from (3.65b) as

Input: $\mathbf{p}_n$ `// starting value for` $\mathbf{p}$
Input: $\mathbf{h}_n$ `// starting value for` $\mathbf{h}$
$\mathbf{p} \leftarrow \mathbf{p}_n$ `// last converged solution is starting value for current time step`
repeat `// main iterative loop`
  foreach *element* **do**
    foreach *Gauss point* **do**
      $\mathbf{h}_{\bar{g}} \leftarrow \mathbf{h}_{gn}$ `// starting value is converged solution at the end of the previous time step`
      while true do `// sub-iterative loop`
        $\mathbf{Q}_g \leftarrow \mathbf{Q}_g\left(\mathbf{r}_g(\mathbf{p}_e), \mathbf{h}_{\bar{g}}, \mathbf{p}_{en}, \mathbf{h}_{gn}\right)$
        $\mathbf{A}_g \leftarrow \frac{\hat{\delta}\mathbf{Q}_g}{\hat{\delta}\mathbf{h}_{\bar{g}}}$
[A]   solve $\mathbf{A}_g \Delta\mathbf{h}_g + \mathbf{Q}_g = \mathbf{0}$ for unknown $\Delta\mathbf{h}_g$
        if *error criterion for* $\|\mathbf{Q}_g\|$ *and* $\|\mathbf{h}_g\|$ *is fulfilled* **then**
          solve $\mathbf{A}_g \widehat{D_{\mathbf{r}_g}\mathbf{h}_g} + \left.\frac{\hat{\delta}\mathbf{Q}_g}{\hat{\delta}\mathbf{r}_g}\right|_{\mathbf{h}_{\bar{g}}=cont.} = \mathbf{0}$ for unknown
            $\widehat{D_{\mathbf{r}_g}\mathbf{h}_g}$ `// already factorized matrix` $\mathbf{A}_g$ `from step [A] is used in this step`
          export $\mathbf{h}_g + \Delta\mathbf{h}_g$ to $\mathbf{h}$ data structure
          break enclosing loop
        end if
        $\mathbf{h}_{\bar{g}} \leftarrow \mathbf{h}_{\bar{g}} + \Delta\mathbf{h}_g$
      end while
      $\mathbf{h}_g \leftarrow \mathbf{h}_{\bar{g}}\Big|_{\frac{D\mathbf{h}_{\bar{g}}}{D\mathbf{r}_g}=\widehat{D_{\mathbf{r}_g}\mathbf{h}_g}}$
      $\mathbf{R}_g \leftarrow \mathbf{R}_g\left(\mathbf{p}_e, \mathbf{h}_g, \mathbf{p}_{en}, \mathbf{h}_{gn}\right)$
      $\mathbf{K}_g \leftarrow \frac{\hat{\delta}\mathbf{R}_g}{\hat{\delta}\mathbf{p}_e}$
      add $w_{gp}\mathbf{R}_g$ to $\mathbf{R}_e$ and $w_{gp}\mathbf{K}_g$ to $\mathbf{K}_e$
    end foreach
    add $\mathbf{R}_e$ to $\mathbf{R}$ and $\mathbf{K}_e$ to $\mathbf{K}$
  end foreach
  solve $\mathbf{K}\Delta\mathbf{p} + \mathbf{R} = \mathbf{0}$ for unknown $\Delta\mathbf{p}$
  $\mathbf{p} \leftarrow \mathbf{p} + \Delta\mathbf{p}$
until *error criterion for* $\|\mathbf{R}\|$ *and* $\|\mathbf{p}\|$ *is fulfilled*
Result: $\mathbf{p}_n$, $\mathbf{h}_n$, $\mathbf{p}$ and $\mathbf{h}$
`// converged solution in time` t_n `and` t_{n+1}

Box 3.5. Automation of Newton–Raphson iterative scheme for primal analysis of time-dependent Gauss point coupled problem (3.64) using global AD exceptions for one time step

$$\mathbf{A}_g = \frac{\hat{\delta}\mathbf{Q}_g}{\hat{\delta}\mathbf{h}_g} \tag{3.68}$$

and the automation of independent tangent operator $\mathbf{K}_g$ using a local AD exception for the definition of an implicit dependency of $\mathbf{h}_g$ on $\mathbf{r}_g$ results from (3.67) as

$$\mathbf{K}_g = \left.\frac{\hat{\delta}\mathbf{R}_g}{\hat{\delta}\mathbf{p}_e}\right|_{\frac{D\mathbf{h}_g}{D\mathbf{r}_g}=-\mathbf{A}_g^{-1}\frac{\hat{\delta}\mathbf{Q}_g}{\hat{\delta}\mathbf{r}_g}} \tag{3.69}$$

and using global AD exception for the same implicit dependency is given by

$$\begin{aligned} \mathbf{h}_g &\leftarrow \mathbf{h}_g\big|_{\frac{D\mathbf{h}_g}{D\mathbf{r}_g}=-\mathbf{A}_g^{-1}\frac{\hat{\delta}\mathbf{Q}_g}{\hat{\delta}\mathbf{r}_g}}, \\ \mathbf{K}_g &= \frac{\hat{\delta}\mathbf{R}_g}{\hat{\delta}\mathbf{p}_e}. \end{aligned} \tag{3.70}$$

Note that equations (3.69), (3.70) are numerically equivalent. Solution and automation of time-independent and time-dependent Gauss point coupled problems are summarized in Table 3.3 and presented in Box 3.5. The automation is identical for time-independent (3.63) and time-dependent (3.64) problems. Box 3.4 only presents a solution procedure for one time or incremental step. The solution for one step can then be used within a general path-following procedure with constant steps (see Box 3.2) or addaptive (see Box 3.3) stepping.

References

Bazaraa, M.S., H.D. Sherali, and C.M. Shetty. 1993. *Nonlinear programming, theory and algorithms*, 2nd ed. Chichester: Wiley.

Boersma, A., and P. Wriggers. 1997. An algebraic multigrid solver for finite element computations in solid mechanics. *Engineering Computations* 14: 202–215.

Braess, D. 2007. *Finite elements: theory, fast solvers, and applications in solid mechanics*. Cambridge: Cambridge University Press.

Brenner, S.C., and L.R. Scott. 2002. *The mathematical theory of finite element methods*. Berlin: Springer.

Ciarlet, P.G. 1988. *Mathematical elasticity I: three-dimensional elasticity*. Amsterdam: North-Holland.

Ciarlet, P.G. 1989. *Introduction to numerical linear algebra and optimization, English Translation of 1982 Edition in Cambridge Texts in Applied Mathematics*. Cambridge: Cambridge University Press.

Crisfield, M.A. 1981. A fast incremental/iterative solution prodedure that handles snap through. *Computers and Structures* 13: 55–62.

Crisfield, M.A. 1991. *Non-linear finite element analysis of solids and structures*, vol. 1. Chichester: Wiley.

Crisfield, M.A., and J. Shi. 1991. A review of solution procedures and path-following techniques in relation to the non-linear finite element analysis of structures. In *Computational Methods in Nonlinear Mechanics*, ed. P. Wriggers, and W. Wagner. Berlin: Springer.

Douglas, C., G. Haase, and U. Langer. 2003. *A tutorial on elliptic PDE solvers and their parallelization*. Philadelphia: SIAM.

Duff, I.S. 2004. Ma57 - a new code for the solution of sparse symmetric definite and indefinite systems. *ACM Transactions on Mathematical Software* 30: 118–154.

Duff, I.S., A.M. Erisman, and J.K. Reid. 1989. *Direct methods for sparse matrices*. Oxford: Clarendon Press.

Elman, H.C., D.J. Silvester, and A.J. Wathen. 2005. *Finite elements and fast iterative solvers with applications in incompressible fluid dynamics*. Oxford: Oxford University Press.

Hackbusch, W. 1994. *Iterative solution of large sparse systems*. New York: Springer.

Isaacson, E., and H.B. Keller. 1966. *Analysis of numerical methods*. London: Wiley.

Johnson, C. 1987. *Numerical solution of partial differential equations by the finite element method*. Cambridge: Cambridge University Press.

Jung, M., and U. Langer. 2001. *Methode der finiten Elemente für Ingenieure*. Teubner.

Keller, H.B. 1977. Numerical solution of bifurcation and nonlinear eigenvalue problems. In *Application of bifurcation theory*, ed. P. Rabinowitz, 359–384. New York: Academic Press.

Kickinger, F. 1996. Algebraic multigrid solver for discrete elliptic second order problems. Technical Report 96-5, Department of Mathematics, Johannes Kepler University, Linz.

Korneev, V.G., U. Langer, and L. Xanthis. 2003. On fast domain decomposition solving procedures for hp-discretizations of 3d elliptic problems. *Computational Methods in Applied Mathematics* 3: 536–559.

Luenberger, D.G. 1984. *Linear and nonlinear programming*, 2nd ed. Reading: Addison-Wesley.

Marsden, J.E., and T.J.R. Hughes. 1983. *Mathematical foundations of elasticity*. Englewood Cliffs: Prentice-Hall.

Meyer, A. 1990. A parallel preconditioned conjugate gradient method using domain decomposition and inexact solvers on each subdomain. *Computing* 45: 217–234.

Michaleris, P., D. Tortorelli, and C. Vidal. 1994. Tangent operators and design sensitivity formulations for transient non-linear coupled problems with applications to elastoplasticity. *International Journal for Numerical Methods in Engineering* 37: 2471–2499.

Ortega, J., and W. Rheinboldt. 1970. *Iterative solution of nonlinear equations in several variables*. New York: Academic Press.

Ramm, E. 1981. Strategies for tracing the nonlinear response near limit points. In *Nonlinear finite element analysis in structural mechanics*, ed. W. Wunderlich, E. Stein, and K.J. Bathe. Berlin: Springer.

Rheinboldt, W. 1984. *Methods for solving systems of nonlinear equations*. Philadelphia: Society for Industrial and Applied Mathematics.

Riks, E. 1972. The application of Newtons method to the problem of elastic stability. *Journal of Applied Mechanics* 39: 1060–1066.

Riks, E. 1984. Some computational aspects of stability analysis of nonlinear structures. *Computer Methods in Applied Mechanics and Engineering* 47: 219–260.

Saad, Y. 2003. Iterative methods for sparse linear systems. *SIAM*.

Schenk, O., and K. Gärtner. 2004. Solving unsymmetric sparse systems of linear equations with pardiso. *Journal of Future Generation Computer Systems* 20: 475–487.

Schwetlick, H., and H. Kretschmar. 1991. *Numerische Verfahren für Naturwissenschaftler und Ingenieure*. Leipzig: Fachbuchverlag.

Schweizerhof, K., and P. Wriggers. 1986. Consistent linearization for path following methods in nonlinear fe-analysis. *Computer Methods in Applied Mechanics and Engineering* 59: 261–279.

Taylor, R.L. 2000. A mixed-enhanced formulation for tetrahedral finite elements. *International Journal for Numerical Methods in Engineering* 47: 205–227.

Vainberg, M.M. 1964. *Variational methods for the study of nonlinear operators*. San Francisco: Holden Day.

Wagner, W. 1991. Zur Behandlung von Stabilitätsproblemen mit der Methode der Finiten Elemente. Technical Report F91/1, Forschungs- und Seminarberichte aus dem Bereich der Mechanik der Universität Hannover.
Wagner, W., and P. Wriggers. 1988. A simple method for the calculation of secondary branches. *Engineering Computations* 5: 103–109.
Wriggers, P. 2008. *Nonlinear finite elements*. Berlin: Springer.
Zienkiewicz, O.C., and R.L. Taylor. 2000a. *The finite element method*, vol. 2, 5th ed. Oxford: Butterworth-Heinemann.

Chapter 4
Automation of Discretization Techniques

Finite elements will be applied to discretize weak forms of the nonlinear continuum or structural problems. Within this approach the primary variables, like displacements or rotations etc., have to be approximated within a finite element by interpolation functions. The emphasis of this chapter is to describe how the automation technique introduced in Chap. 2 can be employed to generate finite element vectors and matrices automatically as well as providing efficient source code for different finite element programs. Hence this chapter is focused on the essential details needed to apply the systems *AceGen* to a standard nonlinear analysis of continua using finite elements. To achieve this goal, first the necessary interpolation functions and related mappings are discussed then two-dimensional continuum elements are derived using *AceGen*. The derivation starts from different configurations and also as well from the functional and weak forms presented in the previous chapter.

Besides the fact that all relevant equations are presented in this chapter it is assumed that the reader has an understanding of finite element methods for linear problems since basic operations like e.g. assembly processes are not discussed. For this the reader has to consult standard text books for finite elements, like e.g. Hughes (1987), Zienkiewicz and Taylor (1991), Bathe (1996), Braess (2007), Oñate (2009).

A continuum body is discretized by an approximation of the geometry of the body B in the initial configuration

$$B \approx B^h = \bigcup_{e=1}^{n_e} \Omega_e . \tag{4.1}$$

This approach leads to a subdivision of B into n_e finite elements. The configuration of one element is described by $\Omega_e \subset B^h$, see Fig. 4.1 for the two-dimensional case. The boundary of the finite element region ∂B^h consist of element boundaries $\partial \Omega_e$ related to the elements Ω_e: $\partial B^h = \left(\bigcup_{e=1}^{n_e} \partial \Omega_e \right) \backslash \partial B^h_{\text{int}}$ where $\partial B^h_{\text{int}}$ stands for interior element boundaries. In general this only provides an approximation of the real geometry of the boundary ∂B.

J. Korelc and P. Wriggers, *Automation of Finite Element Methods*,
DOI 10.1007/978-3-319-39005-5_4

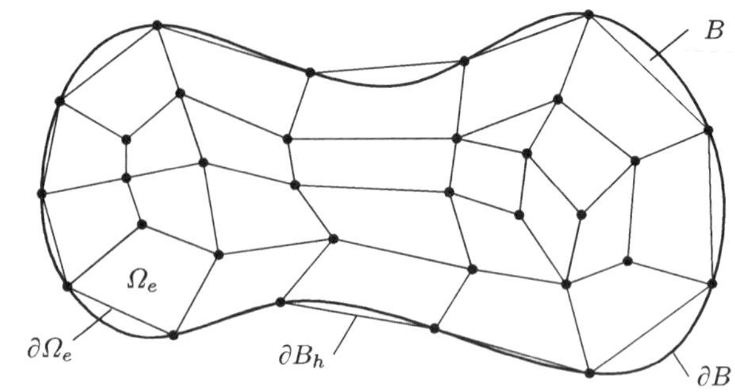

Fig. 4.1 Discretization of body B

An overlapping of finite elements is not allowed as well as gaps are not allowed. Both would violate the compatibility requirements of continuum theory. Hence the assembled elements have to be continuous in the region B.

4.1 General Isoparametric Concept

Within the finite element approach interpolation functions have to be selected for the approximation of the primary field variables. Hence the exact solution of the mathematical model is approximated within a single finite element by

$$\boldsymbol{u}_{\text{exact}}(\boldsymbol{X}) \approx \boldsymbol{u}_h(\boldsymbol{X}) = \sum_{I=1}^{n_{en}} N_I(\boldsymbol{X})\,\mathbf{u}_I = (\mathbf{N}\,\mathbf{u})^T \tag{4.2}$$

In (4.2) the vector $\boldsymbol{X}$ denotes the position vector with respect to the initial configuration in Ω_e, $N_I(\boldsymbol{X})$ are the shape functions which are defined in Ω_e and the unknown nodal quantities of the primary variable are represented by $\mathbf{u}_I$ (these could be e.g. the nodal displacements $\mathbf{u}_I = \{u_I, v_I, w_I\}^T$ for a three-dimensional displacement formulation using the weak form (1.90)). The vector of shape functions $\mathbf{N}$ is given by

$$\mathbf{N} = \{N_1, N_2, \ldots, N_I, \ldots, N_{n_{en}}\} \tag{4.3}$$

where n_{en} is the number of nodal points defining the element. With this definition it is clear that $\mathbf{u} = \{\{u_I, v_I, w_I\} : I = 1, \ldots, n_{en}\}$ is a nodal wise ordered nested set of nodal displacements related to one element that contains $n_{en} \times n_{\text{dim}}$ nodal components related to the spacial dimensions n_{dim} and the number of nodes n_{en}.

An exact solution of the mathematical model is in general unknown and never appears as a part of the finite element equations. Consequently, within the formulation of the finite element equations the index h that indicates an approximated solution can be omitted without loosing the clarity of the formulation. In the subsequent derivations a quantity without an index h always indicates an approximated solution evaluated within a single finite element Ω_e, thus $\boldsymbol{u}_h \equiv \boldsymbol{u}$.

The set of nodal displacements $\mathbf{u}_I$ can be defined in *AceGen* by the command

```
uIO⊢Table[SMSReal[nd$$[i,"at",j]],{i,nen},{j,ndim}]
```
(4.4)

for an arbitrary dimensional problem where `uIO` $\equiv \mathbf{u}$, `ndim` $\equiv n_{\text{dim}}$ and `nen` $\equiv n_{en}$. Once these nodal displacements are defined the interpolation of the displacements `u` $\equiv \boldsymbol{u}$ within an element Ω_e is provided by

```
u⊨Nh.uIO
```
(4.5)

where the shape functions `Nh` $\equiv \mathbf{N}$ have to be defined according to the chosen interpolation within a finite element.

One basic requirement for the choice of the approximation $\boldsymbol{u}$ is the convergence of the finite element solution to the true solution of the underlying partial differential equation. Different possibilities exist to construct interpolation or ansatz functions for geometry and variables within the finite element method. For convergence reasons these functions have to be complete up to the approximation order (e.g. a polynomial including the terms $1, x, y, z, x^2, y^2, z^2, xy, yz, zx$ is complete up to second order), see e.g. Zienkiewicz and Taylor (2000b).

Due to its general applicability the isoparametric concept is mainly used as interpolation scheme for many engineering problems. Within this concept geometry and variables are interpolated by the same ansatz functions. Isoparametric elements allow a very good and sufficiently accurate mapping of arbitrary geometries into a finite element mesh. Furthermore this concept is extremely well suited for nonlinear problems since a discretization of the spatial formulation is easily obtained. This is due to the fact that it makes no difference whether the mapping onto a reference element $\Omega_\square$, needed in the isoparametric formulation, is performed from the initial or the spatial configuration. One further advantage stems from the local orthogonal coordinate system at the reference element $\Omega_\square$ which means that neither co-nor contravariant derivatives have to be computed, even if the bodies under consideration, e.g. shells or beams, depict curved geometries.

The isoparametric concept can be expressed mathematically within a single finite element Ω_e, see also Fig. 4.2, by the interpolation of spatial and initial configuration[1]

$$\begin{aligned} X_e &= \mathbf{N}(\boldsymbol{\Xi})\,\mathbf{X}_c \; Y_e = \mathbf{N}(\boldsymbol{\Xi})\,\mathbf{Y}_c \; Z_e = \mathbf{N}(\boldsymbol{\Xi})\,\mathbf{Z}_c, \\ x_e &= \mathbf{N}(\boldsymbol{\Xi})\,\mathbf{x}_c \; y_e = \mathbf{N}(\boldsymbol{\Xi}), \mathbf{y}_c \; z_e = \mathbf{N}(\boldsymbol{\Xi})\,\mathbf{z}_c. \end{aligned} \tag{4.6}$$

[1] In an Computer algebra system like *Mathematica* the scalar product of two one-dimensional matrices is denoted by a "·" as it is used also in this book for the scalar product of vectors. However since we have here one-dimensional matrices of length n_{en} the scalar product is defined by the multiplication of the transposed vector $\mathbf{N}$ with the vector containing the coordinates.

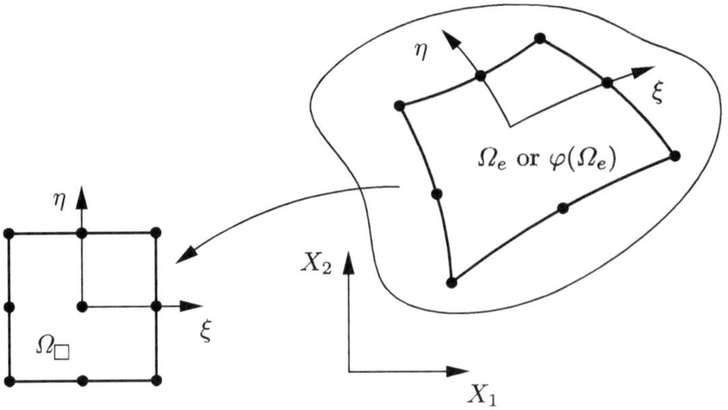

Fig. 4.2 Isoparametric mapping onto the reference configuration

Furthermore the nodal coordinates are contained in vectors

$$\begin{aligned}\mathbf{X}_c &= \{X_1, X_2, \ldots, X_I, \ldots, X_{n_{en}}\}^T \\ \mathbf{Y}_c &= \{Y_1, Y_2, \ldots, Y_I, \ldots, Y_{n_{en}}\}^T \\ \mathbf{Z}_c &= \{Z_1, Z_2, \ldots, Z_I, \ldots, Z_{n_{en}}\}^T\end{aligned} \tag{4.7}$$

or collected in a nodal wise ordered nested set of all coordinates of all element nodes

$$\mathbf{X} = \{\{X_I, Y_I, Z_I\} : I = 1, \ldots, n_{en}\} \tag{4.8}$$

with dimension $n_{en} \times n_{\text{dim}}$ where $\mathbf{X}_I = \{X_I, Y_I, Z_I\}^T$ is a geometrical nodal point of Ith element node. Interpolation of the coordinates within an element Ω_e then follows as

$$X = \{X_e, Y_e, Z_e\}^T = \sum_{I=1}^{n_{en}} N_I(\boldsymbol{\Xi})\mathbf{X}_I = (\mathbf{N}(\boldsymbol{\Xi})\mathbf{X})^T \,. \tag{4.9}$$

where the shape functions $\mathbf{N}(\boldsymbol{\Xi})$ have to be defined according to the chosen interpolation within a finite element. The nodal wise ordered nested set of nodal points $\mathbf{X}_I$ can be defined in *AceGen* by the command

```
𝕏IO⊢Table[SMSReal[nd$$[i,"X",j]],{i,nen},{j,ndim}]
```
(4.10)

where 𝕏IO denotes the set $\mathbf{X}_I$ and the interpolation of the coordinates within an element Ω_e is provided by

```
𝕏⊨Nh.𝕏IO
```
. (4.11)

Usually a polynomial is chosen for the ansatz (interpolation) function N_I. It is defined in the reference configuration $\Omega_\square$ in order to characterize arbitrary element geometries in the initial or spatial configuration. The ansatz functions within a finite element in the initial configuration Ω_e have been replaced in Eq. (4.6) by the shape functions $N_I(\boldsymbol{\Xi})$ defined within the reference element $\Omega_\square$. Thus for each element Ω_e a transformation (4.6) has to be performed, which relates the coordinates $\boldsymbol{X} = \boldsymbol{X}(\boldsymbol{\Xi})$ to the coordinates

$$\boldsymbol{\Xi} = \{\Xi_1, \Xi_2, \Xi_3\}^T = \{\xi, \eta, \zeta\}^T \tag{4.12}$$

of the reference element $\Omega_\square$. This transformation has to fulfill the following requirements:

- For each point within the reference element $\Omega_\square$ there exists one and only one point in Ω_e or $\varphi(\Omega_e)$.
- The geometrical nodal points $\mathbf{X}_I$ or $\mathbf{x}_I$ of $\Omega_\square$ are related to points in Ω_e or $\varphi(\Omega_e)$.
- Each part of the boundary on $\Omega_\square$, which is defined by the nodal points of $\mathbf{X}_I$ or $\mathbf{x}_I$ corresponds to the associated part of the boundary of Ω_e or $\varphi(\Omega_e)$.

With these assumptions isoparametric transformations preserve the element type (e.g. a triangle remains a triangle) in the initial or deformed finite element configurations.

The volume or area $\Omega_\square$ of the reference element will basically never be occupied by the real configuration of an element undergoing a physical deformation process. However the reference configuration $\Omega_\square$ provides an easy way to handle different configurations and can be used for the integration of the element matrices etc. This especially simplifies formulations related to the current configuration since it is absolutely arbitrary whether the transformation is performed from the reference element to the current or the initial configuration.

This transformation process is depicted in Fig. 4.3. The mapping of an element from the initial configuration Ω_e to the current configuration $\varphi(\Omega_e)$ is performed using the approximate deformation map φ_h which is described by φ_e to show its relation to a specific finite element Ω_e. For the mapping the deformations gradient is needed which is here denoted by $\boldsymbol{F}$.
It is easy to see that the mapping in Fig. 4.3 is the discrete version of the continuum mechanical description of the motion of a body, which is shown in Fig. 1.1. Additionally the reference configuration $\Omega_\square$ is introduced here which is needed for the isoparametric mapping. From Fig. 4.3 the following kinematical relations can be deduced that are valid for a single finite element

$$\boldsymbol{F} = \mathbf{j}_e\,\mathbf{J}_e^{-1} \quad \text{and} \quad J_F = \det \boldsymbol{F} = \frac{\det \mathbf{j}_e}{\det \mathbf{J}_e}. \tag{4.13}$$

These show that the deformation gradient is defined by the isoparametric mapping from $\Omega_\square$ to the initial configuration Ω_e and to the current configuration $\varphi(\Omega_e)$. In this mapping the gradients $\mathbf{j}_e$ and $\mathbf{J}_e$ are defined as

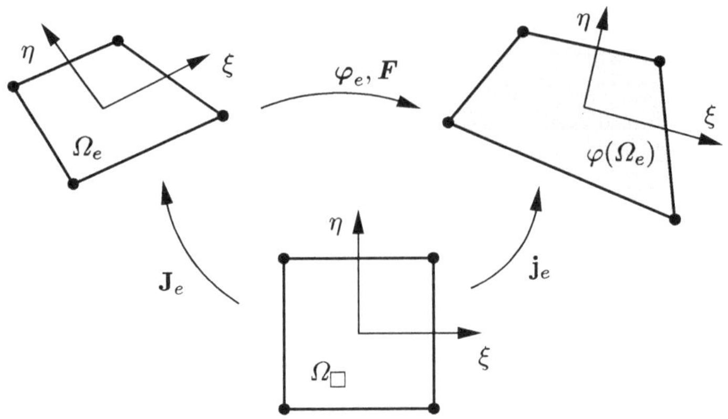

Fig. 4.3 Isoparametric mapping of the deformation of a finite element Ω_e

$$
\begin{aligned}
\mathbf{j}_e &= \mathrm{Grad}_{\Xi}\, \mathbf{x} = \frac{\partial \mathbf{x}}{\partial \boldsymbol{\Xi}}, \\
\mathbf{J}_e &= \mathrm{Grad}_{\Xi}\, \mathbf{X} = \frac{\partial \mathbf{X}}{\partial \boldsymbol{\Xi}}.
\end{aligned}
\tag{4.14}
$$

The command to be used within *AceGen* to compute the Jacobian $\mathbf{J}_e$ and its determinant is provided below

```
Je⊨SMSD[X,Ξ];        Jed⊨Det[Je];
```
. (4.15)

With these relations it is relatively simple to compute gradients related to the initial or current configuration. Exemplarily the gradients of the displacement vector field $\boldsymbol{u}$ in (4.2) can be specified within the element Ω_e by

$$
\begin{aligned}
\mathrm{Grad}\, \boldsymbol{u} &= \frac{\partial \boldsymbol{u}}{\partial \mathbf{X}}, \\
\mathrm{grad}\, \boldsymbol{u} &= \frac{\partial \boldsymbol{u}}{\partial \mathbf{x}}.
\end{aligned}
\tag{4.16}
$$

These gradients in can be written with (4.14) completely in terms of the reference configuration $\Omega_\square$

$$
\begin{aligned}
\mathrm{Grad}\, \boldsymbol{u} &= \frac{\partial \boldsymbol{u}}{\partial \Xi}\,\frac{\partial \Xi}{\partial \mathbf{X}} = \mathbf{J}_e^{-T}\,\frac{\partial \boldsymbol{u}}{\partial \Xi}, \\
\mathrm{grad}\, \boldsymbol{u} &= \frac{\partial \boldsymbol{u}}{\partial \Xi}\,\frac{\partial \Xi}{\partial \mathbf{x}} = \mathbf{j}_e^{-T}\,\frac{\partial \boldsymbol{u}}{\partial \Xi}.
\end{aligned}
\tag{4.17}
$$

The only difference in the formulation of both gradients in (4.17) consists of the exchange of the gradient $\mathbf{j}_e$ by $\mathbf{J}_e$, and vice versa. Hence especially for non linear finite element methods this formulation provides the most flexible approach.[2]

Within the automation process for the generation of finite element matrices the detailed description of the transformation above is not needed. Hence, once the shape functions are defined, the displacement gradient (1.12) can be computed directly by

$$\boldsymbol{H} = \operatorname{Grad} \boldsymbol{u} = \left.\frac{\hat{\delta}\boldsymbol{u}}{\hat{\delta}\boldsymbol{X}}\right|_{\frac{D\Xi}{DX}=\mathbf{J}_e^{-1}} \tag{4.19}$$

where the exception rule takes the special structure of the chain rule used in (4.17) into account. This is reflected in the input for *AceGen* by

```
X⊢SMSFreeze[Nh.XIO]
Je⊨SMSD[X,Ξ]
H⊨SMSD[u,X,"Dependency"→{Ξ,X,SMSInverse[Je]}]
```

. (4.20)

Similarly, the spatial gradient can be computed by

$$\operatorname{grad} u = \left.\frac{\hat{\delta}\boldsymbol{u}}{\hat{\delta}\boldsymbol{x}}\right|_{\frac{D\Xi}{Dx}=\mathbf{j}_e^{-1}} \tag{4.21}$$

and the corresponding *AceGen* input is given by

```
X⊨Nh.XIO
x⊢SMSFreeze[X+u]
je⊨SMSD[x,Ξ]
gradu⊨SMSD[u,x,"Dependency"→{Ξ,x,SMSInverse[je]}]
```

. (4.22)

Note that the `SMSFreeze` functions is applied on initial coordinates $\boldsymbol{X}$ in Eq. (4.20) and on spatial coordinates $\boldsymbol{x}$ in Eq. (4.22). The `SMSFreeze` functions has to be be applied on any variable that is used as independent variable for differentiation.

For continuum elements, shell or beam elements for shear elastic formulations an essential requirement for the ansatz or interpolation functions $N_I(\boldsymbol{\Xi})$ is the C^0 continuity. Furthermore ansatz functions $N_I(\boldsymbol{\Xi})$ have to be used which are complete

[2]In an analogous way the transformations between the gradients of different configurations, see (1.23), are derived e.g. for the gradient of a single scalar shape function N_I

$$\nabla_X N_I = \mathbf{J}_e^{-T}\, \nabla_{\Xi} N_I \quad \text{and} \quad \nabla_x N_I = \mathbf{j}_e^{-T}\, \nabla_{\Xi} N_I \,. \tag{4.18}$$

polynomials in the coordinate space X_1, X_2 and X_3 in which the mechanical problem is formulated. Different possibilities exist to construct such interpolation functions.

Ansatz functions which are also well suited for application within the isoparametric concept are provided by the Lagrangian interpolation functions, see e.g. Zienkiewicz and Taylor (2000b). In case of one dimension

$$N_I(\xi) = \prod_{\substack{J=1 \\ J \neq I}}^{n} \frac{(\xi_J - \xi)}{(\xi_J - \xi_I)} \tag{4.23}$$

is obtained for a Lagrangian polynomial of order $n - 1$. The index I describes the node at which the polynomial has to assume the value "1" while the ansatz function has the value "0" at all other nodes J of the element Ω_e.

For two- or three dimensional interpolation the product form

$$N_J(\xi, \eta) = N_I(\xi)\, N_K(\eta) \quad \text{or} \quad N_J(\xi, \eta, \zeta) = N_I(\xi)\, N_K(\eta)\, N_L(\zeta) \tag{4.24}$$

is used with $J = 1, \dots n_{\text{dim}}$ and $I, K, L = 1, \dots n$. The ansatz functions are defined in the local coordinate system $\boldsymbol{\Xi} = \{\Xi_1, \Xi_2, \Xi_3\}^T = \{\xi, \eta, \zeta\}^T$. Thus a transformation to the physical coordinates X_1, X_2 or X_3 is necessary, see Fig. 4.2 or 4.3, which defines the problems in the physical space. Within the next sections the isoparametric ansatz functions are specified for one-, two- and three-dimensional applications.[3]

4.1.1 One-Dimensional Interpolations

Ansatz functions. Here one-dimensional ansatz or shape functions are discussed which are C^0-continuous. These shape functions are derived from (4.23) by inserting the proper local coordinates into the formula. Since only one coordinate ξ is present the general ansatz for the coordinates and displacement field in one element is with (4.6) and (4.7) given by

$$X_e = \sum_{I=1}^{n_{en}} N_I(\xi)\, X_I, \qquad u_e = \sum_{I=1}^{n_{en}} N_I(\xi)\, u_I. \tag{4.25}$$

X_e denotes the coordinate within the element and u_e is the associated displacement field. The value n_{en} is related to the number of interpolation functions and $n_{en} - 1$ determines the order of interpolation, e.g. $n_{en} = 3$ is related to a quadratic element.

[3]For classical beam and shell theories, due to the fact that the mathematical models are of higher order, different interpolation functions are needed which are e.g. C^1-continuous. The associated formulations and specifications of the interpolation functions will be provided in the sections where the beam and shell theories are described.

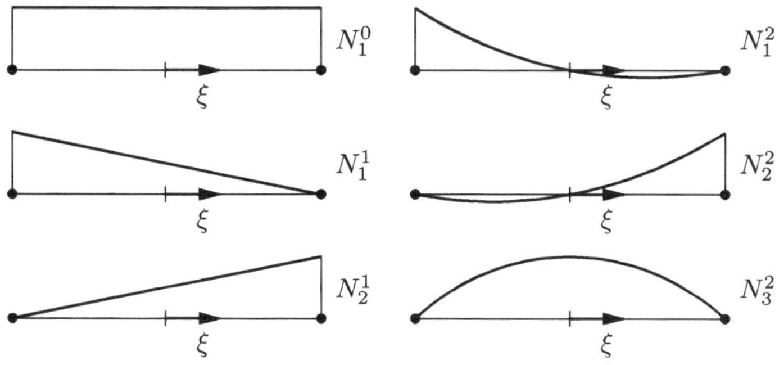

Fig. 4.4 One-dimensional shape functions: constant, linear and quadratic

The local coordinate $\xi \in [-1, 1]$ is used within the one-dimensional reference element, see Fig. 4.4.

The shape functions $N_I(\xi)$ follow from (4.23) and are different with respect to the order of the polynomial approximation. The shape functions are stated in the following up to quadratic order. Here the upper index denotes the order of the polynomial, see Fig. 4.4.

- Constant shape function

$$N_1^0(\xi) = 1 \tag{4.26}$$

- Linear shape functions

$$N_1^1\,(\xi) = \frac{1}{2}\,(1-\xi) \qquad N_2^1\,(\xi) = \frac{1}{2}\,(1+\xi) \tag{4.27}$$

- Quadratic shape functions

$$N_1^2\,(\xi) = \frac{1}{2}\,\xi\,(\xi-1) \qquad N_3^2\,(\xi) = (1-\xi^2) \qquad N_2^2\,(\xi) = \frac{1}{2}\,\xi\,(1+\xi) \tag{4.28}$$

It can be easily shown that these shape functions fulfill the conditions discussed in the previous section. The isoparametric mapping of a function u_e onto the reference element is obtained by using Eq. $(4.25)_1$.

Integration within the reference element. Within the finite element method the weak form contributions have to be integrated within each finite element Ω_e. Using the isoparametric concept these integrations are performed within the reference element $\Omega_\square$. Hence the integrals related to the element Ω_e are transformed

$$\int\limits_{(\Omega_e)} f(X)\,dX = \int\limits_{(\Omega_\square)} f(\xi)\,\frac{dX}{d\xi}\,d\xi = \int\limits_{-1}^{+1} f(\xi)\,J_e(\xi)\,d\xi. \tag{4.29}$$

The integration in the reference element $\Omega_\square$ is defined with respect to the parameter space $[-1 \leq \xi \leq +1]$.

In general the integration is performed numerically since the product $f(\xi)\, J_e(\xi)$ is usually no polynomial but a rational function. Thus the integral in (4.29) is approximated by

$$\int_{-1}^{+1} f(\xi)\, J_e(\xi)\, \delta\xi \approx \sum_{g=1}^{n_g} f(\xi_g)\, J_e(\xi_g)\, w_{gp}. \tag{4.30}$$

w_{gp} are the weighting factors and ξ_g are the coordinates of the evaluation points g.

Due to its accuracy and thus efficiency the Gauss integration is applied. Table C.1 provides the weighting factors and the positions of the evaluations points up to order $n_g = 3$. Note that a polynomial of order $p = 2\, n_g - 1$ is integrated exactly by a Gauss integration with n_g points.

4.1.2 Two-Dimensional Interpolations

Shape functions. Two-dimensional C^0-continuous shape functions are provided for triangles and quadrilaterals with linear and quadratic order of interpolation.

Triangular elements. The most simple two-dimensional element is a triangle with three nodes. Its linear shape function can be constructed directly or by using the isoparametric formulations. In the latter case the same shape functions are used for geometry and field variables. Figure 4.5 depicts the triangular element in its reference configuration $\Omega_\square$, described by the ξ–η-coordinates, and its initial configuration Ω_e using the X_1–X_2 coordinate system.

The shape functions are given by

$$N_1 = 1 - \xi - \eta \qquad N_2 = \xi \qquad N_3 = \eta \tag{4.31}$$

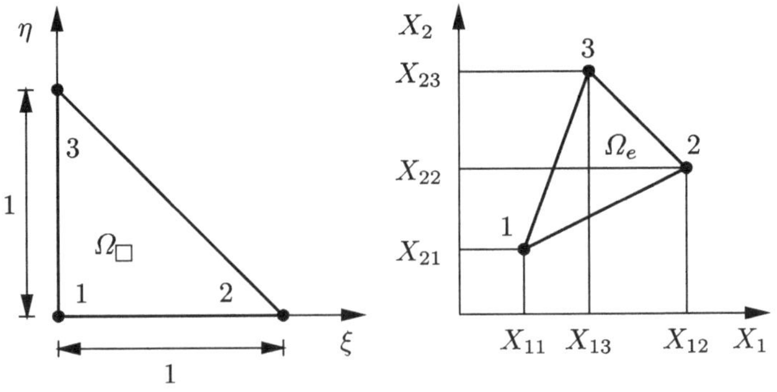

Fig. 4.5 3-node triangular element

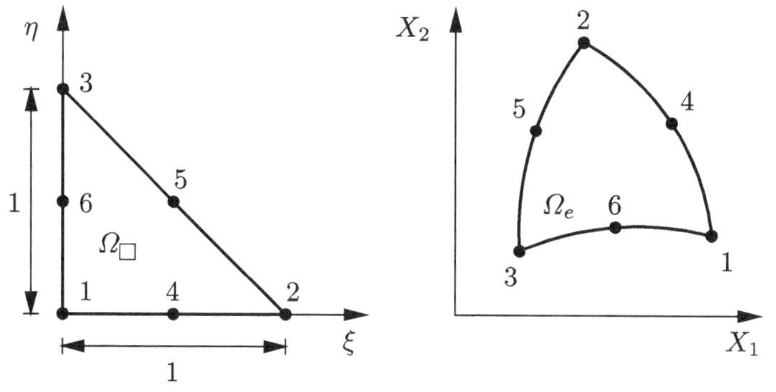

Fig. 4.6 6-node triangular element

which fulfill the condition of being "1" at their node I and "0" at node J, see (4.23). It is easy to verify that the partial derivatives of the shape functions with respect to ξ and η are constant. Hence kinematical quantities, like the strains, are also constant within the element.[4]

A better approximation is provided by the 6-node triangular element, see Fig. 4.6, which is based on quadratic shape functions

$$\begin{aligned} N_1 &= \lambda\,(2\,\lambda - 1), & N_4 &= 4\,\xi\,\lambda, \\ N_2 &= \xi\,(2\,\xi - 1), & N_5 &= 4\,\xi\,\eta, \\ N_3 &= \eta\,(2\,\eta - 1), & N_6 &= 4\,\eta\,\lambda. \end{aligned} \tag{4.32}$$

where the abbreviation $\lambda = 1 - \xi - \eta$ has been introduced.

Quadrilateral elements. The simplest quadrilateral element consists of four nodes. The associated interpolation functions for geometry and field variables are bi-linear and follow from the product form (4.24) using the one-dimensional shape functions (4.27)

$$N_I\,(\xi, \eta) = \frac{1}{2}\,(1 + \xi_I\,\xi)\,\frac{1}{2}\,(1 + \eta_I\,\eta). \tag{4.33}$$

where $\boldsymbol{\Xi}_I = (\xi_I\,, \eta_I)$ are the corner coordinates defined in Fig. 4.7 for the reference element $\Omega_\Box$

[4]A pure displacement triangular element derived on the basis of the linear shape functions has two degrees of freedom per node. Hence the element has in total $2 \times 3 = 6$ unknowns. Of these, three are needed to describe the rigid body motions (two translations and one rotation) and three needed to model the constant strain states.

Besides that the element is very simple and also very robust in nonlinear applications it does not perform very well since its approximation properties are not too good. Due to that the element is very "stiff", meaning its approximation will yield smaller displacements than an analytical solution. This is especially negative when a structure undergoes bending deformations or when incompressible material has to be considered. In such cases higher order elements or special elements have to be applied, see also Chap. 6.

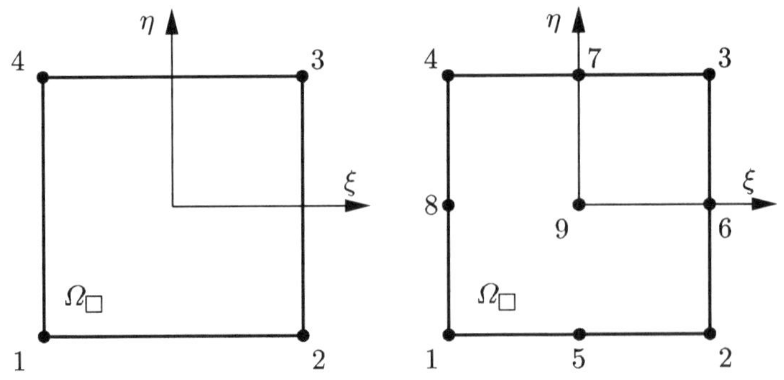

Fig. 4.7 Isoparametric quadrilateral element

$$\Xi_1 = (-1, -1) \quad \Xi_2 = (1, -1) \quad \Xi_3 = (1, 1) \quad \Xi_4 = (-1, 1). \tag{4.34}$$

In the same way the shape functions for the 9-node element in Fig. 4.7 follow from the product form (4.24) using quadratic interpolation functions (4.28). The shape function with respect to the reference element $\Omega_\square$ are given for

- the vertex nodes ($I = 1, 2, 3, 4$):

$$N_I\,(\xi, \eta) = \frac{1}{4}\,(\xi^2 + \xi_I\,\xi)\,(\eta^2 + \eta_I\,\eta), \tag{4.35}$$

- the middle edge nodes ($I = 5, 6, 7, 8$):

$$N_I\,(\xi, \eta) = \frac{1}{2}\,\xi_I^2(\xi^2 + \xi_I\,\xi)\,(1 - \eta^2) + \frac{1}{2}\,\eta_I^2(\eta^2 + \eta_I\,\eta)\,(1 - \xi^2), \tag{4.36}$$

- and the middle node ($I = 9$):

$$N_9\,(\xi, \eta) = (1 - \xi^2)\,(1 - \eta^2). \tag{4.37}$$

This is however not the only possibility to formulate the shape functions for the 9-node element. Often a hierarchical formulation—starting from the 4-node element—is applied, see e.g. Zienkiewicz and Taylor (2000b).

Integration within the reference space. For the computation of the weak form (1.90) an integration is necessary over the area of each finite element Ω_e which in the isoparametric formulation is performed in the parameter space of the reference element $\Omega_\square$, see Fig. 4.7 . Thus the integral has to be transformed from the initial or current configuration to the reference configuration

$$\int\limits_{(\Omega_e)} f(\mathbf{X})\, dA = \int\limits_{(\Omega_\square)} f(\boldsymbol{\Xi})\, \det \mathbf{J}_e(\boldsymbol{\Xi})\, d\,\square = \int\limits_{-1}^{+1}\int\limits_{-1}^{+1} f(\xi\,,\eta)\, \det \mathbf{J}_e\, d\xi\, d\eta. \quad (4.38)$$

The integration over the reference area $\Omega_\square$ is performed numerically since the product $f(\boldsymbol{\Xi})$ det $\mathbf{J}_e(\boldsymbol{\Xi})$ in general does not represent a polynomial but a rational function. This yields for the integral in (4.38) the approximation

$$\int\limits_{-1}^{+1}\int\limits_{-1}^{+1} f(\xi\,,\eta)\ \det\ \mathbf{J}_e\, d\xi\, d\eta \approx \sum_{g=1}^{n_g} f(\xi_g\,,\eta_g)\ \det\ \mathbf{J}_e(\xi_g\,,\eta_g)\, w_{gp}. \quad (4.39)$$

Normally Gauss integration is applied due to its efficiency. The weighting factors w_{gp} and the coordinates of the Gauss or evaluation points ξ_g and η_g are summarized in Table C.2. Table C.2 contains Gauss points and weighting factors up to $n_g = 3 \times 3$. These integration formulae integrate polynomials in $\xi^i\,\eta^k$ exact up to the order $i + k \leq m$. Let us remark that the integration formulae can be derived in the same way as the two-dimensional shape functions by a product form, using one-dimensional Gauss integration in every coordinate direction. The Gauss integration has been chosen within the finite element method due to its accuracy and efficiency. Other integration rules which can be applied for numerical integration of (5.71) can be found in e.g. Dhatt and Touzot (1985).

The transformation of the integrals from the initial configuration to the reference configuration is different for triangular elements. In general

$$\int\limits_{(\Omega_e)} f(\mathbf{X})\, dA = \int\limits_{0}^{1}\int\limits_{0}^{1-\xi} f(\xi\,,\eta)\ \det \mathbf{J}_e\, d\eta\, d\xi \quad (4.40)$$

is obtained where the last integral can again be approximated by the numerical integration (4.39). Table C.3 contains the related evaluation points and weighting functions. These formulae integrate polynomials in the reference space $\xi^k\,\eta^l$ up to order m (with $m \geq k + l$) exact.

Many more rules for the numerical integration of triangular elements exists which have either different evaluation points and weighting factors or a higher approximation order. Such rules can be found in e.g. Zienkiewicz and Taylor (1989) and Dhatt and Touzot (1985).

4.1.3 Three-Dimensional Interpolation

Three dimensional finite elements can have a variety of different shapes. These can be tetrahedral or hexahedral shapes but also mixtures of both types for special

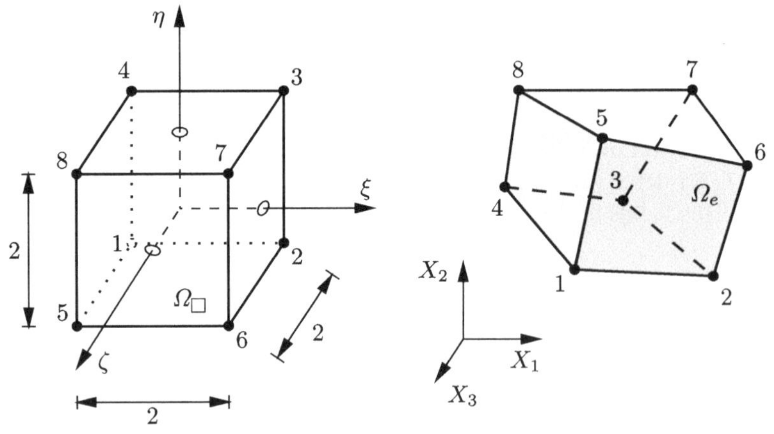

Fig. 4.8 Isoparametric 8-node hexahedral element

geometries like prisms. In this section the formulations are restricted to the tetrahedral or hexahedral shaped elements. Shape functions for other element types can be found e.g. in Dhatt and Touzot (1985). Again the isoparametric formulation is used to be able to generate general shape functions for the discretization of arbitrary three-dimensional geometries.

The shape functions for the three-dimensional hexahedral element are given by

$$N_I(\xi\,,\eta\,,\zeta) = \frac{1}{2}\,(1+\xi_I\,\xi)\,\frac{1}{2}\,(1+\eta_I\,\eta)\,\frac{1}{2}\,(1+\zeta_I\,\zeta), \tag{4.41}$$

which follow from the product form (4.24) together with (4.27). Figure 4.8 describes the associated 8-node hexahedral element in its reference configuration $\Omega_\square$, and its initial configuration Ω_e. The related quadratical shape functions can be derived from (4.24) with (4.28). This yields an interpolation with 27 nodes per element where the functions N_I result from the 27 possible combinations of the local coordinates $\xi\,,\eta\,,\zeta$ with their values -1, 0, $+1$ at the nodes within the reference element $\Omega_\square$

$$N_I(\xi\,,\eta\,,\zeta) = N_I(\xi)\,N_I(\eta)\,N_I(\zeta). \tag{4.42}$$

The shape functions for tetrahedral elements can be defined analogous to the two-dimensional formulation. This leads to the ansatz functions

- 4-node tetrahedral element (linear shape functions)

$$N_1 = 1-\xi-\eta-\zeta, \quad N_2 = \xi, \quad N_3 = \eta, \quad N_4 = \zeta. \tag{4.43}$$

- 10-node tetrahedral element (quadratic shape functions)

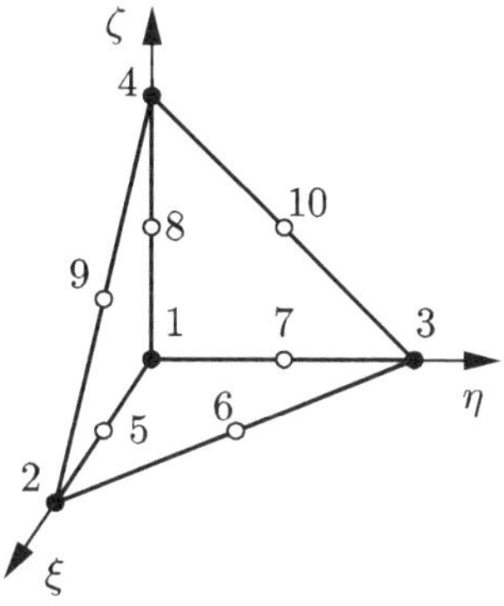

Fig. 4.9 Isoparametric 4 and 10 node tetrahedral

$$\begin{aligned} N_1 &= \lambda\,(2\,\lambda - 1)\,, & N_6 &= 4\,\xi\,\eta, \\ N_2 &= \xi\,(2\,\xi - 1)\,, & N_7 &= 4\,\eta\,\lambda, \\ N_3 &= \eta\,(2\,\eta - 1)\,, & N_8 &= 4\,\zeta\,\lambda, \\ N_4 &= \zeta\,(2\,\zeta - 1)\,, & N_9 &= 4\,\xi\,\zeta, \\ N_5 &= 4\,\xi\,\lambda\,, & N_{10} &= 4\,\eta\,\zeta., \end{aligned}$$

with $\lambda = 1 - \xi - \eta - \zeta$.

The associated node numbers of both elements are depicted in Fig. 4.9.

Computation of the derivatives. The derivatives of the shape functions with respect to the coordinates in the initial and spatial configuration follow from (4.18). This yields for the derivatives with respect to the coordinates in the initial configuration

$$\nabla_X N_I = \begin{Bmatrix} N_{I,1} \\ N_{I,2} \\ N_{I,3} \end{Bmatrix} = \mathbf{J}_e^{-T} \begin{Bmatrix} N_{I,\xi} \\ N_{I,\eta} \\ N_{I,\zeta} \end{Bmatrix}. \tag{4.44}$$

The Jacobi matrix $\mathbf{J}_e$ of element Ω_e which is applied within the derivation is computed with (4.17) from

$$\mathbf{J}_e = \frac{\partial \boldsymbol{X}}{\partial \boldsymbol{\Xi}} = \begin{bmatrix} X_{1,\xi} & X_{1,\eta} & X_{1,\zeta} \\ X_{2,\xi} & X_{2,\eta} & X_{2,\zeta} \\ X_{3,\xi} & X_{3,\eta} & X_{3,\zeta} \end{bmatrix} = \frac{\hat{\delta} X}{\hat{\delta} \boldsymbol{\Xi}} \quad \text{with } \boldsymbol{\Xi} = \{\xi, \eta, \zeta\}^T. \tag{4.45}$$

The components of $\mathbf{J}_e$ are given by

$$X_{m,k} = \sum_{I=1}^{n} N_{I,k}\, X_{m\,I},$$

where the partial derivative denoted by k is the corresponding derivative with respect to ξ, η or ζ. The input for *AceGen* that evaluates $\mathbf{J}_e$ is provided in Box 4.1 for tri-linear shape functions (4.41) where the command `SMSD` provides the differentiation of $\boldsymbol{X}$ with respect to $\boldsymbol{\Xi}$.

```
XIO⊢Table[SMSReal[nd$$[i,"X",j]],{i,nen},{j,ndim}];
Ξn={{-1,-1,-1},{1,-1,-1},{1,1,-1},{-1,1,-1},
   {-1,-1,1},{1,-1,1},{1,1,1},{-1,1,1}};
Nh⊨Table[1/8 (1+ξ Ξn⟦i,1⟧) (1+η Ξn⟦i,2⟧) (1+ζ Ξn⟦i,3⟧),{i,1,8}];
X⊢SMSFreeze[Nh.XIO];
Je⊨SMSD[X,Ξ];

uIO⊢SMSReal[Table[nd$$[i,"at",j],{i,nen},{j,ndim}]];
u⊨Nh.uIO;
H⊨SMSD[u,X,"Dependency"→{Ξ,X,SMSInverse[Je]}];
```

Box 4.1. *AceGen* input segment for the computation of the derivative of the initial coordinates with respect to the reference coordinates

The last three lines in Box 4.1 are an example for the computation of derivatives with respect $\boldsymbol{X}$ of function depending on $\boldsymbol{\Xi}$. Here the displacement gradient $\boldsymbol{H} = \text{Grad}\, \boldsymbol{u}$ is considered. With the definition of the ansatz for the displacement field the derivative follows simply by denoting the dependency of $\boldsymbol{\Xi}$ on $\boldsymbol{X}$ in `SMSD` that involves the inverse of $\mathbf{J}_e$, see (4.44).

4.1.3.1 Integration Within the Parameter Space

As in the two-dimensional case integration of the shape functions and their derivatives over the volume of element Ω_e is carried out in the parameter space of the reference element $\Omega_\Box$. This yields

$$\int\limits_{(\Omega_e)} f(\boldsymbol{X})\, dV = \int\limits_{(\Omega_\Box)} f(\boldsymbol{\Xi}) \det \mathbf{J}_e(\boldsymbol{\Xi})\, d\Box$$
$$= \int\limits_{-1}^{+1}\int\limits_{-1}^{+1}\int\limits_{-1}^{+1} f(\xi,\eta,\zeta) \det \mathbf{J}_e\, d\xi\, d\eta\, d\zeta. \tag{4.46}$$

The integration over $\Omega_\Box$ is performed numerically. Hence the last integral in (4.46) is approximated by

$$\int\limits_{-1}^{+1}\int\limits_{-1}^{+1}\int\limits_{-1}^{+1} f(\xi\,,\eta\,,\zeta) \det \mathbf{J}_e\, d\xi\, d\eta\, d\zeta$$
$$\approx \sum_{g=1}^{n_g} f(\xi_g,\eta_g,\zeta_g) \det \mathbf{J}_e(\xi_g,\eta_g,\zeta_g)\, w_{gp} \tag{4.47}$$

In case of the tri-linear brick element a $2 \times 2 \times 2$ integration with totally 8 Gauss points is necessary. Accordingly for the ansatz with quadratic shape functions a $3 \times 3 \times 3$ integration with totally 27 Gauss points has to be used. Since this is rather time consuming, integration rules are stated in Table C.5 which only need 6 points for an ansatz with tri-linear shape functions and 14 points for an ansatz with quadratic shape functions, see Irons (1971). With this special integration rules the computational effort is reduced by 25 % for linear and by almost 50 % for quadratic ansatz functions.

The integration rules for tetrahedral elements are provided in Table C.6 for linear and quadratic ansatz functions. The coordinates of the sampling points ξ_g , η_g and ζ_g are related to the coordinate system used in Fig. 4.9. Further integration rules for tetrahedral and hexahedral elements can be found in e.g. Zienkiewicz and Taylor (1989) or Dhatt and Touzot (1985).

As an example for the *AceGen* input, the volume load term in (1.98)

$$-\int_B \rho_0\, \boldsymbol{b} \cdot \boldsymbol{u}\, dV = -\int_{-1}^{+1}\int_{-1}^{+1}\int_{-1}^{+1} \rho_0\, \bar{\boldsymbol{b}} \cdot \boldsymbol{u}(\boldsymbol{\Xi})\, \det \mathbf{J}_e(\boldsymbol{\Xi})\, d\xi\, d\eta\, d\zeta$$

```
SMSInitialize["H1BodyLoad","Environment"→"AceFEM"];
SMSTemplate["SMSTopology"→"H1","SMSDefaultIntegrationCode"→7,
  "SMSDomainDataNames"->{"ρ0 -density","bX -force per unit mass X",
    "bY -force per unit mass Y","bZ -force per unit mass Z"},
  "SMSDefaultData"->{1,0,0}];
SMSStandardModule["Tangent and residual"];
nen=SMSNoNodes;ndim=SMSNoDimensions;np=SMSNoDOFGlobal;
{ρ0,bX,bY,bZ}⊢
  SMSReal[Table[es$$["Data",i],{i,Length[SMSDomainDataNames]}]];
SMSDo[Ig,1,SMSInteger[es$$["id","NoIntPoints"]]];(* start Gauss loop *)
Ξ={ξ,η,ζ}⊢Table[SMSReal[es$$["IntPoints",i,Ig]],{i,3}];(*point*)
XIO⊢Table[SMSReal[nd$$[i,"X",j]],{i,nen},{j,ndim}];
Ξn={{-1,-1,-1},{1,-1,-1},{1,1,-1},{-1,1,-1},
   {-1,-1,1},{1,-1,1},{1,1,1},{-1,1,1}};
Nh⊨Table[1/8 (1+ξ Ξn[[i,1]]) (1+η Ξn[[i,2]]) (1+ζ Ξn[[i,3]]),{i,1,8}];
X⊢SMSFreeze[Nh.XIO];Je⊨SMSD[X,Ξ];Jed⊨Det[Je];
uIO⊢SMSReal[Table[nd$$[i,"at",j],{i,nen},{j,ndim}]];
pe=Flatten[uIO];
u⊨Nh.uIO;bb⊨{bX,bY,bZ};
W⊨-ρ0 u.bb;
wgp⊢SMSReal[es$$["IntPoints",4,Ig]];(*weight*)
Rg⊨Jed SMSD[W,pe];
SMSExport[wgp Rg,p$$,"AddIn"→True];
SMSEndDo[];(* end Gauss loop *)
SMSWrite[];
```

Box 4.2. *AceGen* input for the GAUSS loop to compute the body force for hexahedral element is numerically integrated for an element with tri-linear shape functions where $\mathbb{b}\mathbb{b} \equiv \bar{\boldsymbol{b}} = \{b_X, b_Y, b_Z\}^T$ is the applied body force and $\boldsymbol{u}(\boldsymbol{\Xi})$ has to be computed according to (4.6). Box 4.2 contains the Gauss loop structure in *AceGen*.

4.2 Discretization of Potentials and Weak Forms

The one-, two- and three-dimensional ansatz functions can now be used to describe the geometry and to discretize the field variables within the kinematical relations and weak forms stemming from continuum mechanics. Here this discretization process will be shown for the variational equations (1.90) and (1.95) derived in Chap. 3. These variational forms are purely deformation based. Mixed approximations—for e.g. deformation and stresses—will be covered in later Chapters.

The region of interest, which is the volume of the solid to be discretized, will be subdivided in n_e finite elements as shown in Fig. 4.1. Within this process the geometry is approximated using (4.1). An ansatz as given in (4.6) and (4.7) is then selected for the displacement field $\boldsymbol{u}$, the coordinates $\boldsymbol{X}$ and the test function $\boldsymbol{\eta}$ to approximate these quantities within each finite element Ω_e. With these approximations the integral describing the weak forms is given by

$$\int_B (\ldots)\, dV \approx \int_{B_h} (\ldots)\, dV_h = \mathsf{A}_{e=1}^{n_e} \int_{\Omega_e} (\ldots)\, d\Omega = \mathsf{A}_{e=1}^{n_e} \int_{\Omega_\square} (\ldots)\, d\square \tag{4.48}$$

The operator A is chosen here instead of the $\sum$ symbol in order to denote that an assembly process takes place in which all element contributions have not only to be added up but also the kinematical compatibility between the elements has to be fulfilled. The whole process then leads to an algebraic system of nonlinear equations for a given problem, as will be shown later. Hence the assembly process denoted by A stands for fulfillment of the inter element compatibility of the test functions and the displacement field and for the fulfillment of the global displacement boundary conditions. Since the assembly process is the same as the one used within the finite element method for linear problems it is here not described in detail, detailed descriptions of this process can be found in e.g. Hughes (1987); Bathe (1996); Zienkiewicz and Taylor (2000b); Oñate (2009).

The application of the automated approach to finite element problems is now introduced by the discretization of three-dimensional solid problems. The weak form and the functional form for a hyperelastic body will be used to demonstrate the general use of the system *AceGen*. The formulations will be provided for the spatial and initial configurations.

4.2.1 Three-Dimensional Solid Element, Initial Configuration

In this section the matrix formulation for a three-dimensional finite element with 8-nodes is developed with respect to the initial configuration. Only the element vectors and matrices related to the internal strain energy are considered. The external loading terms are not discretized. Within the formulation the Neo-Hooke constitutive equation is applied. Two different formulations will be used to derive the matrices necessary for a nonlinear computation using Newton's method. The first one is based directly on the strain energy function while the second one relies on the weak form. For the theoretical background see Sect. 1.3.

1. Using the variational formulation with the strain energy the only quantities to be described are the strain energy function W, the related kinematics, e.g. $\boldsymbol{C}$ and the finite element ansatz functions $\boldsymbol{u}_h$. Thus the following procedure has to be executed to obtain element residual $\mathbf{R}_e$ and tangent matrix $\mathbf{K}_e$

$$\boldsymbol{u}_h \rightarrow \boldsymbol{C} \rightarrow \int W(\boldsymbol{C})\, dV \rightarrow \mathbf{R}_e \rightarrow \mathbf{K}_e .$$

2. Within the weak form the constitutive equations relating strains and stresses $\boldsymbol{S}$ as well as the strains $\boldsymbol{C}$ have to be formulated based on the finite element ansatz functions. Here the procedure is in general given by

$$\boldsymbol{u}_h \rightarrow \boldsymbol{C} \text{ and } \boldsymbol{S} = 2\,\frac{\partial W}{\partial \boldsymbol{C}} \rightarrow \int \boldsymbol{S} \cdot \frac{1}{2}\,\delta \boldsymbol{C}\, dV \rightarrow \mathbf{R}_e \rightarrow \mathbf{K}_e .$$

In the following we will use both formulations and compare their efficiency with regard to automatic code development.

Discretization based on the strain energy. For the computation of the residuum the variational form (1.98) is applied. Its internal energy part can be written for an element Ω_e as

$$\int_{\Omega_e} W(\boldsymbol{C})\, dV \approx \sum_{g=1}^{n_g} W(\boldsymbol{C}(\xi_g, \eta_g, \zeta_g)) \det \mathbf{J}_e(\xi_g, \eta_g, \zeta_g)\, w_{gp} \tag{4.49}$$

where the $\sum_g$ denotes the numerical integration generally applied to formulate finite element matrices. For the eight node element a $2 \times 2 \times 2$ Gauss integration is sufficient ($n_g = 8$).

The two ingredients needed in this formulation are W and the right Cauchy–Green tensor. The latter has to be computed from the displacement field using the selected finite element ansatz and then can be evaluated at integration point level. To describe the material behaviour the Neo–Hooke strain energy (5.10) is used, leading to

$$W(\boldsymbol{C}) = \frac{\Lambda}{2}(J_F - 1)^2 - \mu \ln J_F + \frac{1}{2}\mu\,(I_C - 3). \tag{4.50}$$

Since the Jacobian J_F can be written as $J_F = \det \boldsymbol{F}$ and the first invariant as $I_C = \operatorname{tr} \boldsymbol{C}$, all terms in (4.50) are known as functions of $\boldsymbol{C}$ and $\boldsymbol{F}$ in Ω_e which both depend upon the displacement field through the selected finite element interpolation.

Now the right Cauchy–Green tensor has to be computed from the displacement interpolation within the element Ω_e. Here isoparametric interpolation is selected for a hexahedral element for displacements and coordinates.[5] In the case of three dimensional elements there are three displacements at each node, see Fig. 4.10. The displacement vector $\boldsymbol{u}$ and initial coordinates $\boldsymbol{X}$ can be expressed within the element with (4.9) using the shape functions N_I

$$\boldsymbol{u} = \{u, v, w\} = \sum_{I=1}^{8} N_I\,(\Xi)\,\mathbf{u}_I = (\mathbf{N}\,\mathbf{u})^T \tag{4.51}$$

$$\boldsymbol{X} = \{X, Y, Z\} = \sum_{I=1}^{8} N_I\,(\Xi)\,\mathbf{X}_I = (\mathbf{N}\,\mathbf{X})^T. \tag{4.52}$$

The shape functions for the three-dimensional hexahedral element can be found in (4.41). The components of the right Cauchy Green strain tensor (1.13), needed in (4.50), follow for a finite element Ω_e as

$$\boldsymbol{C} = \boldsymbol{F}^T\,\boldsymbol{F}, \tag{4.53}$$

where the deformation gradient $\boldsymbol{F}$ is computed in the three-dimensional case from (1.12) using (4.51)

$$\boldsymbol{F} = \mathbf{1} + \operatorname{Grad} \boldsymbol{u} = \begin{bmatrix} 1\,0\,0 \\ 0\,1\,0 \\ 0\,0\,1 \end{bmatrix} + \sum_{I=1}^{8} \begin{bmatrix} u_I\,N_{I,1} & u_I\,N_{I,2} & u_I N_{I,3} \\ v_I\,N_{I,1} & v_I\,N_{I,2} & v_I N_{I,3} \\ w_I\,N_{I,1} & w_I\,N_{I,2} & w_I N_{I,3} \end{bmatrix}. \tag{4.54}$$

[5]The formulation can easily be extended to higher order or triangular elements. In that case only the ansatz functions have to be exchanged, see e.g. Sect. 6.2.1.

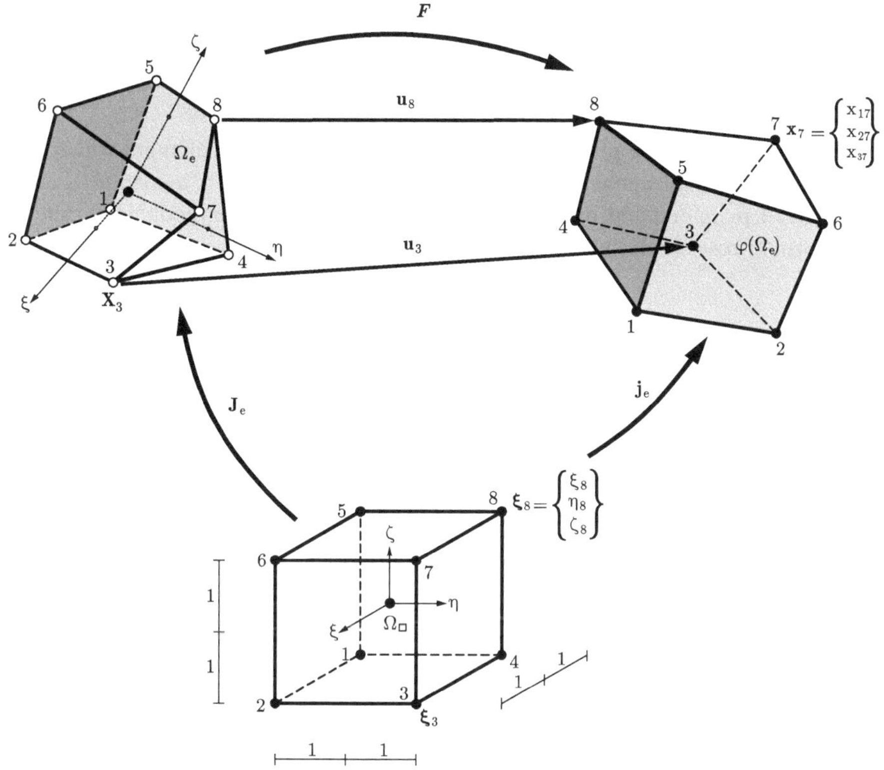

Fig. 4.10 Configurations and some nodal vectors of the eight node H1 element

However when employing the *ADB* form then the deformation gradient $\boldsymbol{F}$ is simply given by

$$\boldsymbol{F} = \mathbf{1} + \left.\frac{\hat{\delta}\boldsymbol{u}}{\hat{\delta}\boldsymbol{X}}\right|_{\frac{D\Xi}{DX}=\mathbf{J}_e^{-1}} . \tag{4.55}$$

Its determinant is denoted by $J_F = \det \boldsymbol{F}$. The right Cauchy–Green tensor and the deformation gradient defined above are evaluated at an integration point (ξ_g, η_g, ζ_g).

The residual of the finite element is now obtained using *AceGen*. Since all quantities in (4.49) depend upon the displacement field the variation

$$\delta \int_{\Omega_e} W(\boldsymbol{C})\, dV \approx \left[\sum_{g=1}^{n_g} \frac{\partial W(\boldsymbol{C}(\mathbf{p}_e))}{\partial \mathbf{p}_e} J_e\, w_{gp} \right] \delta \mathbf{p}_e \tag{4.56}$$

can be computed, where $\mathbf{p}_e$ is the unknown vector related to the element which contains all nodal displacements: $\mathbf{p}_e = \{u_1, v_1, w_1, u_2, v_2, w_2, \ldots, u_8, v_8, w_8\}^T$. The variation of $\mathbf{p}_e$ is denoted by $\boldsymbol{\eta}_e = \delta \mathbf{p}_e$. From (4.56) it can be recognized that the *ADB* form of residual $\mathbf{R}_g$ at an integration point is given by

$$\mathbf{R}_g = \frac{\partial W(\boldsymbol{C}(\mathbf{p}_e))}{\partial \mathbf{p}_e} J_e = \frac{\hat{\delta} W}{\hat{\delta} \mathbf{p}_e} J_e. \tag{4.57}$$

The total size of $\mathbf{R}_g$ is related to the number of entries in $\mathbf{p}_e$ which is 24 for the three-dimensional eight node element. Differentiation of this term with respect to the element displacements $\mathbf{p}_e$ yields the tangent stiffness matrix at an integration point

$$\mathbf{K}_g = \frac{\partial \mathbf{R}_g(\mathbf{p}_e)}{\partial \mathbf{p}_e} = \frac{\hat{\delta} \mathbf{R}_g}{\hat{\delta} \mathbf{p}_e}. \tag{4.58}$$

The residual and tangent matrix can also be expressed component wise for one integration point by

$$R_{gm} = \frac{\hat{\delta} W}{\hat{\delta} p_{em}} J_e \quad \text{and} \quad K_{gmn} = \frac{\hat{\delta} R_{gm}}{\hat{\delta} p_{en}}. \tag{4.59}$$

The component wise form is preferable in the case of large number of element unknowns in order to avoid symbolic generation of large matrices. The element residual and tangent matrix are obtained by numerical integration

$$R_{em} = \sum_{g=1}^{n_g} R_{gm}\, w_{gp} \quad \text{and} \quad K_{emn} = \sum_{g=1}^{n_g} K_{gmn}\, w_{gp}.$$

However both are automatically computed within the *AceGen* procedure. Box 4.3 describes the code for *AceGen* needed to create the element based on the formulation discussed above. The implementation presented in Box 4.3 is not the only way how the three dimensional hexahedral solid element can be implemented in *AceGen*. Alternative implementations are discussed and compared in Sect. 4.3.

```
SMSInitialize["H1-W-B","Environment"→"AceFEM"];
SMSTemplate["SMSTopology"→"H1","SMSSymmetricTangent"→True,
  "SMSDomainDataNames"->{"E -elastic modulus","ν -poisson ratio"},
  "SMSDefaultData"->{21000,0.3},"SMSDefaultIntegrationCode"→7];
nen=SMSNoNodes;ndim=SMSNoDimensions;np=SMSNoDOFGlobal;
SMSStandardModule["Tangent and residual"];
{Em,ν}⊢SMSReal[Table[es$$["Data",i],{i,2}]];{λ,μ}⊨SMSHookeToLame[Em,ν];
XIO⊢Table[SMSReal[nd$$[i,"X",j]],{i,nen},{j,ndim}];
uIO⊢SMSReal[Table[nd$$[i,"at",j],{i,nen},{j,ndim}]];pe=Flatten[uIO];
SMSDo[Ig,1,SMSInteger[es$$["id","NoIntPoints"]]]; (*begin Gauss loop*)
Ξ={ξ,η,ζ}⊢Table[SMSReal[es$$["IntPoints",i,Ig]],{i,3}];
wgp⊢SMSReal[es$$["IntPoints",4,Ig]];
Ξn={{-1,-1,-1},{1,-1,-1},{1,1,-1},{-1,1,-1},
   {-1,-1,1},{1,-1,1},{1,1,1},{-1,1,1}};
Nh⊨Table[1/8 (1+ξ Ξn[[i,1]]) (1+η Ξn[[i,2]]) (1+ζ Ξn[[i,3]]),{i,1,nen}];
```

```
X⊢SMSFreeze[Nh.XIO];u⊨Nh.uIO;Je⊨SMSD[X,Ξ];Jed⊨Det[Je];
H⊨SMSD[u,X,"Dependency"→{Ξ,X,SMSInverse[Je]}];
F⊨IdentityMatrix[3]+H;Ct⊨Fᵀ.F;JF⊨Det[F];
W⊨1/2 λ (JF-1)^2+μ (1/2 (Tr[Ct]-3)-Log[JF]);
SMSDo[m,1,np];
  Rgm⊨Jed SMSD[W,pe,m];
  SMSExport[wgp Rgm,p$$[m],"AddIn"→True];
  SMSDo[n,m,np];
    Kgmn⊨SMSD[Rgm,pe,n];
    SMSExport[wgp Kgmn,s$$[m,n],"AddIn"→True];
  SMSEndDo[];
SMSEndDo[];
```

Segment A. Formulation dependent *AceGen* segment

```
SMSEndDo[];
(*end Gauss loop*)
SMSWrite[];
```

Box 4.3. *AceGen* input for an element based on a strain energy function in the initial configuration B (element "W in B"), here called H1-W-B

Discretization based on the weak form. For the computation of the residuum the internal virtual work expression given in (1.90) has to be specified at element level

$$\int_{\Omega_e} \frac{1}{2} \boldsymbol{S} \cdot \delta \boldsymbol{C}\, dV \approx \sum_{g=1}^{n_g} \frac{1}{2} \boldsymbol{S}(\xi_g, \eta_g, \zeta_g) \cdot \delta \boldsymbol{C}(\xi_g, \eta_g, \zeta_g) \det \mathbf{J}_e(\xi_g, \eta_g, \zeta_g)\, w_{gp} \tag{4.60}$$

Here expressions for the 2nd Piola–Kirchhoff stress $\boldsymbol{S}$ and the virtual right Cauchy–Green strain $\delta \boldsymbol{C}$ are needed at integration point level.

For a three-dimensional solid the 2nd Piola–Kirchhoff stress[6] and the right Cauchy–Green strain can be written as

$$\boldsymbol{S} = \begin{bmatrix} S_{11} & S_{12} & S_{13} \\ S_{12} & S_{22} & S_{23} \\ S_{13} & S_{23} & S_{33} \end{bmatrix} \quad \text{and } \boldsymbol{C} = \begin{bmatrix} C_{11} & C_{12} & C_{13} \\ C_{12} & C_{22} & C_{23} \\ C_{13} & S_{23} & C_{33} \end{bmatrix} \tag{4.61}$$

Both, $\boldsymbol{S}$ and $\boldsymbol{C}$ are symmetric. Since *AceGen* performs operations internally directly on the components of tensors, one has to specify this fact explicitly in the *AceGen* input. It is then considered by *AceGen* automatically. The 2nd Piola–Kirchhoff stress follows from the Neo Hooke constitutive Eq. (5.11) within the element Ω_e[7]

$$\boldsymbol{S} = 2\,\frac{\partial W}{\partial \boldsymbol{C}} = \Lambda\, J_F\,(J_F - 1)\,\boldsymbol{C}^{-1} + \mu\,(\mathbf{1} - \boldsymbol{C}^{-1}). \tag{4.62}$$

Note that $J_F^2 = \det \boldsymbol{C}$. Since the inverse and the determinant of $\boldsymbol{C}$ are computed directly within the *AceGen* system it only remains to compute $\boldsymbol{C}$ in terms of the chosen finite element ansatz. This has been stated already in (4.53).

```
𝕏⊢SMSFreeze[ℕh.𝕏IO];𝕦⊨ℕh.𝕦IO;
Je⊨SMSD[𝕏,Ξ];Jed⊨Det[Je];
ℍ⊨SMSD[𝕦,𝕏,"Dependency"→{Ξ,𝕏,SMSInverse[Je]}];
𝔽⊨IdentityMatrix[3]+ℍ;JF⊨Det[𝔽];
ℂt⊨𝔽ᵀ.𝔽;ℂei⊨SMSInverse[ℂt];
𝕊⊨λ JF (JF-1) ℂei+μ (IdentityMatrix[3]-ℂei);
SMSDo[m,1,np];
  δℂt⊨SMSD[ℂt,𝕡e,m];
  Rgm⊨Jed 1/2 Tr[δℂt.𝕊];
  SMSExport[wgp Rgm,p$$[m],"AddIn"→True];
  SMSDo[n,m,np];
    Kgmn⊨SMSD[Rgm,𝕡e,n];
    SMSExport[wgp Kgmn,s$$[m,n],"AddIn"→True];
  SMSEndDo[];
SMSEndDo[];
```

Box 4.4. A segment of *AceGen* input for an element based on the weak form with respect to the initial configuration B (element "G in B"), here called H1-G-B

Having defined the weak form equations one can now follow the general procedure for the derivation of the *ADB* form of the arbitrary weak form equations presented in Sect. 2.6.2. The variation of the right Cauchy–Green strain is computed within an element as

[6] For a stress calculation within an actual design the 2nd Piola–Kirchhoff stresses have to be transformed to real stresses, e.g. the Cauchy stresses using (3.72).

[7] The first part of Eq. (4.62), partial derivative of W, is easier to use within *AceGen* when the strain energy function is known.

$$\delta \boldsymbol{C} = D\boldsymbol{C}\,\delta\mathbf{p}_e = \frac{\partial \boldsymbol{C}(\mathbf{p}_e)}{\partial \mathbf{p}_e}\delta\mathbf{p}_e.$$

with the nodal vector of the displacements $\mathbf{p}_e = \{u_1, v_1, w_1, u_2, v_2, w_2, \ldots, u_8, v_8, w_8\}^T$. Hence the residual can now be expressed component wise at an integration point by[8]

$$R_{gm} = \frac{1}{2}\boldsymbol{S} \cdot \frac{\partial \boldsymbol{C}(\mathbf{p}_e)}{\partial p_{em}} J_g = \frac{1}{2}\boldsymbol{S} \cdot \frac{\hat{\delta}\boldsymbol{C}}{\hat{\delta} p_{em}} J_g \tag{4.63}$$

The above equation can also be written in index notation

$$R_{gm} = \frac{1}{2} S_{ik} \frac{\partial C_{ik}}{\partial p_{em}} J_g.$$

The element residual is then defined at an integration point by

$$\mathbf{R}_g^T = \{R_{g1}, R_{g2}, \ldots, R_{g24}\}.$$

Again the 24 × 24 tangent stiffness matrix is computed at each integration point by

$$K_{gmn} = \frac{\partial R_{gm}(\mathbf{p}_e)}{\partial p_{en}} = \frac{\hat{\delta} R_{gm}}{\hat{\delta} p_{en}}. \tag{4.64}$$

With the last equations all necessary relations and vectors are known that are needed for a successful implementation of this element using the automated approach. The segment of the input for *AceGen* is depicted in Box 4.4. The complete *AceGen* input for element H1-G-B can be obtained by replacing the Segment A in Box 4.3 containing the formulation dependent part of *AceGen* input by the input segment given in Box 4.4.

The approach discussed in Sect. 2.6.2 suggests to explore the formulation of a "pseudo" potential that generates the weak form. Using the exceptions within the operator `SMSD` that generates the derivatives it is possible to derive from a pseudo potential the correct formulation that is equivalent to using directly the weak form. In this approach the residual (4.63) stemming from the weak form (4.60) is reformulated as

$$R_{gm} = \frac{1}{2}\boldsymbol{S}(\mathbf{p}_e) \cdot \frac{\partial \boldsymbol{C}(\mathbf{p}_e)}{\partial p_{em}} J_e = \left.\frac{\hat{\delta} W^P(\mathbf{p}_e)}{\hat{\delta} p_{em}}\right|_{\boldsymbol{S}=\text{const.}} J_e \tag{4.65}$$

with the pseudo potential

$$W^P = \frac{1}{2}\boldsymbol{S}(\boldsymbol{p}_e) \cdot \boldsymbol{C}(\boldsymbol{p}_e) = \frac{1}{2}\, tr\left(\boldsymbol{S}(\boldsymbol{p}_e)^T \boldsymbol{C}(\boldsymbol{p}_e)\right). \tag{4.66}$$

---

[8]The scalar product can also be written as $\boldsymbol{S} \cdot \frac{\partial \boldsymbol{C}}{\partial p_m} = \mathrm{tr}\,[\boldsymbol{S}^T \frac{\partial \boldsymbol{C}}{\partial p_m}]$ by using the trace operator, see Appendix B.1.3 and B.1.5.

Hence the input for *AceGen* can be simplified in comparison to the formulation[9] used in Box 4.7, especially the definition of the virtual strain $\delta \boldsymbol{C}$. The complete *AceGen* input for element H1-WP-B can be obtained by replacing the Segment A in Box 4.3 by the input segment given in Box 4.5.

```
X⊢SMSFreeze[Nh.XIO];u⊨Nh.uIO;Je⊨SMSD[X,Ξ];Jed⊨Det[Je];
H⊨SMSD[u,X,"Dependency"→{Ξ,X,SMSInverse[Je]}];
F⊨IdentityMatrix[3]+H;JF⊨Det[F];
Ct⊨Fᵀ.F;Cei⊨SMSInverse[Ct];
S⊢SMSFreeze[λ JF (JF-1) Cei+μ (IdentityMatrix[3]-Cei)];
WP⊨1/2 Tr[Ct.Transpose[S]];
SMSDo[m,1,np];
  Rgm⊨Jed SMSD[WP,pe,m,"Constant"→S];
  SMSExport[wgp Rgm,p$$[m],"AddIn"→True];
  SMSDo[n,m,np];
    Kgmn⊨SMSD[Rgm,pe,n];
    SMSExport[wgp Kgmn,s$$[m,n],"AddIn"→True];
  SMSEndDo[];
SMSEndDo[];
```

Box 4.5. A segment of *AceGen* input for the formulation of weak form with respect to the initial configuration B using the psudo potential W^P (element "W^P in B"), here called H1-WP-B

4.2.2 *Three-Dimensional Solid Element, Current Configuration*

The same element as in the last section will be derived, but now based on the current configuration. Within the formulation the Neo Hooke constitutive equation is used. Again two different formulations will be employed to derive the matrices necessary for a nonlinear computation using Newton's method. The first one is based directly on the strain energy function while the second one relies on the weak form.

Discretization based on the strain energy. For the computation of the residuum the strain energy in (1.98) is applied. Its internal energy part can be written for an element Ω_e in the current configuration $\varphi(\Omega_e)$ as

[9]It is of course also possible to formulate the pseudo potential W^P using the strain energy function W, see (4.50). Then the 2nd Piola–Kirchhoff stress is computed from

$$S = 2\,\frac{\partial W}{\partial \boldsymbol{C}}$$

and can be introduced as stress in the pseudo potential.

$$\int_{\varphi(\Omega_e)} \frac{1}{J_F} W(\boldsymbol{b})\, dv \approx \sum_{g=1}^{n_g} \frac{1}{J_F} W(\boldsymbol{b}(\xi_g, \eta_g, \zeta_g)) \det \mathbf{j}_e(\xi_g, \eta_g, \zeta_g)\, w_{gp} \tag{4.67}$$

Here the two ingredients needed are W and the left Cauchy–Green tensor $\boldsymbol{b}$. Note that the strain energy W in (4.50) has to be divided by the determinant of the deformation gradient since $W = \rho_0 \Psi$ which has to be transformed completely to the deformed configuration $\rho \Psi = \frac{\rho_0}{J_F} \Psi = \frac{W}{J_F}$, see also the arguments in Sect. 5.1. The left Cauchy–Green tensor $\boldsymbol{b}$ has to be computed from the displacement field using the selected finite element ansatz. For W the Neo–Hooke strain energy (5.10) with the parameters related to (5.11) is used, leading to[10]

$$W(\boldsymbol{b}) = \frac{\Lambda}{2}(J_F^2 - 1) - \mu \ln J_F + \frac{1}{2}\mu(I_b - 3). \tag{4.68}$$

Since the Jacobian J_F can be written as $J_F = \sqrt{det(\boldsymbol{b})}$ and the first invariant as $I_b = \operatorname{tr} \boldsymbol{b}$ all terms in (4.68) are known as functions of $\boldsymbol{b}$ in $\varphi(\Omega_e)$.

Now the left Cauchy–Green tensor has to be computed from the displacement interpolation within the element $\varphi(\Omega_e)$. For the quadrilateral element isoparametric interpolation is selected. Thus displacements and coordinates are interpolated in the same way as in the implementation based on the initial configuration, see (4.51).[11] The components of the left Cauchy Green strain tensor (1.18), needed in (4.68), follow in the three-dimensional case for a finite element Ω_e as

$$\boldsymbol{b} = \boldsymbol{F}\boldsymbol{F}^T \quad \text{with} \quad \boldsymbol{F} = \sum_{I=1}^{8} \begin{bmatrix} x_{1I}\, N_{I,1} & x_{1I}\, N_{I,2} & x_{1I} N_{I,3} \\ x_{2I}\, N_{I,1} & x_{2I}\, N_{I,2} & x_{2I} N_{I,3} \\ x_{3I}\, N_{I,1} & x_{3I}\, N_{I,2} & x_{3I} N_{I,3} \end{bmatrix} \tag{4.69}$$

where the deformation gradient $\boldsymbol{F}$ is computed using the coordinate transformations related to the isoparametric formulation, see e.g. (4.17). The nodal coordinates $x_{\alpha I} = X_{\alpha I} + u_{\alpha I}$ are related to the current configuration $\bar{\varphi}_e$. The deformation gradient can also be obtained from $\boldsymbol{F}^{-1} = \mathbf{1} - \nabla \mathbf{u}$, see (1.25), when all quantities have to be derived from the current configuration.

The internal residual of the finite element is now obtained using *AceGen*. Since all quantities in (4.49) depend upon the displacement field the variation

$$\delta \int_{\varphi(\Omega_e)} \frac{1}{J_F} W(\boldsymbol{b})\, dv \approx \left(\sum_{g=1}^{n_g} \frac{1}{J_F} \frac{\partial W(\boldsymbol{b})}{\partial \mathbf{p}_e} j_e\, w_{gp} \right) \delta \mathbf{p}_e \tag{4.70}$$

[10]Note that a dependency of W on the left Cauchy–Green tensor cannot be used for anisotropic materials.

[11]The formulation can easily be extended to higher order or triangular elements. In that case only the ansatz functions have to be exchanged.

can be computed. Here j_e denotes the determinant of the Jacobi matrix $\mathbf{j}_e$ related to the transformation of the element volume from the current configuration to the reference configuration, see (4.17) and Fig. 4.10. The unknown vector $\mathbf{p}_e$ is related to the element and contains all components of the nodal displacement vectors: $\mathbf{p}_e = \{u_1, v_1, w_1, u_2, v_2, w_2, \ldots, u_8, v_8, w_8\}^T$. From (4.70) it can be recognized that the *ADB* form of the element residual $\mathbf{R}_g$ is given by

$$\mathbf{R}_g = \frac{1}{J_F} \frac{\hat{\delta} W(\boldsymbol{b})}{\hat{\delta} \mathbf{p}_e} j_e \tag{4.71}$$

The total size of $\mathbf{R}_g$ is related to the number of nodes, here 8, and the degree of freedoms per node, here 3, and thus has 24 entries. Eight Gauss points are sufficient to compute the element residual

$$\mathbf{R}_e = \sum_{g=1}^{8} \mathbf{R}_g \, w_{gp}. \tag{4.72}$$

Differentiation of the residual in (4.71) with respect to the element nodal displacements $\mathbf{p}_e$ yields the tangent stiffness matrix at a Gauss point

$$\mathbf{K}_g = \frac{\partial \mathbf{R}_g(\mathbf{p}_e)}{\partial \mathbf{p}_e} = \frac{\hat{\delta} \mathbf{R}_g(\mathbf{p}_e)}{\hat{\delta} \mathbf{p}_e}. \tag{4.73}$$

The input for *AceGen* is depicted in Box 4.6 where the relation $\boldsymbol{F}^{-1} = \mathbf{1} - \operatorname{grad} \boldsymbol{u}$ is used to obtain the deformation gradient.

```
X⊨Nh.XIO;u⊨Nh.uIO;x⊢SMSFreeze[X+u];
je⊨SMSD[x,Ξ];jed⊨Det[je];
gradu⊨SMSD[u,x,"Dependency"→{Ξ,x,SMSInverse[je]}];
invF⊨IdentityMatrix[3]-gradu;F⊨SMSInverse[invF];
b⊨F.Fᵀ;JF⊨Det[F];
W⊨1/2 λ (JF-1)^2+μ (1/2 (Tr[b]-3)-Log[JF]);
SMSDo[m,1,np];
  Rgm⊨jed/JF SMSD[W,pe,m];
  SMSExport[wgp Rgm,p$$[m],"AddIn"→True];
  SMSDo[n,m,np];
    Kgmn⊨SMSD[Rgm,pe,n];
    SMSExport[wgp Kgmn,s$$[m,n],"AddIn"→True];
  SMSEndDo[];
SMSEndDo[];
```

Box 4.6. A segment of *AceGen* input for H1 element based on a strain energy function in the spatial configuration $\varphi(B)$ (element "W in $\varphi(B)$")

Discretization based on the weak form. For the computation of the residuum the internal virtual work expression given in (1.90) has to be specified at element level

$$\int_{\varphi(\Omega_e)} \boldsymbol{\sigma} \cdot (\nabla^S \boldsymbol{\eta})\, dv \approx \sum_{g=1}^{n_g} \boldsymbol{\sigma}(\boldsymbol{\Xi}_g) \cdot (\nabla^S \boldsymbol{\eta})(\boldsymbol{\Xi}_g) \det \mathbf{j}_e(\boldsymbol{\Xi}_g)\, w_{gp} \tag{4.74}$$

Here expressions for the Cauchy stress $\boldsymbol{\sigma}$ and the virtual strain $(\nabla^S \boldsymbol{\eta})$ are needed. The push forward of the constitutive Eq. (5.11) leads to

$$\boldsymbol{\sigma} = \Lambda(J_F - 1)\, \mathbf{1} + \frac{\mu}{J_F}\, (\boldsymbol{b} - \mathbf{1}), \tag{4.75}$$

which immediately can be evaluated at integration point level when the left Cauchy–Green tensor $\boldsymbol{b}$ and the determinant of the deformation gradient J_e are known. Another way to compute $\boldsymbol{\sigma}$ is based on Eq. (5.6)

$$\boldsymbol{\sigma} = 2\, J_F^{-1}\, \boldsymbol{b}\, \frac{\partial W(\boldsymbol{b})}{\partial \boldsymbol{b}}$$

which yields the Cauchy stresses for a given isotropic strain energy function W.

Using the displacement interpolation as well as the interpolation for the variations, the virtual strain is computed at integration point level as

$$\bar{\nabla}^S \boldsymbol{\eta} = \frac{1}{2} \left(\operatorname{grad} \boldsymbol{\eta} + \operatorname{grad} \boldsymbol{\eta}^T \right) \tag{4.76}$$

where the differentiation has to be performed with respect to the current coordinates x_1, x_2, x_3. However this formulation is not well suited for the generalized automated finite element method since the scheme used in (4.63) cannot be applied. Hence another formulation is advantageous in which the variation of spatial gradient is computed from

$$\operatorname{grad} \boldsymbol{\eta} = \left. \frac{\partial \operatorname{grad} \boldsymbol{u}_e}{\partial \mathbf{p}_e} \right|_{\mathbf{j}_e = \text{const.}} \delta \mathbf{p}_e \tag{4.77}$$

where $\mathbf{j}_e$ is the Jacobian of the transformation from the current configuration $\varphi(B)$ to the reference configuration $\square$, given in (4.17). The spatial gradient $\operatorname{grad} \boldsymbol{u}_e$ is given in Eq. (4.21). By the latter form the differentiation is performed on the displacement part of the gradient, leaving the configuration of the spatial gradient 'frozen'.

By using the variation of the spatial gradient (4.77) the *ADB* form of contributions to the weak form can now be written component wise for each Gauss point g as

$$R_{gm} = \boldsymbol{\sigma} \cdot \left. \frac{\hat{\delta}\left(\frac{1}{2}(\operatorname{grad} \boldsymbol{u}_e + \operatorname{grad} \boldsymbol{u}_e^T)\right)}{\hat{\delta} p_{em}} \right|_{\mathbf{j}_e = \text{const.}} j_e. \tag{4.78}$$

Again the tangent stiffness matrix is computed at each integration point by

$$K_{gmn} = \frac{\hat{\delta} R_{gm}(\mathbf{p}_e)}{\hat{\delta} p_{en}}. \tag{4.79}$$

The corresponding input segment for *AceGen* can be found in Box 4.7. The segment is used to replace the Segment A in Box 4.3 in order to obtain complete *AceGen* input.

```
𝕏⊨Nh.𝕏IO;𝕦⊨Nh.𝕦IO;𝕩⊢SMSFreeze[𝕏+𝕦];
je⊢SMSFreeze[SMSD[𝕩,Ξ]];jed⊨Det[je];
gradu⊨SMSD[𝕦,𝕩,"Dependency"→{Ξ,𝕩,SMSInverse[je]}];
invF⊨IdentityMatrix[3]-gradu;F⊨SMSInverse[invF];
𝕓⊨F.Fᵀ;JF⊨Det[F];
σ⊨λ (JF-1) IdentityMatrix[3]+μ/JF (𝕓-IdentityMatrix[3]);
SMSDo[m,1,np];
  δgradu⊨SMSD[gradu,pe,m,"Constant"→je];
  ∇Sηm⊨1/2 (δgradu+δgraduᵀ);
  Rgm⊨jed Tr[∇Sηm.σᵀ];
  SMSExport[wgp Rgm,p$$[m],"AddIn"→True];
  SMSDo[n,m,np];
    Kgmn⊨SMSD[Rgm,pe,n];
    SMSExport[wgp Kgmn,s$$[m,n],"AddIn"→True];
  SMSEndDo[];
SMSEndDo[];
```

Box 4.7. A segment of *AceGen* input for an element based on the weak form in the spatial configuration $\varphi(B)$ (element "G in $\varphi(B)$")

4.2.3 Comparison of Formulations

In the last sections different versions of a standard three-dimensional H1-element were derived for finite hyperelastic deformations. The discretization was based on a strain energy function W and a weak form G when formulated with respect to the initial and current configuration. In continuum theory all element formulations are equivalent and have to yield the same results. This is also true for all element formulations developed in the last section. Their results, when applied to an engineering problem, do not differ up to computer precision.

Table 4.1 presents the size of the generated C source code file, the time required by *AceGen* to derive the formulas and to generate the element code and the time required by *AceFEM* to evaluate the element tangent matrices and assembly the global tangent matrix for a model problem depicted in Fig. 4.11.

It is known that finite element discretizations based on the weak form of the current configuration are very efficient when classical coding techniques are employed, see e.g. Simo and Armero (1992); Zienkiewicz and Taylor (2000a); Wriggers (2008). This observation is not true for elements generated by *AceGen*. Table 4.1 depicts

Table 4.1 Different formulations of the H1-element

Formulation	Eq.	*AceGen* input	Code size [kB]/[norm.]	*AceGen* time [s]/[norm.]	**K** and **R** time [norm.]
W in B	(4.49)	4.3	14.1/1.00	7/1.0	1.00
G in B	(4.60)	4.4	15.4/1.07	8/1.1	1.76
W^P in B	(4.65)	4.5	17.1/1.22	9/1.3	1.62
W in $\varphi(B)$	(4.67)	4.6	56.4/3.44	43/6.1	5.84
G in $\varphi(B)$	(4.74)	4.7	22.0/1.66	17/2.4	3.18

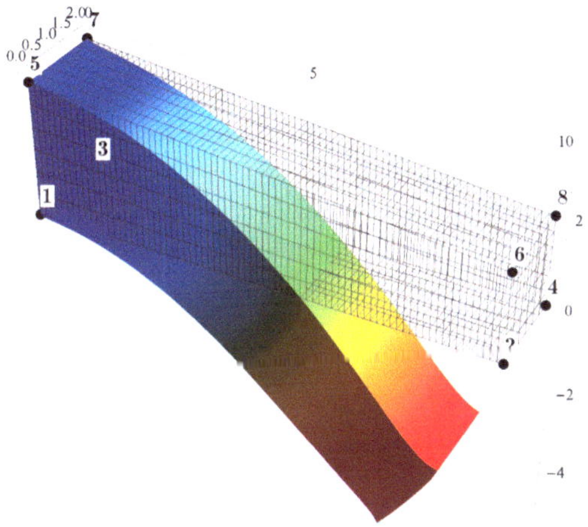

$T_1(0,0,0)$
$T_2(10,0,0)$
$T_3(0,2,0)$
$T_4(10,2,0)$
$T_5(0,0,3)$
$T_6(10,0,2)$
$T_7(0,2,3)$
$T_8(10,2,2)$
$X = 0 : u = v = w = 0$
$X = 10 : w = -5$
$E = 1000$
$\nu = 0.3$
$n_e = 2700$
$n_{tp} = 10\,976$

Fig. 4.11 Test example used to assess the performance of alternative formulations and implementations of the H1 element

a significant increase of the generated code size for formulations based on the current configuration $\varphi(B)$. Since the code size relates strongly to the execution time of such element as well as to the time required by *AceGen* to derive formulas and generate the element code, it is clear that elements formulated with respect to the initial configuration B are more efficient. The most efficient formulations follows directly from the potential W (4.49) leading to the smallest code size and the shortest derivation time. The code size is three times smaller than the code generated from the potential (4.67) in the current configuration $\varphi(B)$. Thus formulations with respect to the initial configuration should be preferred when using *AceGen* .

The results for the element based on a strain energy function in the spatial configuration $\varphi(B)$ (element "W in $\varphi(B)$") require additional explanation. With the relation $J_e = \frac{j_e}{J_F}$ the *ADB* formulation (4.71) of the element based on a strain energy function in the spatial configuration leads directly to the *ADB* formulation of the H1 element based on a strain energy function in the initial configuration (4.57). Consequently, one would expect that the elements "W in B" and "W in $\varphi(B)$" would have similar codes size and numerical efficiency. However, the "W in $\varphi(B)$" element is by far the largest and the slowest. The efficiency of the results is in this case influenced by the ability of *AceGen* to perform global optimization of the derived set of formulas. It is an imperative to produce codes in reasonable time, otherwise debugging of the *AceGen* inputs would be impossible and the whole automation procedure impractical for the development of the new numerical models. Only the most simple global relations are tested., while the higher level relations that would, if applied, transform the "W in $\varphi(B)$" element into "W in B" element are not applied. Consequently, the convoluted way how the deformation gradient is obtained from $\boldsymbol{F}^{-1} = \mathbf{1} - \text{grad}\,\boldsymbol{u}$ is not automatically resolved into simple $\boldsymbol{F} = \mathbf{1} + \text{Grad}\,\boldsymbol{u}$ and thus results in a significant amount of redundant code.

Test example: A three dimensional solid is investigated that acts like a clamped tapered beam. The eight coordinates $T_i\ ,\ i = 1, 8$ describing the geometry of the tapered beam are depicted in Fig. 4.11. At $X = 0$ the beam is clamped by setting all displacement variables to zero. At the end of the beam at $X = 10$ a vertical displacement of $w = -5$ is described. The constitutive date are $E = 1000$ for the Young's modulus and $\nu = 0.3$ for the Poisson ratio. These values are equivalent to the Lame constants: $\mu = E\,/\,(2(1+\nu)) = 384.62$ and $\Lambda = E\,\nu\,/\,((1+\nu)(1-2\nu)) = 576.92$. The mesh consists of 6 elements in Y- and Z-direction and 75 elements in X-direction leading in total to $n_e = 2700$ finite elements and $n_{tp} = 10976$ number of unknowns.

4.3 Alternative Implementations

Additionally to the different continuum formulations of the same element we can also introduce different implementations of the same formulation of the same element. Implementation of the same formulation can differ only in *AceGen* input or it can differ also in solution algorithm and consequently in the efficiency of the generated code. Some representative possibilities are discussed in this section. Different implementations of the H1 element based on a strain energy function in the initial configuration (element "W in B") are provided as an example.

An important factor that has to be considered when the particular implementation of the same element is chosen is the physical size of the generated code. If n_p unknown parameters are used in a numerical procedure at element level, then an explicit form of the element residual and the tangent matrix will have at least $n_p + n_p^2$ terms. Thus, explicit code for all terms can be generated only if the number of unknowns is small. In a loop is a sequence of formulas which is specified once but can be carried out several times in succession. Thus, loops are important instruments that can be used to reduce the physical size of the generated code. A formula that is evaluated in a loop is called a "characteristic formula". Characteristic formulas can be produced by automatic differentiation with respect to variables with an index as described in Sect. 2.6.3.

The *AceGen* input of the "W in B" element presented in Box 4.3 contains three nested count-controlled loops:

1. loop over integration points with I_g as a counter,
2. loop over components of the residual $\mathbf{R}$ with m as a counter,
3. loop over components of the tangent matrix $\mathbf{K}$ with n as a counter.

Consequently, one characteristic formula was generated for an arbitrary element of the residual vector and one characteristic formula for an arbitrary element of the tangent matrix. Each of these loops can be unrolled to increase the execution speed. Loop unrolling is a loop transformation technique that attempts to optimize a program's execution speed at the expense of its size (space-time trade-off). Loops are first re-written as a repeated sequence of independent formulas. These formulas are then evaluated for the particular value of the counter and additionally optimized by *AceGen* . Unrolling of loops is combined with the generation of characteristic formulas. This means the set of unknowns of the element $\mathbf{p}_e = \{p_{e1}, p_{e2}, \dots, p_{en_p}\}^T$ has to be divided into disjoint subsets in a way that matches the chosen loop unrolling. Among almost countless combinations six specific ones are selected and combined in this section:

a completely loop free implementation,
b implementation with integration loop only,
c parameter wise implementation (element "*W* in *B*"),
d nodal wise implementation,
e implementation with **K** loop unrolled,
f implementation with **R** loop unrolled.

The general characteristics of the considered implementations are presented in Table 4.2. The table presents the number of *Do* loop structures and the number of explicitly derived elements of the tangent and residual used for the evaluation of the tangent matrix and the residual. The schematic position of the explicitly derived elements is also depicted in Fig. 4.12 where every black box stands for one explicitly derived element of the residual vector or the tangent matrix.

Completely loop free implementation. Within the completely loop free implementation all three loops are unrolled and explicit formulas are generated for all elements of the residual and the tangent matrix at all Gauss points. Since the element is integrated with $2 \times 2 \times 2$ GAUSS integration that means that in total $8 \times 24 = 192$ formulas are generated for the evaluation of the residual and $8 \times 24 \times 24 = 4\,608$ formulas for the evaluation of the tangent matrix. The corresponding *AceGen* input is presented in Box 4.8.

Table 4.2 Characteristics of different implementations of the H1-element

Implementation	*AceGen* input	Integ. loop	**R** loop	**K** loop	No. of **R** formulas	No. of **K** formulas
a. Loop free	4.8	No	No	No	8×24	$8 \times 24 \times 24$
b. Integration loop	4.9	Yes	No	No	24	24×24
c. Parameter wise	4.3	Yes	Yes	Yes	1	1
d. Node wise	4.10	Yes	Yes	Yes	3	3×3
e. **K** loop unrolled	4.11	Yes	Yes	No	1	24
f. **R** loop unrolled	4.12	Yes	No	Yes	24	24

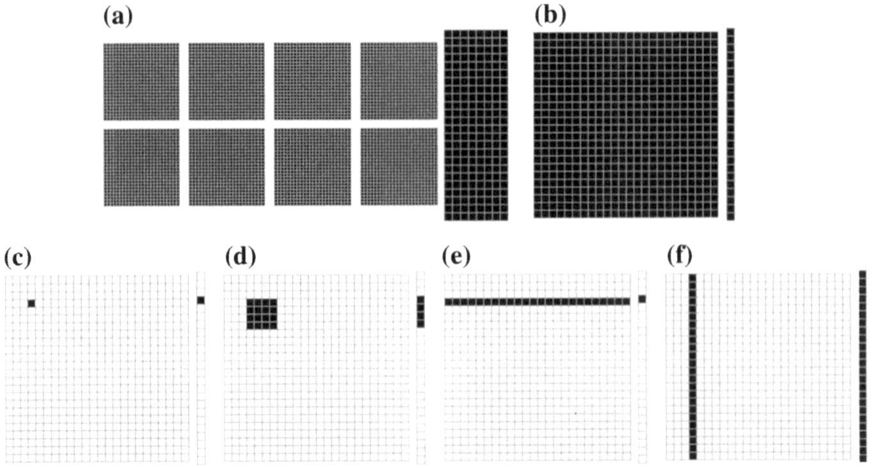

Fig. 4.12 Schematic position of characteristic or explicit formulas of tangent matrix and residual for considered implementations of H1 element

```
SMSInitialize["All-Loops-Free","Environment"→"AceFEM",
  "VectorLength"→10 000];
SMSTemplate["SMSTopology"→"H1","SMSSymmetricTangent"→True,
  "SMSDomainDataNames"->{"E -elastic modulus","ν -poisson ratio"},
  "SMSDefaultData"->{21 000,0.3}];
nen=SMSNoNodes;ndim=SMSNoDimensions;np=SMSNoDOFGlobal;
SMSStandardModule["Tangent and residual"];
{Em,ν}⊢SMSReal[Table[es$$["Data",i],{i,2}]];{λ,μ}⊨SMSHookeToLame[Em,ν];
𝕏IO⊢Table[SMSReal[nd$$[i,"X",j]],{i,nen},{j,ndim}];
𝕦IO⊢SMSReal[Table[nd$$[i,"at",j],{i,nen},{j,ndim}]];𝕡e=Flatten[𝕦IO];
Ξn={{-1,-1,-1},{1,-1,-1},{1,1,-1},{-1,1,-1},
  {-1,-1,1},{1,-1,1},{1,1,1},{-1,1,1}};ξg⊨1/Sqrt[3];
{ℝe,𝕂e}=Sum[
    Ξ={ξ,η,ζ}⊢SMSFreeze[ξg Ξn⟦Ig⟧];wgp=1;
    Nh⊨Table[1/8 (1+ξ Ξn⟦i,1⟧) (1+η Ξn⟦i,2⟧) (1+ζ Ξn⟦i,3⟧),{i,1,nen}];
    𝕏⊢SMSFreeze[Nh.𝕏IO];𝕦⊨Nh.𝕦IO;𝕁e⊨SMSD[𝕏,Ξ];Jed⊨Det[𝕁e];
    ℍ⊨SMSD[𝕦,𝕏,"Dependency"→{Ξ,𝕏,SMSInverse[𝕁e]}];
    𝔽⊨IdentityMatrix[3]+ℍ;ℂt⊨𝔽ᵀ.𝔽;JF⊨Det[𝔽];
    W⊨1/2 λ (JF-1)^2+μ (1/2 (Tr[ℂt]-3)-Log[JF]);
    ℝg⊨Jed SMSD[W,𝕡e];
    𝕂g⊨SMSD[ℝg,𝕡e];
    wgp {ℝg,𝕂g}
    ,{Ig,1,8}];(*explicit sum over gauss points*)
SMSExport[ ℝe,p$$];
SMSExport[ Table[𝕂e⟦i,j⟧,{i,1,np},{j,i,np}],
  Table[s$$[i,j],{i,1,np},{j,i,np}]];
SMSWrite[];
```

Box 4.8. *AceGen* input for an H1 element that generates the code that is completely loops free

Implementation with an integration loop only. Only the integration loop is used in this case to reduce the amount of generated code. here 24 formulas are generated for the evaluation of the residual and $24 \times 24 = 576$ formulas for the evaluation of the tangent matrix. The corresponding segment of the *AceGen* input is presented in Box 4.9. The complete *AceGen* input can be obtained by replacing the Segment A in Box 4.3 by the input segment given in Box 4.9.

```
𝕏⊢SMSFreeze[ℕh.𝕏IO];𝕦⊨ℕh.𝕦IO;𝕁e⊨SMSD[𝕏,Ξ];Jed⊨Det[𝕁e];
ℍ⊨SMSD[𝕦,𝕏,"Dependency"→{Ξ,𝕏,SMSInverse[𝕁e]}];
𝔽⊨IdentityMatrix[3]+ℍ;ℂt⊨𝔽ᵀ.𝔽;JF⊨Det[𝔽];
W⊨1/2 λ (JF-1)^2+μ (1/2 (Tr[ℂt]-3)-Log[JF]);
ℝg⊨Jed SMSD[W,𝕡e];
SMSExport[wgp ℝg,p$$,"AddIn"→True];
𝕂g⊨SMSD[ℝg,𝕡e];
SMSExport[wgp Table[𝕂g⟦i,j⟧,{i,1,np},{j,i,np}],
  Table[s$$[i,j],{i,1,np},{j,i,np}],"AddIn"→True];
```

Parameter wise implementation. The "parameter wise implementation" is the one used in the previous Section to generate different formulations of the hyperelastic H1 element. The formulation was denoted "W in B" and the corresponding *AceGen* input is presented in Box 4.3. All the unknowns form one set, thus accordingly to derivations presented in Sect. 2.6.3, $n_s = 1$ and $\mathbf{p}_{e1} = \mathbf{p}_e$. One characteristic formula is generated for an arbitrary element of the residual vector and one characteristic formula for an arbitrary element of the tangent matrix as presented in Fig. 4.12c.

Nodal wise implementation. The *AceGen* input presented in Box 4.3 of the element element "W in B" is using abstract handling of nested sets (e.g. deformation gradient $\boldsymbol{F}$ or set of nodal coordinates $\mathbf{X}$) and abstract operations performed on nested sets available in *Mathematica*. While at the level of *AceGen* input the quantities such as $\boldsymbol{F}$ are treated as a single object, *AceGen* in fact always stores and optimizes nested sets in component wise form. Standard finite element codes that are written in FORTRAN or C language can manipulate sets only at the component level, unless additional object oriented libraries are used. The component based arrangement of nodal quantities is needed in that case. The components of the displacements (u, v, w) and coordinates (X, Y, Z) can be expressed within the element using the shape functions $\mathbf{N}$ as

$$\begin{aligned} u_e &= \mathbf{N}\,\mathbf{u}_c, \quad v_e = \mathbf{N}\,\mathbf{v}_c \quad \text{and} \quad w_e = \mathbf{N}\,\mathbf{v}_c, \\ X_e &= \mathbf{N}\,\mathbf{X}_c, \quad Y_e = \mathbf{N}\,\mathbf{Y}_c \quad \text{and} \quad Z_e = \mathbf{N}\,\mathbf{Z}_c \end{aligned} \tag{4.80}$$

where nodal coordinates and displacements are contained in vectors

$$\begin{aligned} &\mathbf{u}_c = \{u_1, u_2, \ldots, u_8\}^T, \ \mathbf{v}_c^T = \{v_1, v_2, \ldots, v_8\}^T \text{ and} \\ &\mathbf{w}_c = \{w_1, w_2, \ldots, w_8\}^T, \\ &\mathbf{X}_c = \{X_1, X_2, \ldots, X_8\}^T, \ \mathbf{Y}_c^T = \{Y_1, Y_2, \ldots, Y_8\}^T \text{ and} \\ &\mathbf{Z}_c = \{Z_1, Z_2, \ldots, Z_8\}^T. \end{aligned} \tag{4.81}$$

The latter arrangement can be seen in the *AceGen* input provided in Box 4.10. The unknowns are divided into three subsets with equal length ($n_s = 8$) as follows

$$\mathbf{p}_{e1} = \mathbf{u}_e, \ \mathbf{p}_{e2} = \mathbf{v}_e, \ \mathbf{p}_{e3} = \mathbf{w}_e. \tag{4.82}$$

Three characteristic formulas are generated for an arbitrary element of the residual vector and 3×3 characteristic formulas for an arbitrary element of the tangent matrix as presented in Fig. 4.12d. The nodal wise implementation closely resembles the way how the finite element codes are derived manually.

```
SMSInitialize["Node-Wise-Loops","Environment"→"AceFEM"];
SMSTemplate["SMSTopology"→"H1","SMSSymmetricTangent"→True,
  "SMSDomainDataNames"->{"E -elastic modulus","ν -poisson ratio"},
  "SMSDefaultData"->{21 000,0.3},"SMSDefaultIntegrationCode"→7];
nen=SMSNoNodes;ndim=SMSNoDimensions;np=SMSNoDOFGlobal;
SMSStandardModule["Tangent and residual"];
{Em,ν}⊢SMSReal[Table[es$$["Data",i],{i,2}]];
{λ,μ}⊨SMSHookeToLame[Em,ν];
{Xc,Yc,Zc}⊢Table[SMSReal[nd$$[i,"X",j]],{j,ndim},{i,nen}];
{uc,vc,wc}⊢Table[SMSReal[nd$$[i,"at",j]],{j,ndim},{i,nen}];
SMSDo[Ig,1,SMSInteger[es$$["id","NoIntPoints"]]];
Ξ={ξ,η,ζ}⊢Table[SMSReal[es$$["IntPoints",i,Ig]],{i,3}];
wgp⊢SMSReal[es$$["IntPoints",4,Ig]];
Ξn={{-1,-1,-1},{1,-1,-1},{1,1,-1},{-1,1,-1},
   {-1,-1,1},{1,-1,1},{1,1,1},{-1,1,1}};
Nh⊨Table[1/8 (1+ξ Ξn⟦i,1⟧) (1+η Ξn⟦i,2⟧) (1+ζ Ξn⟦i,3⟧),{i,1,nen}];
X⊢SMSFreeze[{Nh.Xc,Nh.Yc,Nh.Zc}];u⊨{Nh.uc,Nh.vc,Nh.wc};
Je⊨SMSD[X,Ξ];Jed⊨Det[Je];
H⊨SMSD[u,X,"Dependency"→{Ξ,X,SMSInverse[Je]}];
F⊨IdentityMatrix[3]+H;Ct⊨Fᵀ.F;JF⊨Det[F];
W⊨1/2 λ (JF-1)^2+μ (1/2 (Tr[Ct]-3)-Log[JF]);
SMSDo[m,1,nen];
  Rgm⊨Jed {SMSD[W,uc,m],SMSD[W,vc,m],SMSD[W,wc,m]};
  SMSExport[wgp Rgm,Table[p$$[3 (m-1)+i],{i,3}],"AddIn"→True];
  SMSDo[n,m,nen];
    Kgmn⊨{SMSD[Rgm,uc,n],SMSD[Rgm,vc,n],SMSD[Rgm,wc,n]}ᵀ;
    SMSExport[wgp Kgmn,
      Table[s$$[3 (m-1)+i,3 (n-1)+j],{i,3},{j,3}],"AddIn"→True];
  SMSEndDo[];
SMSEndDo[];
SMSEndDo[];
SMSWrite[];
```

Box 4.10. *AceGen* input for an element with the node wise arrangement of the unknowns

Implementations with K or R loop unrolled. In the parameter wise implementation the loop over the elements of the row of the tangent matrix is nested within the loop over the elements of the residual. Each of this loops can be independently unrolled.

Within the implementation with the **K** loop unrolled the inner loop is unrolled as presented in Fig. 4.12e and in Box 4.11. The set of unknowns is formed into one subset ($n_s = 1$ and $\mathbf{p}_{e1} = \mathbf{p}_e$), thus a characteristic formula is generated for an arbitrary (mth) element of the residual vector, while in the inner loop explicit expressions are generated for all 8 elements of an arbitrary (with an index m) row of the tangent matrix.

```
X⊢SMSFreeze[Nh.XIO];u⊨Nh.uIO;Je⊨SMSD[X,Ξ];Jed⊨Det[Je];
H⊨SMSD[u,X,"Dependency"→{Ξ,X,SMSInverse[Je]}];
F⊨IdentityMatrix[3]+H;Ct⊨Fᵀ.F;JF⊨Det[F];
W⊨1/2 λ (JF-1)^2+μ (1/2 (Tr[Ct]-3)-Log[JF]);
SMSDo[m,1,np];
  Rgm⊨Jed SMSD[W,pe,m];
  SMSExport[wgp Rgm,p$$[m],"AddIn"→True];
  Kgm⊨SMSD[Rgm,pe];
  SMSExport[wgp Kgm,Table[s$$[m,i],{i,np}],"AddIn"→True];
SMSEndDo[];
```

Box 4.11. *AceGen* input segment for an element with the tangent loop unrolled

Alternatively only the **R** loop can be unrolled. Explicit expressions are generated for all 8 elements of the residual vector. For each element of the residual vector, a characteristic expression is generated for an arbitrary element (nth element) of the corresponding row of the tangent matrix. Thus, a vector of 8 characteristic expressions forms a characteristic column of the tangent matrix as presented in Fig. 4.12f and in Box 4.12.

```
X⊢SMSFreeze[Nh.XIO];u⊨Nh.uIO;Je⊨SMSD[X,Ξ];Jed⊨Det[Je];
H⊨SMSD[u,X,"Dependency"→{Ξ,X,SMSInverse[Je]}];
F⊨IdentityMatrix[3]+H;Ct⊨Fᵀ.F;JF⊨Det[F];
W⊨1/2 λ (JF-1)^2+μ (1/2 (Tr[Ct]-3)-Log[JF]);
Rg⊨Jed SMSD[W,pe];
SMSExport[wgp Rg,p$$,"AddIn"→True];
SMSDo[n,1,np];
  Kgn⊨SMSD[Rg,pe,n];
  SMSExport[wgp Kgn,Table[s$$[i,n],{i,np}],"AddIn"→True];
SMSEndDo[];
```

Box 4.12. *AceGen* input segment for an element with the residual loop unrolled

The complete *AceGen* input can be obtained by replacing the Segment A in Box 4.3 by the input segment given in Box 4.11 or Box 4.12.

4.3.1 Comparison of Implementations

In Table 4.3 the code size, the *AceGen* time needed to generate C code and the numerical efficiency of the considered implementations are compared. While comparing the implementations several criteria have to be considered:

Table 4.3 Code size and efficiency of considered implementations of the H1-element

Implementation	*AceGen* input	*AceGen* time [s]	Number of formulas	C code size [kB]	**K** and **R** time [norm.]
a. Loop free	4.8	2337	3884	302.5	1.01
b. Integration loop	4.9	80	552	39.3	0.78
c. Parameter wise	4.3	7	225	14.1	1.00
d. Node wise	4.10	7	225	9.7	0.83
e. **K** loop unrolled	4.11	8	224	14.4	0.97
f. **R** loop unrolled	4.12	9	228	14.5	1.00

- the size of the generated code,
- the time needed to generate the code,
- the numerical efficiency of the generated code and
- from the practical reasons also the simplicity of the *AceGen* input.

The simplicity of the *AceGen* input influences the time needed to write the input and most important the time needed to find and correct all the mistakes in it. Numerical efficiency is measured by the time needed by *AceFEM* to evaluate element tangent matrices and assembly the global tangent matrix of the problem depicted in Fig. 4.11. The time is normalized against the time needed for parameter wise implementation. All the implementations are equivalent and the results, when applied to an engineering problem, do not differ up to computer precision.

The actual unrolling of the loops does not by itself measurably increase the numerical efficiency since the time needed to administer the loops (to increase the counters and test the exit condition) is negligible when compared to the time needed to evaluate the complex formulas involved. If the formulas obtained by unrolling, and thus evaluated for the integer value of the loop counter, do not simplify significantly when compared to the same formulas evaluated for symbolic value of the counter, then the two implementation would have comparable numerical efficiency.

The "parameter wise", the "**K** loop unrolled" and the "**R** loop unrolled" implementations have almost the same code size and numerical efficiency as well as *AceGen* time, however of those three implementations the "parameter wise" implementation has the most simple *AceGen* input. The "integration loop only" implementation leads to the fastest code with 22 % shorter evaluation time, but it also takes 11.2 times longer to be generated by *AceGen* and results in 2.9 times larger code size when compared to "parameter wise" implementation. The completely "loops free" implementation leads to 21.5 times larger C code file and it takes 333.8 times longer to be derived. In principle the "loop free" implementation should lead to the fastest code, however the shear number of generated formulas (3884 formulas are generated by *AceGen* during the derivation of the code) and the size of the generated code prevents the *AceGen* and the C code compiler to perform an efficient code optimization in real time.

The node wise implementation explores the fact that the three scalar fields u, v and w are interpolated independently and simplifies the resulting formulas accordingly. The result is 17 % shorter evaluation time than the "parameter wise" implementation. The "node wise" implementation also leads to the 30 % smaller code size than the "parameter wise" implementation. In fact leads to the smallest code size of all implementations. It is interesting that the node wise implementation from all implementations resembles most closely the way how the finite element codes are derived manually. The only drawback of the "node wise" implementation is a rather complex *AceGen* input.

Taking into account all above considerations, the "parameter wise" implementation presented in Box 4.3 has the best overall performances and it will be used throughout this book as a primal way for the implementation of various finite elements. The *AceGen* input presented in Box 4.3 is abstract and requires only small changes when the input is used as a basis for development of more general form of elements that also includes the use of different numbers of degrees of freedoms at nodal points, like in mixed methods.

References

Bathe, K.J. 1996. *Finite element procedures*. Englewood Cliffs: Prentice-Hall.

Braess D. 2007. *Finite elements: theory, fast solvers, and applications in solid mechanics*. Cambridge: Cambridge University Press.

Dhatt, G., and G. Touzot. 1985. *The finite element method displayed*. Chichester: Wiley.

Hughes, T.R.J. 1987. *The finite element method*. Englewood Cliffs, New Jersey: Prentice Hall.

Irons, B. 1971. Quadrature rules for brick based finite elements. *International Journal for Numerical Methods in Engineering* 3: 293–294.

Oñate, E. 2009. *Structural analysis with the finite element method. linear statics: volume 1: basis and solids*. Springer: Heidelberg.

Simo, J.C., and F. Armero. 1992. Geometrically non-linear enhanced strain mixed methods and the method of incompatible modes. *International Journal for Numerical Methods in Engineering* 33: 1413–1449.

Wriggers, P. 2008. *Nonlinear finite elements*. Berlin: Springer.

Zienkiewicz, O.C., and R.L. Taylor. 1989. *The finite element method*, vol. 1, 4th ed. London: McGraw Hill.

Zienkiewicz, O.C., and R.L. Taylor. 1991. *The finite element method*, vol. 2, 4th ed. London: McGraw Hill.

Zienkiewicz, O.C., and R.L. Taylor. 2000a. *The finite element method*, vol. 1, 5th ed. Oxford: Butterworth-Heinemann.

Zienkiewicz, O.C., and R.L. Taylor. 2000b. *The finite element method*, vol. 2, 5th ed. Oxford: Butterworth-Heinemann.

Chapter 5
Materials

Constitutive equations have to be formulated in continuum mechanics that characterizes the material response of a solid body. The constitutive theory describes, in relation to the nature of the task, either the microscopic or the macroscopic behaviour of a material. For most materials like steel, concrete or rubber a macroscopic description is sufficient. Then a functional dependence of stresses or heat flux with respect to the motion or temperature has to be considered. Since real materials can exhibit very complex behaviour, approximations have to be applied within the derivation process of constitutive equations. These however have to be extensive enough to cover all effects observed in experimental investigations. Furthermore basic principles from mechanics have to be obeyed to obtain theoretically sound constitutive equations, for more detail see e.g. Truesdell and Noll (1965), Bertram (1989).

5.1 Elastic Materials

5.1.1 General Formulation

Purely elastic material behaviour is discussed in this section under the assumption of so called Green elasticity also named hyper elasticity, see e.g. Ogden (1984), Chap. 4. This description is valid for many materials—like e.g. foam or rubbers—which undergo finite deformations. In case of small strains these constitutive equations reduce to the classical law of Hooke known form the linear theory of elasticity.

The constitutive equation for the second Piola–Kirchhoff stress tensor can be derived from the potential Ψ in case of a hyper elastic material. $W = \rho_0 \, \Psi$ describes the strain energy stored in the body (for this reason it is called strain energy function). The derivative of the strain energy function with respect to the right Cauchy–Green tensor yields

J. Korelc and P. Wriggers, *Automation of Finite Element Methods*,
DOI 10.1007/978-3-319-39005-5_5

$$
\boldsymbol{S} = 2\,\rho_0\,\frac{\partial \Psi(\boldsymbol{C})}{\partial \boldsymbol{C}} = 2\,\frac{\partial W(\boldsymbol{C})}{\partial \boldsymbol{C}}; \quad S_{AB} = 2\,\rho_0\,\frac{\partial \Psi(C_{CD})}{\partial C_{AB}} = 2\,\frac{\partial W(C_{CD})}{\partial C_{AB}}. \tag{5.1}
$$

The exclusive dependence of Ψ on $\boldsymbol{C}$ is substantiated by the general principles as discussed in the last section. For a more detailed treatment see e.g. Truesdell and Noll (1965), Malvern (1969) and Ogden (1984).

The material behaviour is independent on directions for technically important classes of isotropic materials which includes e.g. steel, aluminium, rubber or concrete. With this assumption it is possible to specify the general function W in (5.1). By introducing isotropy groups a function can be derived which only depends upon the invariants of the strain tensors, see e.g. Ogden (1984). Using the invariants of the right Cauchy–Green deformation tensor $\boldsymbol{C} = \boldsymbol{F}^T\,\boldsymbol{F}$ it follows

$$
W(\boldsymbol{C}) = W\,(I_C,\ II_C,\ III_C). \tag{5.2}
$$

The invariants $I_C,\ II_C,\ III_C$[1] are defined in Appendix B.1.5. By using the invariants in their explicit form the strain energy (5.2) can also be expressed as

$$
W(\boldsymbol{C}) = W\,(\mathrm{tr}\,\boldsymbol{C},\ \mathrm{tr}\,(\mathrm{Cof}\,\boldsymbol{C}),\ \det\boldsymbol{C}). \tag{5.3}
$$

The description of the elastic material by the isotropic tensor function (5.2), see e.g. Ogden (1984), leads—by using the chain rule—to an expression for the second Piola–Kirchhoff stresses in terms of the right Cauchy–Green tensor

$$
\boldsymbol{S} = 2\left[\left(\frac{\partial W}{\partial I_C} + I_C\,\frac{\partial W}{\partial II_C}\right)\boldsymbol{1} - \frac{\partial W}{\partial II_C}\,\boldsymbol{C} + III_C\,\frac{\partial W}{\partial III_C}\,\boldsymbol{C}^{-1}\right]. \tag{5.4}
$$

In this equation the relations

$$
\frac{\partial I_C}{\partial \boldsymbol{C}} = \boldsymbol{1}, \quad \frac{\partial II_C}{\partial \boldsymbol{C}} = I_C\,\boldsymbol{1} - \boldsymbol{C}, \quad \frac{\partial III_C}{\partial \boldsymbol{C}} = III_C\,\boldsymbol{C}^{-1} \tag{5.5}
$$

were applied. They describe the derivatives of the invariants of a tensor with respect to the tensor itself.

The constitutive equation (5.4) can also be related directly to the current configuration. In that case the Cauchy stress tensor can be expressed in terms of the left Cauchy–Green tensors.

---

[1]By expressing the invariants in terms of principal stretches

$$
I_C = \lambda_1^2 + \lambda_2^2 + \lambda_3^2\,, \quad II_C = \lambda_1^2\,\lambda_2^2 + \lambda_2^2\,\lambda_3^2 + \lambda_3^2\,\lambda_1^2\,, \quad III_C = \lambda_1^2\,\lambda_2^2\,\lambda_3^2
$$

the strain energy function assumes the form

$$
W(\boldsymbol{C}) \equiv W\,(\lambda_1^2,\ \lambda_2^2,\ \lambda_3^2).
$$

This approach is not considered within this book.

With Eq. (1.72) the Cauchy stress $\boldsymbol{\sigma} = J_F^{-1}\,\boldsymbol{F}\,\boldsymbol{S}\,\boldsymbol{F}^T$ is obtained and hence by considering (5.1)

$$\boldsymbol{\sigma} = 2\,J_F^{-1}\,\rho_0\,\boldsymbol{F}\,\frac{\partial\psi(\boldsymbol{C})}{\partial\boldsymbol{C}}\,\boldsymbol{F}^T$$

follows. Based on (5.4) it can be shown, using also (1.48), that this equation is equivalent to

$$\boldsymbol{\sigma} = 2\,\boldsymbol{b}\,\rho\,\frac{\partial\Psi(\boldsymbol{b})}{\partial\boldsymbol{b}} = 2\,J_F^{-1}\,\boldsymbol{b}\,\frac{\partial W(\boldsymbol{b})}{\partial\boldsymbol{b}}. \tag{5.6}$$

where $\boldsymbol{b} = \boldsymbol{F}\,\boldsymbol{F}^T$ is the left Cauchy–Green tensor.[2]

Compressible materials like foams can be described by introducing a term that describes the volumetric change within the material. Thus a function depending on the determinant J_F of the deformation gradient $\boldsymbol{F}$ is best suited. Hence the function $g(J_F)$ can be used to substitute the third invariant of $\boldsymbol{C}$ ($III_C = J_F^2$) leading to

$$W(I_C, II_C, J_F) = \hat{W}(I_C, II_C) + g(J_F) \tag{5.7}$$

Mathematical investigations regarding existence of solutions restrict the choice of the constitutive parameters, see e.g. Marsden and Hughes (1983), Ogden (1984) and Ciarlet (1988). One restriction produces the effect that the strain energy function (5.7) yields at the undeformed configuration $\boldsymbol{F} = \mathbf{1}$ the constitutive tensor of the classical linear theory. The second restriction is related to the existence of solutions in finite elasticity, see Ogden (1972). Parameters determined from experiments have to obey both restrictions. To further fulfill also polyconvexity which ensures existence of solutions in finite elasticity further relations have to be met by the constitutive parameters, see Marsden and Hughes (1983). This demand is stronger than the second restriction. When fitting experimental results to material parameters of hyperelastic strain energy functions the above conditions have to be met.

5.1.2 Neo-Hooke Material

The compressible Neo-Hooke material is one of the simplest hyperelastic materials. It has the form

$$W(I_C, J_F) = g(J_F) + \frac{1}{2}\,\mu\,(I_C - 3). \tag{5.8}$$

[2]This relation is only valid for isotropic materials. Thus for anisotropic materials formulation (5.1) has to be used, see e.g. Schröder (2010).

The function $g(J_F)$ in (5.7) and (5.8) has to be convex for compressible materials. Furthermore the growth conditions

$$\lim_{J_F \to +\infty} W \to \infty \qquad \text{and} \qquad \lim_{J_F \to 0} W \to \infty \tag{5.9}$$

have to be fulfilled by the strain energy.[3] These growth conditions also play a role in the mathematical treatment of finite elasticity, e.g. for questions regarding existence and uniqueness of solutions, see e.g. Marsden and Hughes (1983), Ciarlet (1988).

In Ciarlet (1988) a special ansatz was chosen for the compressible part in (5.8) to fulfil the growth conditions

$$g(J_F) = c\,(J_F^2 - 1) - d \ln J_F - \mu \ln J_F \quad \text{with} \quad c > 0, d > 0. \tag{5.10}$$

Function $g(J_F)$ has to be fitted to experimental data. Different choices for the function $g(J_F)$ and a related discussion can be found in e.g. Doll and Schweizerhof (2000).

By inserting relation (5.10) for $g(J_F)$ into (5.8) a constitutive relation for the second Piola–Kirchhoff stress tensor is obtained. With the special choice of $c = \Lambda\,/\,4$ and $d = \Lambda\,/\,2$ it follows

$$\boldsymbol{S} = \frac{\Lambda}{2}\,(J_F^2 - 1)\,\boldsymbol{C}^{-1} + \mu\,(\mathbf{1} - \boldsymbol{C}^{-1}). \tag{5.11}$$

The constitutive parameters Λ, μ are known as the Lamé constants. It should be remarked that the constitutive relation for the second Piola–Kirchhoff stress tensor does not have anything in common with a linear relation between $\boldsymbol{S}$ and $\boldsymbol{E}$ which is often used in engineering applications of structural members.[4]

Equation (5.11) can be referred to the current configuration by expressing the second Piola–Kirchhoff stress tensor by the Cauchy stress tensor via $\boldsymbol{\sigma} = J_F^{-1}\,\boldsymbol{F}\,\boldsymbol{S}\,\boldsymbol{F}^T$. After some algebraic manipulations relation

[3] The latter conditions can be physically interpreted in the way that stresses have to approach $-\infty$ for a volume going to zero and $+\infty$ for a volume approaching ∞.

[4] Nonlinear elastic material behaviour is often described in engineering literature by a linear relation between the second Piola–Kirchhoff stress tensor and the Green–Lagrange strain tensor (St. Venant material)

$$\boldsymbol{S} = \Lambda\,\mathrm{tr}\,\boldsymbol{E}\,\mathbf{1} + 2\,\mu\,\boldsymbol{E}. \tag{5.12}$$

This constitutive equation corresponds to Hooke's laws of the infinitesimal theory of elasticity with the Lamé constants Λ and μ (these constants can be converted to the modulus of elasticity $E = \frac{(3\Lambda + 2\mu)\,\mu}{\Lambda + \mu}$ and Poisson's ratio $\nu = \frac{\Lambda}{2\,(\Lambda + \mu)}$). Generally it can be shown that this constitutive equation is restricted to deformations with large displacements and finite rotations but small strains. St. Venant's law depicts major deficiencies in the compressible range: in the limit case of the compression of a body to volume "0" the stress $\boldsymbol{\sigma}$ approaches zero instead of $\lim_{J_F \to 0} \boldsymbol{\sigma} \to -\infty$. Thus the St. Venant material equation is not applicable for general simulations of solids within the finite deformation range. However it can be successfully used for large deflection analysis of thin structural members like beams or shells.

Table 5.1 Hyperelastic Materials

Materials	Function
Neo Hooke	$W(I_C, J_F) = \frac{1}{2}\mu(I_C - 3) + g(J_F)$
Mooney Rivlin	$W(I_C, II_C) = c_1(I_C - 3) + c_2(II_C - 3)$
St. Venant	$W(\boldsymbol{E}) = \frac{\Lambda}{2} I_E^2 + \mu\,\boldsymbol{E} \cdot \boldsymbol{E}$
General polyconvex	$W(\boldsymbol{C}) = \frac{\alpha}{2}(\mathrm{tr}\,\boldsymbol{C})^2 + \frac{\beta}{2}[\mathrm{tr}(cof\,\boldsymbol{C})]^2 + \gamma\, g(\det \boldsymbol{C})$
Generalized material	$W = c_1(I_C - 3)^\alpha + c_2(II_C - 3)^\beta + c_3(III_C - 1)^\gamma$

$$\boldsymbol{\sigma} = \frac{\Lambda}{2}\left(J_F - \frac{1}{J_F}\right)\mathbf{1} + \frac{\mu}{J_F}(\boldsymbol{b} - \mathbf{1}) \tag{5.13}$$

is obtained where all terms are living in the spatial configuration.

Table. 5.1 summarizes some isotropic strain energy functions for hyperelastic materials.

5.1.3 Split in Isochoric and Volumetric Parts

In finite elasticity of foam or rubber materials the deformation is split into a volumetric part represented by J_F and an isochoric part described by $\hat{\boldsymbol{C}}$, see (1.20), since both parts can depict different material behaviour, see e.g. Lubliner (1985). This split is also useful when quasi-incompressible materials are described by special numerical formulations—like mixed methods—since the split permits a different treatment of the incompressible part.

One possibility for the formulation of the constitutive equation is given by an additive split of the strain energy function in its volumetric and isochoric parts: $W(\hat{\boldsymbol{C}}, J_F) = \hat{W}(\hat{\boldsymbol{C}}) + U(J_F)$. This leads for the strain energy function introduced in (5.8) to

$$W(\hat{\boldsymbol{C}}, J_F) = U(J_F) + \frac{1}{2}\mu(I_{\hat{C}} - 3). \tag{5.14}$$

Here the term $U(J_F)$ is different from $g(J_F)$: $U(J_F) = \frac{K}{4}(J_F^2 - 1) - \frac{K}{2}\ln J_F$ since the third term in the sum in (5.10) disappears. Furthermore the Lamé constant Λ has to be exchanged by the bulk modulus K.[5]

The 2nd Piola–Kirchhoff stresses are computed via the chain rule

$$\boldsymbol{S} = 2\,\frac{\partial W}{\partial \boldsymbol{C}} = 2\,\frac{\partial \hat{W}}{\partial \hat{\boldsymbol{C}}}\,\frac{\partial \hat{\boldsymbol{C}}}{\partial \boldsymbol{C}} + 2\,\frac{\partial U}{\partial J_F}\,\frac{\partial J_F}{\partial \boldsymbol{C}}. \tag{5.15}$$

[5]Note that the volumetric-isochoric split does not lead to the same constitutive behaviour as described by (5.8), especially for $\nu < 0.5$. However in the limit of incompressible material behaviour formulations (5.8) and (5.14) are equivalent.

It can be shown, see e.g. Wriggers (2008), that the 2nd Piola–Kirchhoff stress tensor can be split in an isochoric and volumetric part

$$\boldsymbol{S} = \boldsymbol{S}_{ISO} + \boldsymbol{S}_{VOL}. \tag{5.16}$$

For the special choice of the strain energy function (5.14) the volumetric and isochoric stresses

$$\boldsymbol{S}_{ISO} = \mu\, J_F^{-\frac{2}{3}} \left(\mathbf{1} - \frac{1}{3}\,\mathrm{tr}\boldsymbol{C}\,\boldsymbol{C}^{-1}\right), \qquad \boldsymbol{S}_{VOL} = \frac{K}{2}\,(J_F^2 - 1)\,\boldsymbol{C}^{-1}. \tag{5.17}$$

can be specified.

The transformation to the current configuration yields for the Kirchhoff stress tensor introduced in (1.73) $\boldsymbol{\tau} = \boldsymbol{F}\,\boldsymbol{S}\,\boldsymbol{F}^T$. By defining the operator $\mathrm{dev}(\bullet) = (\bullet) - \frac{1}{3}\,\mathrm{tr}(\bullet)\,\mathbf{1}$ equation the Kirchhoff stress tensor can be written as

$$\boldsymbol{\tau} = J_F\, p\,\mathbf{1} + \mathrm{dev}\,\widehat{\boldsymbol{\tau}} = \tau_{vol}\,\mathbf{1} + \boldsymbol{\tau}_{iso}. \tag{5.18}$$

This relation depicts clearly the split of the stress tensor into a volumetric and an isochoric part. The following definitions were used in (5.18)

$$p = \frac{\partial U}{\partial J_F} \quad \text{and} \quad \widehat{\boldsymbol{\tau}} = \hat{\boldsymbol{F}}\,2\,\frac{\partial \hat{W}}{\partial \hat{\boldsymbol{C}}}\,\widehat{\boldsymbol{F}}^T. \tag{5.19}$$

It is easily seen from equation (5.6) that the second term of the last equation can also be formulated as $\widehat{\boldsymbol{\tau}} = \widehat{\boldsymbol{b}}\,2\,\partial\hat{W} \,/\, \partial\widehat{\boldsymbol{b}}$. Here the definition $\widehat{\boldsymbol{b}} = J_F^{-\frac{2}{3}}\,\boldsymbol{b}$ has to be applied analogously to (1.20), see also Miehe (1994).

5.1.4 *Anisotropic Strain Energy Functions*

When a certain direction within the material has a different constitutive behaviour than other directions the associated material is called anisotropic. Lately generalizations of polyconvex strain energy functions to the anisotropic range were derived for different classes of anisotropy, for details see Schröder et al. (2008), Schröder (2010).

Here a transversly isotropic strain energy function will be provided that has in general to be added to an isotropic strain energy as given in Sect. 5.1.2. The strain energy function is based on the introduction of a structural tensor that defines a preferred direction $\boldsymbol{a}$. This tensor has the form

$$\boldsymbol{M} = \boldsymbol{a} \otimes \boldsymbol{a}. \tag{5.20}$$

Then the transversely isotropic strain energy part W^{ti} can be formulated for a general polyconvex case as

$$W^{ti}(\boldsymbol{C}, \operatorname{cof} \boldsymbol{C}, \det \boldsymbol{C})$$
$$= C \left[\frac{1}{\alpha+1} (\operatorname{tr}[\boldsymbol{C}\, \boldsymbol{M}])^{\alpha+1} + \frac{1}{\beta+1} (\operatorname{tr}[\operatorname{cof} \boldsymbol{C}\, \boldsymbol{M}])^{\beta+1} + \frac{1}{\gamma} (\det \boldsymbol{C})^{-\gamma} \right] \quad (5.21)$$

where the constitutive parameters C, α, β and γ have to be chosen such that $C > 0$, $\alpha > 0$, $\beta > 0$ and $\gamma > 1$, see Schröder et al. (2008). In that case W^{ti} is polyconvex and stress free in the initial configuration.

Note that the total strain energy will in general consist of an isotropic and an anisotropic part

$$W^{total} = W^{iso} + W^{ti}$$

where the isotropic part W^{iso} can be selected from Table 5.1, for more details see e.g. Schröder et al. (2008) and Schröder (2010).

5.1.5 Automation of Formulation of Elastic Materials

Accordingly to the classification of nonlinear computational problems elastic problems fall into the most basic category of time-independent problems. The general formulation of time-independent problems is presented in Sect. 3.1.3 and automation of time-independent problems in Sect. 3.2.2. A general algorithm for the automation of iterative scheme for primal analysis of time-independent problems is given in Algorithm 3.1. The only problem dependent part in Algorithm 3.1 is the element residual vector $\mathbf{R}_e$. Various possibilities for the discretization and the automation of formulation of the element residuals based on variational potentials or the weak forms of the equilibrium equations has been extensively studied in Sect. 4.2 and it will not be repeated here.

5.2 Elasto-Plastic Materials, Small Deformations

Many materials which are widely used in technical applications depict nonlinear behaviour even when only small deformations are present. Metals are described by elasto-plastic constitutive equations, polymers depict viscous effects furthermore damage effects occur in many cyclic loading processes. For a detailed treatment of such constitutive equations and their physical interpretation see e.g. Hill (1950), Prager (1955), Desai and Siriwardane (1984), Lubliner (1990), Khan and Huang (1995) and de Souza Neto et al. (2008).

A large class of nonlinear materials can be described by the assumption of elasto-plastic behaviour. Among these are materials like steel, aluminium, concrete but also geo-materials as rock and soils. Often the associated constitutive models are very complex. In the following sections several equations will be discussed which describe rate-independent constitutive behaviour for the general case of isotropic and kinematic hardening of metals.

5.2.1 General Formulation

The phenomenological model of the theory of plasticity has to be based on the fact that plastic flow of a material is an irreversible process. It is described in a solid by an additional strain measure and additional variables, known as plastic strains and hardening parameters.

An example for such constitutive equations is the classical model of elasto-plasticity with isotropic (expansion of flow surface) and kinematic (shift of origin of flow surface) hardening. Figure 5.1 illustrates the behaviour of that material for the one-dimensional case. Until point A the material reacts elastically that means the loading and unloading path is the same. Plastic deformations occur once the stress reaches the flow stress σ_A in A after that the stress does not increase as much as in the elastic range. Observe that the flow stress is larger in point B than in point A which is due to hardening. In case of unloading in point B the stress response follows a straight line which is parallel to the elastic tangent between the origin and point A. Hence unloading is related to elastic behaviour. However plastic flow occurs again at point C. Here the flow stress has a smaller absolute value than at point B. This observation is related to a movement of the origin of the elastic zone which is called kinematic hardening. A possible softening which is related to a decrease of the flow stress characterizes part BD in the stress-strain diagram.

In the following a three-dimensional generalization will be given for the phenomenological behaviour described above. In the subsequent development of the mathematical model of elasto-plastic constitutive behaviour that only small deformations occur.

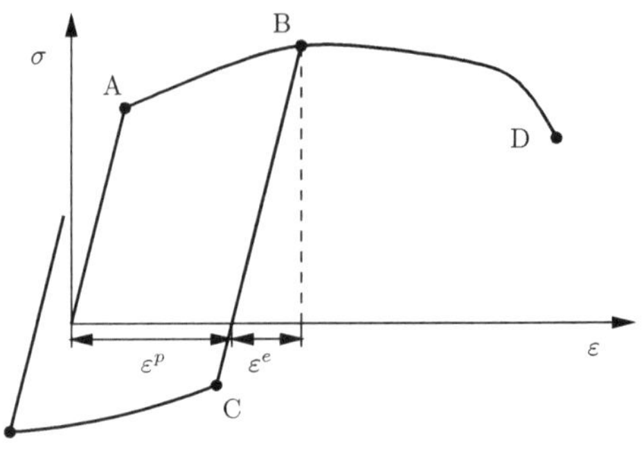

Fig. 5.1 Elasto-plastic constitutive behaviour

Plasticity models can be formulated in a standard framework. Hence for a large class of elasto-plastic materials it is possible to derive a generalized description of the constitutive equations. This will also reflect the fact that there exist many materials of technical interest which can be either described by associative or non-associative flow rules or need multi-surface flow conditions for a proper modeling. This includes most plasticity models e.g. metals, concrete or geo-materials like sand, clay or rock. For an indepth treatment of such material laws see e.g. Desai and Siriwardane (1984), Khan and Huang (1995), Hofstetter and Mang (1995) and de Souza Neto et al. (2008).

The linear strain tensor $\varepsilon = \frac{1}{2}(\text{Grad}\boldsymbol{u} + \text{Grad}\boldsymbol{u}^T)$ can be split additively into an elastic and a plastic part when only small strains are present, see Fig. 5.1,

$$\varepsilon = \varepsilon^e + \varepsilon^p . \tag{5.22}$$

For the general case of an elasto-plastic material with m independent flow surfaces the subsequent equations and evolution laws are obtained based on the introduced notation and $\boldsymbol{\alpha}$ as a set of hardening variables. Associative and non-associative plasticity is distinguished in the following

- Stress $\boldsymbol{\sigma}$ and back stress $\boldsymbol{q}$

$$\begin{aligned} \boldsymbol{\sigma} &= \frac{\partial W(\varepsilon^e, \boldsymbol{\alpha})}{\partial \varepsilon^e} \\ \boldsymbol{q} &= -\frac{\partial W(\varepsilon^e, \boldsymbol{\alpha})}{\partial \boldsymbol{\alpha}} \end{aligned} \tag{5.23}$$

- flow conditions/yield criteria[6] (restrictions for elastic domain), $1 \leqslant s \leqslant m$

$$f_s(\boldsymbol{\sigma}, \boldsymbol{q}) \leqslant 0 \tag{5.24}$$

- Flow rule and evolution equation for hardening

1. associative plasticity

$$\begin{aligned} \dot{\varepsilon}^p &= \sum_{s=1}^{m} \lambda_s \frac{\partial f_s(\boldsymbol{\sigma}, \boldsymbol{q})}{\partial \boldsymbol{\sigma}} \\ \dot{\boldsymbol{\alpha}} &= \sum_{s=1}^{m} \lambda_s \frac{\partial f_s(\boldsymbol{\sigma}, \boldsymbol{q})}{\partial \boldsymbol{q}} \end{aligned} \tag{5.25}$$

[6]Here the flow conditions depend upon the stress $\boldsymbol{\sigma}$ and not solely on the deviatoric stresses $\boldsymbol{s}$ as in classical J_2-plasticity. The reason for this is that inelastic processes of general non metallic materials or of metals in which damage has to be considered are pressure sensitive and hence the flow condition has to depend upon the full stress tensor.

2. non-associative plasticity

$$
\begin{aligned}
\dot{\varepsilon}^p &= \sum_{s=1}^{m} \lambda_s \, \boldsymbol{r}_s(\boldsymbol{\sigma}, \boldsymbol{q}) \\
\dot{\boldsymbol{\alpha}} &= \sum_{s=1}^{m} \lambda_s \, \boldsymbol{h}_s(\boldsymbol{\sigma}, \boldsymbol{q})
\end{aligned}
\tag{5.26}
$$

- Loading-/unloading conditions in Kuhn–Tucker form

$$
\lambda_s \geqslant 0, \qquad f_s(\boldsymbol{\sigma}, \boldsymbol{q}) \leqslant 0, \qquad \lambda_s \, f_s(\boldsymbol{\sigma}, \boldsymbol{q}) = 0 \tag{5.27}
$$

All relations hold for models of plasticity with m flow surfaces. The special case of only one flow surface is naturally included by setting $m = 1$. Tensors $\boldsymbol{r}_s$ and $\boldsymbol{h}_s$ describe the flow direction and the change of hardening for non-associative plasticity in (5.26).

Remark 3.4 Inelastic processes result in mechanical dissipation which can only grow when the plastic deformation increases. Based on the local principle of maximum dissipation some basic properties can be deduced which are needed for the description of plastic flow. The local dissipation is defined by the difference between the stress power and the time derivative of the free energy function. Thus Eqs. (1.54) and (5.64) lead to

$$
\mathcal{D} = \boldsymbol{\sigma} \cdot \dot{\boldsymbol{\varepsilon}} - \frac{D}{Dt} \, \psi(\boldsymbol{\varepsilon}^e, \boldsymbol{\alpha}). \tag{5.28}
$$

The evaluation of the time derivative yields together with (5.22) an expression for the plastic dissipation

$$
\mathcal{D} = \boldsymbol{\sigma} \cdot \dot{\boldsymbol{\varepsilon}}^p + \boldsymbol{q} \cdot \dot{\boldsymbol{\alpha}} \geqslant 0. \tag{5.29}
$$

The principle of maximum plastic dissipation can be formulated as, see e.g. Hill (1950) or Lubliner (1990),

$$
\mathcal{D} = \max_{\boldsymbol{\tau}, \boldsymbol{p}} \; \boldsymbol{\tau} \cdot \dot{\boldsymbol{\varepsilon}}^p + \boldsymbol{p} \cdot \dot{\boldsymbol{\alpha}} \quad \forall \quad \{ (\boldsymbol{\tau}, \boldsymbol{p}) \, | \, f(\boldsymbol{\tau}, \boldsymbol{p}) \leqslant 0 \}. \tag{5.30}
$$

It leads to the inequality

$$
(\boldsymbol{\sigma} - \boldsymbol{\tau}) \cdot \dot{\boldsymbol{\varepsilon}}^p + (\boldsymbol{q} - \boldsymbol{p}) \cdot \dot{\boldsymbol{\alpha}} \geqslant 0, \tag{5.31}
$$

where $\boldsymbol{\tau}$ and $\boldsymbol{p}$ characterize arbitrary stresses which are contained in the elastic domain. From this variational inequality it follows that the flow surfaces have to be convex, see e.g. Lubliner (1990). However this does not exclude flow surfaces with corners which occur frequently in real materials. Such cases require a special mathematical treatment within the framework of non-convex analysis which was substantiated in Moreau (1976), see also Simo (1998).

Based on the variational inequality (5.31) with the constraint $f \leqslant 0$ a saddle point problem is formulated which yields as result the flow rule. The actual stress state follows then as extremal value of the functional

$$L(\boldsymbol{\tau}, \boldsymbol{p}, \lambda) = -\boldsymbol{\tau} \cdot \dot{\boldsymbol{\varepsilon}}^p - \boldsymbol{p} \cdot \dot{\boldsymbol{\alpha}} + \lambda \, f(\boldsymbol{\tau}, \boldsymbol{p}) \longrightarrow EXTREMUM. \tag{5.32}$$

The constraint $f \leqslant 0$ is incorporated in this relation by the method of Lagrange multipliers. The solution of this saddle point problem at $\boldsymbol{\tau} = \boldsymbol{\sigma}$ and $\boldsymbol{p} = \boldsymbol{q}$ leads to

$$\dot{\boldsymbol{\varepsilon}}^p = \lambda \frac{\partial f}{\partial \boldsymbol{\sigma}}, \quad \dot{\boldsymbol{\alpha}} = \lambda \frac{\partial f}{\partial \boldsymbol{q}}, \quad f(\boldsymbol{\tau}, \boldsymbol{p}) = 0, \tag{5.33}$$

together with the Kuhn–Tucker conditions, see e.g. Luenberger (1984). Equation (5.33) represent the associative flow rules (5.25), which determine plastic flow for $f = 0$.

5.2.2 *Integration of Constitutive Equations for Small Inelastic Deformations*

The constitutive equations discussed in Sect. 5.2.1 describe elasto-plastic, material behaviour for small deformations (geometrically linear theory). The inelastic behaviour is governed by time dependent evolution equations which can be scalar or vector valued differential equations. These have to be solved by numerical integration algorithms.

The rate equations which have to be integrated are ordinary differential equations which lead in general to an initial value problem of the form

$$\dot{e}(t) = f\,[\,e(t)\,] \qquad \text{with} \quad e(0) = e_0. \tag{5.34}$$

The integration $(I - ALGO)$ is usually performed by a generalized midpoint rule

$$e = e_n + \Delta t \; f\,(e_{n+\theta}) \tag{5.35}$$

which yields an equation for the variable e at time t depending upon quantities at time t_n with the notation $e_n = e(t_n)$ and $e = e(t)$. In (5.35) the definition $e_{n+\theta} = (1-\theta)\, e_n + \theta\, e, \quad 0 \leqslant \theta \leqslant 1$ was used. The approximation (5.35) leads to the following integration algorithms

- for $\theta = 0$ to the explicit Euler scheme,
- for $\theta = 1$ to the implicit Euler scheme and
- for $\theta = \frac{1}{2}$ to the *midpoint-rule*.

The scheme (5.35) is for $\theta = \frac{1}{2}$ of second order accuracy. Otherwise the integration schemes are first order accurate. Further investigations with respect to consistency,

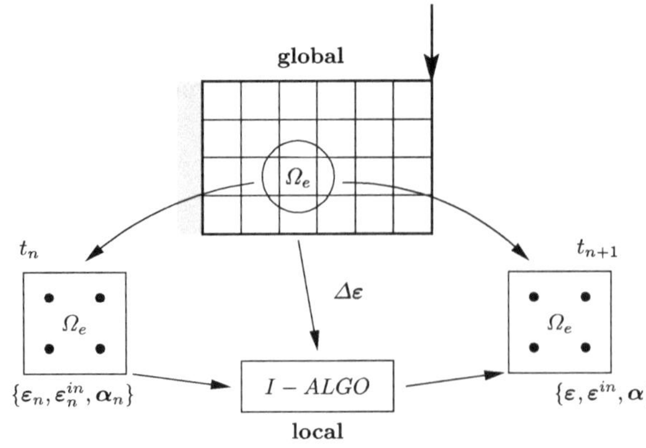

Fig. 5.2 Algorithm for the integration of elasto-plastic material

stability and accuracy of algorithms based on (5.35) can be found in the mathematical literature, see e.g. Gear (1971) or Stoer and Bulirsch (1990) but also in Simo (1998). Consistency and stability are properties which are essential for establishing convergence of the numerical solution for arbitrary small time steps.

The constitutive equations have to be fulfilled at every point of the solid. It is efficient to preserve this local character within the integration schemes. In case of the spatial discretization, using finite elements, the algorithm will be partitioned in such a way that the integration of the constitutive equations ($I - ALGO$) is performed locally at element level. Hence the constitutive equations have to be fulfilled at each integration point (Gauss point) within an element Ω_e. Besides this, the weak form of the equilibrium must be obeyed. From these two requirements follows an iteration which has to be performed within each time step of the solution. Thus we have here the case of a time-dependent Gauss point coupled problem, see Sect. 3.1.4. A schematic description of the associated algorithm for plasticity can be found in Fig. 5.2.

5.2.3 Integration of General Elasto-Plastic Materials

For Eqs. (5.23) to (5.27) a general integration algorithm will be constructed based on the above ideas. For simplicity we will restrict ourselves to materials which are described by one flow surface. Algorithms for the more general case are e.g. described in Simo (1998) and de Souza Neto et al. (2008).

Since the differential equations which describe elasto-plastic deformations are stiff in the mathematical sense, see e.g. Nagtegaal (1982), Simo and Taylor (1985) and Simo and Hughes (1998), implicit integration methods, like the backward Euler scheme, have to be applied which leads for the case of non-associated plasticity to a finite difference approximation of Eqs. (5.23) to (5.27).

$$\sigma = \frac{\partial W(\varepsilon^e, \alpha)}{\partial \varepsilon^e}, \tag{5.36}$$

$$q = -\frac{\partial W(\varepsilon^e, \alpha)}{\partial \alpha}, \tag{5.37}$$

$$\frac{1}{\Delta t}(\varepsilon^p - \varepsilon^p_n) = \lambda\, r(\sigma, q), \tag{5.38}$$

$$\frac{1}{\Delta t}(\alpha - \alpha_n) = \lambda\, h(\sigma, q), \tag{5.39}$$

$$f(\sigma, q) \leqslant 0. \tag{5.40}$$

This constitutes a set of implicit equations for the unknown stresses and strains,[7] note that $\varepsilon^e = \varepsilon - \varepsilon^p$. The equation system has to be fulfilled at each Gauss point of a spatial finite element discretization.

The solution of the equation set (5.36) to (5.40) is not trivial since the inequality constraint in (5.40) has to be fulfilled. A so called operator split algorithm has proven to be most efficient for this task. The idea of this procedure is to freeze the plastic variables at the beginning of a time step from t_n to t. Thus $\varepsilon^{p\,tr} = \varepsilon^p_n$ and $\alpha^{tr} = \alpha_n$. Based on this frozen state trial strains and trial hardening variables are given by

$$\varepsilon^{e\,tr} = \varepsilon - \varepsilon^p_n, \qquad \alpha^{tr} = \alpha_n. \tag{5.41}$$

The superscript $()^{tr}$ denotes *trial* quantities which being fixed for a moment, but may change during the iteration associated with the algorithm. The frozen plastic variables are inserted in Eqs. (5.36) and (5.37)

$$\sigma^{tr} = \frac{\partial W(\varepsilon^{e\,tr}, \alpha^{tr})}{\partial \varepsilon^e} \tag{5.42}$$

$$q^{tr} = -\frac{\partial W(\varepsilon^{e\,tr}, \alpha^{tr})}{\partial \alpha}. \tag{5.43}$$

The trial values of stress, σ^{tr}, and backstress, q^{tr}, are used to check the status (elastic or plastic) within a time step by inserting the values into the yield function

$$f(\sigma^{tr}, q^{tr}) \begin{cases} \leqslant 0 \Rightarrow \text{elastic} \\ > 0 \Rightarrow \text{plastic} \end{cases} \tag{5.44}$$

In case that f denotes an elastic state at t then all plastic variables are updated at the end of the time step by the trial quantities:

$$\varepsilon^p = \varepsilon^{p\,tr}, \qquad \alpha = \alpha^{tr} \tag{5.45}$$

[7]The equation system can be solved either for the stresses σ and the back stress q or the plastic strains ε^p and the hardening variable α at step $n+1$. In the following the solution for the strain variables is preferred. Algorithms that yield the stress variables can be found in e.g. Simo (1998), Simo and Hughes (1998), Wriggers (2008) and de Souza Neto et al. (2008).

and the local algorithm is terminated for that time step. In case that f denotes a plastic state at t the stress state has to be corrected such that it fulfills the yield condition (5.40).

The time increment Δt is eliminated from (5.39) and (5.40) by introducing the rate of the plastic multiplier $\dot{\gamma} = \lambda$. This leads with $\gamma - \gamma_n = \lambda \, \Delta t$, (5.36) and (5.40) to

$$\begin{aligned} \mathbf{Q}_\varepsilon &= \boldsymbol{\varepsilon}^p - \boldsymbol{\varepsilon}^p_n - (\gamma - \gamma_n)\, \boldsymbol{r}(\boldsymbol{\sigma}, \boldsymbol{q}) = \mathbf{0} \\ \mathbf{Q}_\alpha &= \boldsymbol{\alpha} - \boldsymbol{\alpha}_n - (\gamma - \gamma_n)\, \boldsymbol{h}(\boldsymbol{\sigma}, \boldsymbol{q}) = \mathbf{0} \\ Q_f &= f(\boldsymbol{\sigma}, \boldsymbol{q}\,) = 0 \end{aligned} \tag{5.46}$$

Here the stress $\boldsymbol{\sigma}$ and back stress $\boldsymbol{q}$ depend on the strains and hardening variables at time t: $\boldsymbol{\sigma} = \boldsymbol{\sigma}(\varepsilon, \boldsymbol{\alpha})$ and $\boldsymbol{q} = \boldsymbol{q}(\varepsilon, \boldsymbol{\alpha})$. The three equations constitute a nonlinear system of equations at each Gauss point

$$\mathbf{Q}_g = \begin{Bmatrix} \text{vec}(\mathbf{Q}_\varepsilon) \\ \text{vec}(\mathbf{Q}_\alpha) \\ Q_f \end{Bmatrix} \tag{5.47}$$

for the unknowns $\boldsymbol{\varepsilon}^p, \boldsymbol{\alpha}$ and γ that form corresponding Gauss point vectors of unknowns[8]

$$\mathbf{h}_g = \begin{Bmatrix} \text{vec}(\boldsymbol{\varepsilon}^p) \\ \text{vec}(\boldsymbol{\alpha}) \\ \gamma \end{Bmatrix}. \tag{5.48}$$

For the solution of (5.47) Newton's method is applied (iteration index i) leading to the algorithm

$$\begin{aligned} \mathbf{A}_g^{(i)} \, \Delta\mathbf{h}_g^{(i+1)} &= -\mathbf{Q}_g^{(i)} \\ \mathbf{h}_g^{(i+1)} &= \mathbf{h}_g^{(i)} + \Delta\mathbf{h}_g^{(i+1)}. \end{aligned} \tag{5.49}$$

Matrix $\mathbf{A}$ follows from the linearization of the residuals defined in (5.46)

$$\mathbf{A}_g^{(i)} = \frac{\partial \mathbf{Q}_g^{(i)}}{\partial \mathbf{h}_g^{(i)}} \tag{5.50}$$

At the end of the Newton iterations the strains and hardening variables as well as the plastic multiplier are know at a single Gauss point.

Besides the fulfillment of the inequality constraints imposed by plasticity, the global weak form (equilibrium) has to be fulfilled, see also the introductory remarks and Fig. 5.2.

[8]The vectorization operator vec($\mathbf{x}$) converts tensors into column vectors while accounting for the symmetry and skipping the zero entries. It relates to the Voigt notation used in classical finite element formulations.

It is well known that several constitutive formulations depict softening behaviour, and hence can lead to localization. These models can have non-unique solutions and thus are not easily treated by numerical algorithms. Especially implicit schemes have their problems with such constitutive models. Thus for geo-materials explicit schemes using sub-stepping were constructed in e.g. Sloan (1987), Sloan et al. (2001) and Sheng and Sloan (2001). Another approach which combines implicit and explicit methods for plasticity and general damage models can be found in Oliver et al. (2006).

5.2.4 *Automation of Formulation of Small Strain Elasto-Plasticity*

We start with the small strain variant of the standard weak form of equilibrium equations (1.90)

$$\int_B \boldsymbol{\sigma} \cdot \delta\boldsymbol{\varepsilon}\, dV - \int_B \rho_0\, \bar{\boldsymbol{b}} \cdot \delta\boldsymbol{u}\, dV - \int_{\partial B_\sigma} \bar{\boldsymbol{t}} \cdot \delta\boldsymbol{u}\, dA = 0. \tag{5.51}$$

For the computations of the residuum, the internal virtual work expression is specified at the element level as described in Sect. 4.2.1

$$\int_{\Omega_e} \boldsymbol{\sigma} \cdot \delta\boldsymbol{\varepsilon}\, dV \approx \sum_{g=1}^{n_g} \boldsymbol{\sigma}(\xi_g, \eta_g, \zeta_g) \cdot \delta\boldsymbol{\varepsilon}(\xi_g, \eta_g, \zeta_g)\, J_e(\xi_g, \eta_g, \zeta_g)\, w_{gp} \tag{5.52}$$

where $\boldsymbol{\sigma}$ is given by (5.36) and the variation of small strain tensor by

$$\delta\boldsymbol{\varepsilon} = \frac{\partial \boldsymbol{\varepsilon}}{\partial \mathbf{p}_e}\, \delta\mathbf{p}_e. \tag{5.53}$$

By inserting (5.53) into (5.52) and replacing the partial derivative with the computational derivative, the Gauss point contribution $\mathbf{R}_g$ is given by

$$\mathbf{R}_g = J_e \frac{\partial W(\boldsymbol{\varepsilon}^e, \boldsymbol{\alpha})}{\partial \boldsymbol{\varepsilon}^e} \cdot \frac{\partial \boldsymbol{\varepsilon}}{\partial \mathbf{p}_e} = J_e \left. \frac{\hat{\delta} W}{\hat{\delta} \boldsymbol{\varepsilon}^e} \right|_{\mathbf{h}_g = const.} \cdot \frac{\hat{\delta} \boldsymbol{\varepsilon}}{\hat{\delta} \mathbf{p}_e}. \tag{5.54}$$

This vector consequently contributes to the element residual vector $\mathbf{R}_e$ and hence to the global residual vector $\mathbf{R}$ (also internal load or stress divergence term), such that $\int_B \boldsymbol{\sigma} \cdot \delta\boldsymbol{\varepsilon}\, dV = \mathbf{R} \cdot \delta\mathbf{p}$.

With the use of identity $\frac{\partial W}{\partial \boldsymbol{\varepsilon}^e} = \frac{\partial W}{\partial \boldsymbol{\varepsilon}}$ the *ADB formulation of small strain plasticity* is given by

$$\mathbf{R}_g = J_e \left. \frac{\hat{\delta} W}{\hat{\delta} \mathbf{p}_e} \right|_{\mathbf{h}_g = const.} \tag{5.55}$$

As a side effect of the iterative solution of equations (5.47) at a Gauss-point, there exists an implicit (algorithmic) dependency of $\mathbf{h}_g$ on ε. The AD exception in equation (5.55) hides this dependency from the automatic differentiation procedure and ensures correct evaluation of the weak form equations.

The finite element discretization of the weak form and the time discretization of the constitutive equations lead to the following set of coupled nonlinear equations at the current time instant $t = t_{n+1}$,

$$\mathbf{R}(\mathbf{p}, \mathbf{h}, \mathbf{h}_n) = \mathbf{0}, \tag{5.56}$$

$$\mathbf{Q}_g(\varepsilon(\mathbf{p}_e), \mathbf{h}_g, \mathbf{h}_{gn}) = \mathbf{0}, \quad g = 1, 2, \ldots, n_{tg}, \tag{5.57}$$

where $\mathbf{p}$ and $\mathbf{h}$ denote the current unknown global vectors of generalized displacements and internal (history) variables, respectively. Equation $\mathbf{R} = \mathbf{0}$ in (5.56) is the discrete counterpart of the weak form (5.51), while equations $\mathbf{Q}_g = \mathbf{0}$ in (5.57) represent the set of incremental constitutive equations. As presented in Sect. 5.2.3 the Gauss-point equations (5.57) are solved for $\mathbf{h}_g$ at fixed $\mathbf{p}_e$ using the Newton method. The Newton method is also applied to solve the global equilibrium equation (5.56) in an outer loop thus leading to a nested iteration-subiteration solution scheme for the unknowns $\mathbf{p}$ and $\mathbf{h}$. According to the classification of nonlinear computational problems in (3.1) the elasto-plastic problems fall into the category of *time-dependent Gauss point coupled problems.*

The general formulation of time-dependent Gauss point coupled problems is presented in Sect. 3.3.4 and automation of time-depen-dent Gauss point coupled problems is provided in Sect. 3.3.5. The Gauss point residual (5.57) depends explicitly only on a small strain tensor ε, thus a set of intermediate variables $\mathbf{r}_g$ that is a part of general formulation can be identified as a set of mutually independent variables that form the small strain tensor

$$\mathbf{r}_g = \text{var}(\varepsilon). \tag{5.58}$$

The local tangent matrix $\mathbf{A}_g$ and the Gauss point contribution to the global tangent matrix $\mathbf{K}_g$ then follow directly from the general Eqs. (3.68) and (3.69) as

$$\mathbf{A}_g = \frac{\hat{\delta} \mathbf{Q}_g}{\hat{\delta} \mathbf{h}_g}, \tag{5.59}$$

$$\mathbf{K}_g = \left. \frac{\hat{\delta} \mathbf{R}_g}{\hat{\delta} \mathbf{p}_e} \right|_{\frac{D\mathbf{h}_g}{D\mathbf{r}_g} = -\mathbf{A}_g^{-1} \frac{\hat{\delta} \mathbf{Q}_g}{\hat{\delta} \mathbf{r}_g}}. \tag{5.60}$$

Evaluation of the Gauss-point contribution to the element tangent matrix (5.60) requires proper consideration of the implicit dependency $\mathbf{h}_g(\mathbf{r}_g(\varepsilon(\mathbf{p}_e)))$ introduced by the local iterative procedure (5.49). The "missing" implicit derivative $\frac{\partial \mathbf{h}_g}{\partial \mathbf{r}_g} = -\mathbf{A}_g^{-1}\frac{\partial \mathbf{Q}_g}{\partial \mathbf{r}_g}$ is in (5.60) introduced as an AD exception. It can be observed, contrary to elasticity, that the element tangent matrix depends upon the chosen integration algorithm. Due to this reason the tangent is often called consistent tangent in the literature since it is consistent with the elasto-plastic algorithm, see e.g. Simo and Taylor (1985), Simo and Hughes (1998), de Souza Neto et al. (2008) and Korelc (2009). The tangent matrix (5.60) is equivalent to the one produced by the standard consistent linearization procedure introduced by Simo and Taylor (1985).

Input: $\mathbf{p}_e$ `// set of element DOF at time` t_{n+1}
Input: $\mathbf{h}_{gn}$ `// set of Gauss point unknowns at time` t_n
$\varepsilon \leftarrow \varepsilon(\mathbf{p}_e)$, $\mathbf{r}_g \leftarrow \text{vec}(\varepsilon)$
[f] $f^{tr} \leftarrow \text{fQW}(\text{"f"}, \mathbf{h}_{gn})$
if $f^{tr} < 0$ **then**
  $\mathbf{h}_g \leftarrow \mathbf{h}_{gn}$
  export $\mathbf{h}_g$ to $\mathbf{h}$ data structure
else
  $\mathbf{h}_{\bar{g}} \leftarrow \mathbf{h}_{gn}$ `// starting value for local loop`
  while true do `// sub-iterative loop`
[Q] $\mathbf{Q}_g \leftarrow \text{fQW}(\text{"Q"}, \mathbf{h}_{\bar{g}})$
    $\mathbf{A}_g \leftarrow \frac{\hat{\delta}\mathbf{Q}_g}{\hat{\delta}\mathbf{h}_g}$
[LU]   solve $\mathbf{A}_g \Delta\mathbf{h}_g + \mathbf{Q}_g = \mathbf{0}$ for unknown $\Delta\mathbf{h}_g$
    if *error criterion for* $\|\mathbf{Q}_g\|$ *and* $\|\mathbf{h}_g\|$ *is fulfilled* **then**
      solve $\mathbf{A}_g \widehat{D_{\mathbf{r}_g}\mathbf{h}}_g + \left.\frac{\hat{\delta}\mathbf{Q}_g}{\hat{\delta}\mathbf{r}_g}\right|_{\mathbf{h}_{\bar{g}}=const.} = \mathbf{0}$ for unknown $\widehat{D_{\mathbf{r}_g}\mathbf{h}}_g$ `// using factorized` $\mathbf{A}_g$ `from step [LU]`
      export $\mathbf{h}_{\bar{g}} + \Delta\mathbf{h}_g$ to $\mathbf{h}$ data structure
      break enclosing loop
    end if
    $\mathbf{h}_{\bar{g}} \leftarrow \mathbf{h}_{\bar{g}} + \Delta\mathbf{h}_g$
  end while
  $\mathbf{h}_g \leftarrow \mathbf{h}_{\bar{g}}\Big|_{\frac{D\mathbf{h}_{\bar{g}}}{D\mathbf{r}_g}=\widehat{D_{\mathbf{r}_g}\mathbf{h}}_g}$ `// AD exception ensures consistent linearization of` $\mathbf{R}_g$
end if
[W] $W \leftarrow \text{fQW}(\text{"W"}, \mathbf{h}_g)$
$\mathbf{R}_g \leftarrow J_g \left.\frac{\hat{\delta}W(\varepsilon,\mathbf{h}_g)}{\hat{\delta}\mathbf{p}_e}\right|_{\mathbf{h}_g=const.}$ `// AD exception ensures that` $\mathbf{R}_g$ `corresponds to the discretized weak form`
$\mathbf{K}_g \leftarrow \frac{\hat{\delta}\mathbf{R}_g}{\hat{\delta}\mathbf{p}_e}$
Result: $\mathbf{R}_g$, $\mathbf{K}_g$, and $\mathbf{h}_g$

Box 5.1. *ADB* form of Gauss point contribution to the element residual and tangent matrix for elasto-plastic problems

A general automation of the iterative scheme for the primal analysis of a time-dependent Gauss point coupled problem using global AD exceptions is presented in Algorithm 3.5. Here, the general algorithm is applied for small strain elasto-plastic problems. The global part of the solution algorithm is the same, thus only the *ADB* formulation of Gauss point contribution to the element residual and tangent matrix is derived and presented in Algorithm 5.1.

Various material model dependent quantities have to be evaluated in Algorithm 5.1 at three points labeled by [f], [Q] and [W]. In order to avoid repeating the same or similar symbolic input at several points (this is a well known source of errors in programming) the advantage of simultaneous optimization of expressions is exploited. An auxiliary symbolic function fQW is introduced that holds expressions that will be evaluated several times. The function is evaluated symbolically during the *AceGen* session and it just adds expressions to already derived expression. It **does not** results in a Fortran or C subroutine with the same name. The simultaneous expression optimization procedure, built in *AceGen*, can recognize that the same expression has been evaluated several times and replaces later occurrences with the reference to the first definition of the same expression stored in the *AceGen* internal data base. The auxiliary function fQW$(task, \mathbf{h}_g)$ evaluates the yield function f, the Gauss point equations $\mathbf{Q}_g$ or the elastic strain energy function W depending on the value of parameter $task$. The function is evaluated for the current value of the Gauss point unknowns $\mathbf{h}_g$ at the point where the function is called. Note that the total strain ε is known from the solution of the global equation system for load step t_{n+1}. The function is evaluated for the given total strain ε, a set of Gauss point unknowns $\mathbf{h}_{gn}$ at time t_n and a set of material parameters $\mathbf{d}_B$ (e.g. E, $Y0$, etc.) that are defined as global symbols of the *Mathematica* session. Within the *Mathematica* session all the global symbols are automatically defined, also within the calling functions, thus ε, $\mathbf{h}_{gn}$ and $\mathbf{d}_B$ do not have to be named in a parameter list of the fQW function.

Algorithm 5.1 closely follows the general idea of an operator split for the integration of the constitutive model within a load step as described in Sect. 5.2.3. Thus the integration scheme is split into three stages.

1. In the first stage the flow rule $f^{tr} =$ fQW("f", $\mathbf{h}_{gn}$) is evaluated using the trial values of Gauss point unknowns $\mathbf{h}_g = \mathbf{h}_g^{tr} = \mathbf{h}_{gn}$. Thus, $\varepsilon^{p\,tr} = \varepsilon_n^p$ and $\varepsilon^e = \varepsilon - \varepsilon^{p\,tr}$ have to be used when evaluating the stresses using (5.36). If the flow rule is fulfilled then the stress is in the elastic region. In that case the Gauss point unknowns are updated by $\mathbf{h}_g = \mathbf{h}_{gn}$.
2. In case that the flow rule is not fulfilled the stresses have to be returned to the yield surface. This is done by solving the nonlinear equation system (5.47) where the equations $\mathbf{Q}_g =$ fQW("Q", $\mathbf{h}_{\bar{g}}$) are computed for the iterative value of the Gauss point unknowns $\mathbf{h}_g = \mathbf{h}_{\bar{g}}$. As discussed before the constitutive equations have to be properly linearized in order to obtain the consistent tangent for the global solution procedure. The tangent relation needs the differentiation of the Gauss point unknowns with respect to the total strains. This derivative can simply be computed using the equation system of the local Newton method. By solving a linear system

$$\mathbf{A}_g \frac{\partial \mathbf{h}_{\bar{g}}}{\partial \mathbf{r}_g} + \left.\frac{\hat{\delta} \mathbf{Q}_g}{\hat{\delta} \mathbf{r}_g}\right|_{\mathbf{h}_{\bar{g}}=const.} = \mathbf{0} \tag{5.61}$$

for the unknown vector $\frac{\partial \mathbf{h}_{\bar{g}}}{\partial \mathbf{r}_g} = \widehat{D_{\mathbf{r}_g}\mathbf{h}_g}$ the partial derivative of the Gauss point unknowns $\mathbf{h}_g$ with respect to the strains ε is obtained. Since $\mathbf{Q}_g$ is evaluated within the iterative loop, the additional AD exception $\mathbf{h}_{\bar{g}} = const.$ in (5.61) prevents eventual differentiation of the loop itself.

3. Finally the elastic strain energy function $W = \mathrm{fQW}(\text{“W”}, \mathbf{h}_g)$ is evaluated for the converged value $\mathbf{h}_g$ of the Gauss point unknowns and internal load vector is evaluated using *ADB formulation of small strain plasticity* given by Eq. (5.55).

The AD exception that ensures consistent linearization is in Algorithm 5.1 attached to the definition of $\mathbf{h}_g$ rather than to the definition of Gauss point tangent matrix. With this, the computation of the AD exception is related to the plastic branch only. The AD exception that ensures proper evaluation of the internal load vector (5.55) hides the implicit dependency $\mathbf{h}_g(\varepsilon)$. Since this dependency does not exist in the elastic case, the AD exception has no effect on the elastic branch of the algorithm.

Several advantages of the presented formulation can be observed with respect to the standard formulation of elasto-plastic problems:

- Equation (5.55) unifies the elastic and plastic state, thus only two calls to AD procedure are needed (one to evaluate $\mathbf{R}_g$ and one to evaluate $\mathbf{K}_g$) making it optimal for the automatic differentiation and automatic code generation. Algorithm 5.1 in total requires a minimum of 4 calls to the AD procedure.
- The stress and the strain tensors do not appear explicitly in (5.55), thus the question of choosing the optimal stress-strain pair does not arise at all. The only free parameters of the formulation are the strain energy function, the yield condition, the evolution equations and the discretization of the domain and displacements.
- The described formulation can be employed (with the appropriate modifications) to derive small strain as well as finite strain plasticity models, such as multi-surface plasticity, non-associate plasticity models, compressible plasticity models, etc. Also various finite element discretization techniques (standard displacement elements, enhanced strain elements, under-integrated formulations, etc.) can be incorporated.
- Algorithm 5.1 is valid for arbitrary solid and structural elements. The only limitation is that the strain tensor ε is the only quantity that directly depends on a set of global unknowns $\mathbf{p}$ at time t_{n+1} and that no quantity depends on a set of global unknowns $\mathbf{p}_n$ at time t_n. This limitations are warranted by the basic principles of continuum mechanics.

5.2.5 Example: von Mises Plasticity

A often applied plasticity model for metals is named after von Mises. It assumes incompressible plastic deformations that can be experimentally justified for metals.

In such case it is appropriate to introduce deviatoric measures. This yields for the total strains with (1.21) $\boldsymbol{e}_D = \boldsymbol{\varepsilon} - \frac{1}{3}\,\mathrm{tr}\,\boldsymbol{\varepsilon}\,\mathbf{1}$. An analogous relation can be written for the stress tensor which then defines the stress deviator

$$\boldsymbol{s} = \boldsymbol{\sigma} - \frac{1}{3}\,\mathrm{tr}\,\boldsymbol{\sigma}\,\mathbf{1}. \tag{5.62}$$

The stresses $\boldsymbol{\sigma}$ and the internal variables $\boldsymbol{q}$ related to the hardening parameters $\boldsymbol{\alpha}$ can be computed by a derivative of the strain energy function ψ

$$\boldsymbol{\sigma} = \rho_0\,\frac{\partial \psi(\boldsymbol{\varepsilon}^e,\,\boldsymbol{\alpha})}{\partial \boldsymbol{\varepsilon}^e}, \qquad \boldsymbol{q} = -\rho_0\,\frac{\partial \psi(\boldsymbol{\varepsilon}^e,\,\boldsymbol{\alpha})}{\partial \boldsymbol{\alpha}} \tag{5.63}$$

For many applications the assumption of a function ψ is valid which is decoupled in $\boldsymbol{\varepsilon}^e$ and $\boldsymbol{\alpha}$. This leads to

$$\rho_0\,\psi(\boldsymbol{\varepsilon}^e,\,\boldsymbol{\alpha}) = W^e(\boldsymbol{\varepsilon}^e) + W^p(\boldsymbol{\alpha}), \tag{5.64}$$

where $W^e(\boldsymbol{\varepsilon}^e)$ is the elastic strain energy function, see e.g. (5.8), and $W^p(\boldsymbol{\alpha})$ denotes a potential function for the hardening variables. The elastic strain energy function W^e has the explicit form $W^e = \frac{1}{2}\,\boldsymbol{\varepsilon}^e \cdot \mathbb{C}^e[\boldsymbol{\varepsilon}^e]$ for small strains. The partial derivative of this function with respect to the elastic strains yields with

$$\boldsymbol{\sigma} = \rho_0\,\frac{\partial \psi(\boldsymbol{\varepsilon}^e,\,\boldsymbol{\alpha})}{\partial \boldsymbol{\varepsilon}^e} = \mathbb{C}^e[\boldsymbol{\varepsilon}^e], \tag{5.65}$$

the classical Hooke's law of the theory of linear elasticity. Note that Eq. (5.65) can also be written in terms of the deviatoric quantities which leads for isotropic elastic response to

$$\boldsymbol{s} = 2\,\mu\,\boldsymbol{e}^e_D \quad \text{and} \quad p = K\,\mathrm{tr}\varepsilon^e = K\,\mathrm{div}\,\boldsymbol{u}\,. \tag{5.66}$$

Here μ is the shear modulus and K the bulk modulus (with $K = \lambda + \frac{2}{3}\,\mu$). λ and μ are also called Lame constants.

We assume the same structure, given in Eq. (5.65), for W_v, see e.g. Lubliner (1990). Hence $W_v = \frac{1}{2}\,\hat{H}\hat{\alpha}^2 + \frac{1}{3}\,H\,\|\boldsymbol{\alpha}\|^2$ is defined where $\hat{\alpha}$ are the isotropic and $\boldsymbol{\alpha}$ kinematic hardening variables. This definition leads with (5.63) to

$$\boldsymbol{q} = -\frac{2}{3}\,H\,\boldsymbol{\alpha} \quad \text{and} \quad \hat{q} = -\hat{H}\,\hat{\alpha}, \qquad q_{ij} = -\frac{2}{3}\,H\,\alpha_{ij}. \tag{5.67}$$

The elastic domain of the deformation is restricted by the yield condition. Mathematically an inequality constraint has to be formulated which depends upon the stresses and the internal hardening variables. The flow condition must be able to describe two different phenomena as already pointed out in Fig. 5.1 for the one-dimensional case. This is the enlargement of the elastic domain (isotropic hardening) and the shift of the permissible elastic range (kinematic hardening), depicted schematically in Fig. 5.3 for times $t_=$ and t. The flow condition or yield criterion can be written in general form as

$$f(\boldsymbol{\sigma}, \boldsymbol{q}, \hat{q}) \leqslant 0 \tag{5.68}$$

for isotropic and kinematic hardening. In case of the classical von Mises plasticity the flow condition f depends only upon the second invariant of the stress deviator. Hence it can be expressed as

$$f(\boldsymbol{s}, \boldsymbol{q}, \hat{q}) = \sqrt{(\boldsymbol{s} - \boldsymbol{q}) \cdot (\boldsymbol{s} - \boldsymbol{q})} - k(\hat{q}) \leqslant 0. \tag{5.69}$$

For linear isotropic and kinematic hardening

$$f(\boldsymbol{s}, \boldsymbol{q}, \hat{q}) = \| \boldsymbol{s} - \boldsymbol{q} \| - \sqrt{\frac{2}{3}} (Y_0 - \hat{q}) \leqslant 0 \tag{5.70}$$

is obtained explicitly. The generalized stress measure $\boldsymbol{q}$ which is related to kinematic hardening is called back stress. A stress point lies for $f < 0$ in the elastic domain. The stress point is located on the boundary of the flow surface for $f = 0$. This can result in plastic deformations. Values $f > 0$ of the flow condition are not admissible, see Fig. 5.3.

Irreversibility of the plastic flow process is expressed by a flow rule. For most metals an associated flow rule can be used in which the direction of flow is given by the partial derivative of the flow condition with respect to the deviatoric stresses

$$\dot{\boldsymbol{e}}^p_D = \lambda \frac{\partial f}{\partial \boldsymbol{s}}. \tag{5.71}$$

This equation describes the evolution of the deviatoric plastic strains. The direction of plastic flow is given by $\frac{\partial f}{\partial \boldsymbol{s}}$, λ is a scalar which determines the size of the plastic strain increment.

Evolution equations have also to be formulated for the hardening variables. This leads to

$$\dot{\boldsymbol{\alpha}} = \lambda \frac{\partial f}{\partial \boldsymbol{q}}, \qquad \dot{\hat{\alpha}} = \lambda \frac{\partial f}{\partial \hat{q}}. \tag{5.72}$$

Using the flow condition (5.70) the derivative of f yields

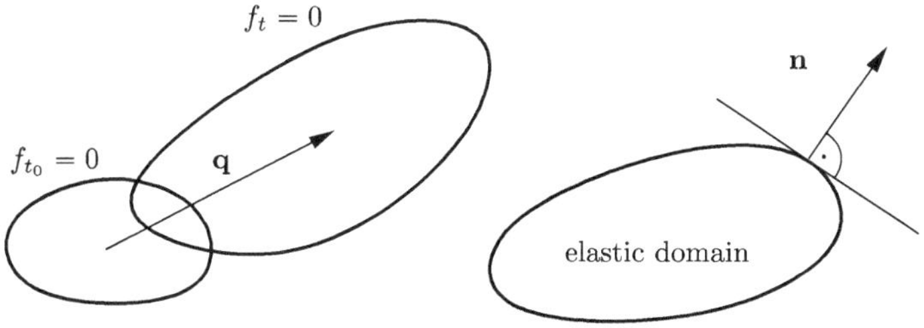

Fig. 5.3 Flow condition and admissible region

$$\frac{\partial f}{\partial \boldsymbol{s}} = \frac{\boldsymbol{s} - \boldsymbol{q}}{\| \boldsymbol{s} - \boldsymbol{q} \|} =: \boldsymbol{n} \quad \text{and} \quad \frac{\partial f}{\partial \boldsymbol{q}} = -\boldsymbol{n} \tag{5.73}$$

which defines the flow direction. Hence the evolution equations

$$\dot{\boldsymbol{e}}_D^p = \lambda \, \boldsymbol{n}, \quad \dot{\boldsymbol{\alpha}} = -\lambda \, \boldsymbol{n} \quad \text{and} \quad \dot{\hat{\alpha}} = \lambda \sqrt{\frac{2}{3}}. \tag{5.74}$$

can be written. Equation (5.74) yields the equivalent plastic strain increment $\dot{\hat{\alpha}} = \sqrt{\frac{2}{3}} \, \|\dot{\boldsymbol{e}}_D^p\|$ since $\|\dot{\boldsymbol{e}}_D^p\| = \lambda$ is valid. Time integration leads to the equivalent (or effective) strain

$$\hat{\alpha} = \int_0^t \sqrt{\frac{2}{3}} \, \|\dot{\boldsymbol{e}}_D^p\| d\tau, \tag{5.75}$$

which provides a measure for plastic distortion, see e.g. Hill (1950) or Lubliner (1990).

In (5.74) the parameter λ describes the magnitude of plastic flow. Generally three cases have to be distinguished when a stress point lies on the flow surface $f = 0$

$$\begin{array}{lll} \dot{f} < 0 \Longrightarrow \lambda = 0 & \text{elastic unloading,} \\ \dot{f} = 0 \Longrightarrow \lambda = 0 & \text{neutral loading,} \\ \dot{f} = 0 \Longrightarrow \lambda > 0 & \text{plastic flow.} \end{array}$$

These different cases can be summarized in the so called Kuhn–Tucker conditions

$$\lambda \geqslant 0, \qquad f \leqslant 0, \qquad \lambda \, f = 0. \tag{5.76}$$

Furthermore the consistency condition

$$\lambda \, \dot{f} = 0, \quad \text{if} \quad f = 0 \tag{5.77}$$

is contained in (5.76). With this, all evolution equations for plastic flow and hardening parameters as well as the flow conditions of elasto-plastic flow are known. An algorithm which can be used to integrate the evolution equations is presented in Sect. 5.2.2 and its automation in Sect. 5.2.4.

5.2.6 Formulation of a von Mises Small Strain Elasto-Plastic Element

Formulation. The general procedure described in Sect. 5.2.3 and its automation described in Sect. 5.2.4 will now be applied for the simple case of von Mises plasticity with linear isotropic hardening.[9] In that case only the first and the last equation in (5.46) have to be considered within the algorithm. The equation set that has to be considered is a simplified version of Eqs. (5.63), (5.70), (5.74) and (5.76). Thus the equations

$$
\begin{aligned}
\varepsilon^e &= \varepsilon - \varepsilon^p \\
W &= \frac{\lambda}{2}\,(\mathrm{tr}\varepsilon^e)^2 + \mu\,\varepsilon^e\cdot\varepsilon^e \\
\boldsymbol{\sigma} &= \frac{\partial W}{\partial \varepsilon^e} \\
\boldsymbol{s} &= \boldsymbol{\sigma} - \frac{1}{3}\mathrm{tr}\,\boldsymbol{\sigma}\,\mathbf{1} \\
\dot{\varepsilon}^p &= \lambda\,\frac{\partial f}{\partial \boldsymbol{\sigma}} = \lambda\,\boldsymbol{n} \\
f(\boldsymbol{s}) &= \sqrt{\boldsymbol{s}\cdot\boldsymbol{s}} - \sqrt{\frac{2}{3}}(Y_0 + \hat{H}\hat{\alpha}) \leqslant 0 \\
\dot{\hat{\alpha}} &= \sqrt{\frac{2}{3}}\lambda
\end{aligned}
\tag{5.78}
$$

govern the constitutive behaviour of von Mises elasto-plasticity with linear isotropic hardening. Implicit integration of 5th equation together with 7th yields

$$
\varepsilon^p - \varepsilon^p_n = (\hat{\alpha} - \hat{\alpha}_n)\sqrt{\frac{3}{2}}\frac{\partial f}{\partial \boldsymbol{\sigma}} \tag{5.79}
$$

with the time increment Δt eliminated by introducing a new variable $\hat{\alpha} - \hat{\alpha}_n = \sqrt{\frac{2}{3}}\lambda\,\Delta t$. The Gauss point nonlinear system (5.46) can now be written as

$$
\mathbf{Q}_\varepsilon = \varepsilon^p - \varepsilon^p_n - (\hat{\alpha} - \hat{\alpha}_n)\sqrt{\frac{3}{2}}\frac{\partial f}{\partial \boldsymbol{\sigma}} = \mathbf{0} \tag{5.80}
$$

$$
Q_f = f = \sqrt{\boldsymbol{s}\cdot\boldsymbol{s}} - \sqrt{\frac{2}{3}}(Y_0 + \hat{H}\hat{\alpha}) = 0. \tag{5.81}
$$

[9]It is well known that the equations describing the von Mises plasticity model lead to a single nonlinear equation for the plastic multiplier, see Simo (1998), Simo and Hughes (1998) and de Souza Neto et al. (2008). Hence the more general approach followed in this section seem to be an overkill. However when comparing the efficiency of both approaches it turns out that the more general method is as efficient—when using *AceGen* —as the special elimination procedure leading to only one equation. Thus we follow in this book the approach discussed in Sect. 5.2.3 that can easily be adopted to more general plasticity formulations.

for the unknown plastic strain tensor ε^p and the equivalent plastic strain $\hat{\alpha}$ from which a set of Gauss point equations

$$\mathbf{Q}_g = \{Q_{\varepsilon 11},\ Q_{\varepsilon 22},\ Q_{\varepsilon 33},\ Q_{\varepsilon 12},\ Q_{\varepsilon 13},\ Q_{\varepsilon 23},\ f\}^T \tag{5.82}$$

for the set of Gauss point unknowns

$$\mathbf{h}_g = \{\varepsilon^p_{11}, \varepsilon^p_{22}, \varepsilon^p_{33}, \varepsilon^p_{12}, \varepsilon^p_{23}, \varepsilon^p_{31}, \hat{\alpha}\}^T \tag{5.83}$$

is extracted. This equation system has in total 7 unknowns (6 components of the symmetric plastic strain and the equivalent plastic strain). The formulation is summarized in Box 5.2 where an automation of the quantities that require operation of differentiation is also indicated.

Given: $\varepsilon, \varepsilon^p_n, \hat{\alpha}_n$ **Find:** $\varepsilon^p, \hat{\alpha}$

1 $\varepsilon^e = \varepsilon - \varepsilon^p$

2 $W = \frac{\lambda}{2}(\mathrm{tr}\varepsilon^e)^2 + \mu\,\varepsilon^e \cdot \varepsilon^e$

3 $\boldsymbol{\sigma} = \frac{\partial W}{\partial \varepsilon^e}$ .... automation $\longrightarrow$ .... $\boldsymbol{\sigma} = \frac{\hat{\delta} W}{\hat{\delta} \varepsilon^e}$

4 $\boldsymbol{s} = \boldsymbol{\sigma} - \frac{1}{3}\mathrm{tr}\,\boldsymbol{\sigma}\,\mathbf{1}$

5 $f = \sqrt{\boldsymbol{s}\cdot\boldsymbol{s}} - \sqrt{\frac{2}{3}}(Y_0 + H\hat{\alpha})$

6 $\boldsymbol{n} = \frac{\partial f}{\partial \boldsymbol{\sigma}}$ .... automation $\longrightarrow$ .... $\boldsymbol{n} = \frac{\hat{\delta} f}{\hat{\delta} \boldsymbol{\sigma}}$

7 $\mathbf{Q}_\varepsilon = \varepsilon^p - \varepsilon^p_n - (\hat{\alpha} - \hat{\alpha}_n)\sqrt{\frac{3}{2}}\,\boldsymbol{n} = \mathbf{0}$

8 $\mathbf{Q}_g = \{Q_{\varepsilon 11},\ Q_{\varepsilon 22},\ Q_{\varepsilon 33},\ Q_{\varepsilon 12},\ Q_{\varepsilon 13},\ Q_{\varepsilon 23},\ f\}^T$

9 $\mathbf{h}_g = \{\varepsilon^p_{11}, \varepsilon^p_{22}, \varepsilon^p_{33}, \varepsilon^p_{12}, \varepsilon^p_{23}, \varepsilon^p_{31}, \hat{\alpha}\}^T$

Box 5.2. Summary of small strain elasto-plastic material model

The formulation follows the general procedure for automation of small strain plasticity problems presented in Sect. 5.2.4. The input for the element is split into two parts, the part that is dependent on the material model and the part that is dependent on the kinematical model. The second part also describes the local loop using the operator split described above and the computation of residual and tangent.

AceGen description of material model. First we need an auxiliary function fQW that contains the part of the *AceGen* input that is dependent on the elasto-plastic model. The fQW function is related to the definition of the strain energy function, plastic flow condition and elasto-plastic constitutive equations. The elasto-plastic auxiliary function fQW for a three-dimensional von Mises small strain material model with linear isotropic hardening is presented in Box 5.3.

```
fQW[task_,hg_]:=Block[{},
        ┌─────────┬─────────┬─────────┐
        │ hg[[1]] │ hg[[4]] │ hg[[5]] │
    εp= │ hg[[4]] │ hg[[2]] │ hg[[6]] │;αh=hg[[7]];
        │ hg[[5]] │ hg[[6]] │ hg[[3]] │
        └─────────┴─────────┴─────────┘
  SMSFreeze[εe,ε-εp,"Symmetric"→True];
  {Em,ν,Y0,Hh}=dB[[1;;4]];{λ,μ}⊨SMSHookeToLame[Em,ν];
  W⊨Simplify[λ/2 Tr[εe]^2+μ Tr[εe.εe]];If[task=="W",Return[W]];
  SMSFreeze[σ,SMSD[W,εe,"Symmetric"→True],"Symmetric"→True];
  s=σ-1/3 IdentityMatrix[3] Tr[σ];
  fg⊨SMSSqrt[Tr[s.s]]-Sqrt[2/3] (Y0+Hh αh);
  If[task=="f",Return[fg]];
  n=Simplify[SMSD[fg,σ,"Symmetric"→True]];
         ┌─────────────┬─────────────┬─────────────┐
         │ hgnIO[[1]]  │ hgnIO[[4]]  │ hgnIO[[5]]  │
    εpn= │ hgnIO[[4]]  │ hgnIO[[2]]  │ hgnIO[[6]]  │;αhn=hgnIO[[7]];
         │ hgnIO[[5]]  │ hgnIO[[6]]  │ hgnIO[[3]]  │
         └─────────────┴─────────────┴─────────────┘
  Qε=εp-εpn-(αh-αhn) Sqrt[3/2] n;
  If[task=="Q",Return[Append[
     {Qε[[1,1]],Qε[[2,2]],Qε[[3,3]],Qε[[1,2]],Qε[[1,3]],Qε[[2,3]]},fg]]];
 ]
```

Box 5.3. *AceGen* input for elasto-plastic auxiliary function fQW for three-dimensional VON MISES small strain material model with linear isotropic hardening.

AceGen description of the Gauss point coupled problem. The material model, independent *AceGen* input, will be presented in several steps. General description of the *AceGen* input for the generation of the elastic three-dimensional element was given in Sect. 2.7. Only the differences related to the elasto-plastic formulation are additionally explained here.

Step 1: *AceGen* and template initialization

```
SMSInitialize["ElastoPlastic","Environment"→"AceFEM"];
nhg=7;lhg=nhg+1;
SMSTemplate["SMSTopology"→"H1","SMSSymmetricTangent"→True,
   "SMSDomainDataNames"->{"E","ν","Y0","H"},
   "SMSDefaultData"->{21000,0.3,24,0},
   "SMSNoTimeStorage"→lhg es$$["id","NoIntPoints"],
   "SMSPostIterationCall"→True
  ];
nen=SMSNoNodes;ndim=SMSNoDimensions;np=SMSNoDOFGlobal;
ndB=4;
```

(5.84)

At the beginning of the session the `SMSInitialize` function starts *AceGen* and the `SMSTemplate` function initializes constants that are needed to create the interface to the chosen FE environment. Additionally to the constants `"Environment"`, `"SMSTopology"`, `"SMSDomainDataNames"`, `"SMSDefaultData"` and `"SMSSymmetricTangent"` (already described in Sect. 2.7) we need to specify:

`"SMSNoTimeStorage"` constant specifies the length of the history data vector per element. In total eight history variables, the six strain components, the plastic multiplier and a state indicator have to be stored at every Gauss point, thus the length of history data vector per integration point is $l_{hg} = 8$ and the length of history data vector per element is $l_{dhe} = l_{hg}\, n_g \equiv$ `lhg es$$["id", "NoIntPoint"]` with the number of integration points `es$$["id","NoIntPoint"]` as an input parameter of the subroutine. The state indicator is set to "0" when the Gauss point is in elastic regime and to "1" when the Gauss point is in plastic regime.

`"SMSPostIterationCall"` option specifies that after the convergence of the global system of equations has been achieved, all the local problems have to be solved again in order to achieve the same accuracy for the global problem and the local problems.

Step 2: Standard "Tangent and residual" user subroutine

```
SMSStandardModule["Tangent and residual"];
XIO⊢Table[SMSReal[nd$$[i,"X",j]],{i,nen},{j,ndim}];
uIO⊢SMSReal[Table[nd$$[i,"at",j],{i,nen},{j,ndim}]];
dB⊢SMSReal[Table[es$$["Data",i],{i,ndB}]];
pe=Flatten[uIO];

(*start of Gauss loop*)
SMSDo[Ig,1,SMSInteger[es$$["id","NoIntPoints"]]];
Ξ={ξ,η,ζ}⊢Table[SMSReal[es$$["IntPoints",i,Ig]],{i,3}];
wgp⊢SMSReal[es$$["IntPoints",4,Ig]];
Ξn={{-1,-1,-1},{1,-1,-1},{1,1,-1},{-1,1,-1},
   {-1,-1,1},{1,-1,1},{1,1,1},{-1,1,1}};
Nh⊨Table[1/8 (1+ξ Ξn⟦i,1⟧) (1+η Ξn⟦i,2⟧) (1+ζ Ξn⟦i,3⟧),
   {i,1,nen}];
X⊢SMSFreeze[Nh.XIO];u⊨Nh.uIO;Je⊨SMSD[X,Ξ];Jed⊨Det[Je];
H⊨SMSD[u,X,"Dependency"→{Ξ,X,SMSInverse[Je]}];
```

Here the standard "Tangent and residual" user subroutine is declared, the loop over Gauss points is started, geometry and displacements are discretized and displacement gradient is evaluated for a three-dimensional 8 noded, izoparametric element. For details see Sect. 4.1.

Step 3: Gauss point data

```
Ihg⊢SMSInteger[(Ig-1) lhg];
hgnIO⊢Table[SMSReal[ed$$["hp",Ihg+ i]],{i,nhg}];
```

The data related to the solution of the locally coupled problem is:

`ed$$["ht",i]` is the ith component of the element history data vector at time t_{n+1},
`ed$$["hp",i]` is the ith component of the element history data vector at time t_n,

The history data vector is organized Gauss point wise, thus the position of the first element of the $\mathbf{h}_g$ vector (ε_{11}^p) within the element history data vector is $I_{hg} = (I_g - 1)\, l_{hg}$ where I_g is the Gauss point index.

Step 4: Strain tensor

```
SMSFreeze[ϵ,1/2 (H+Transpose[H]),"Symmetric"→True];
rg=SMSVariables[ϵ];
```
(5.85)

Here the small strain tensor is defined. The `SMSVariables` function performs vectorization of the strain tensor as defined in (5.58). Note that the `SMSFreeze` function has to be applied here on components of the small strain tensor, because the components will appear in the next step as independent variables for differentiation in Eq. (5.61).

Step 5: Operator split procedure

```
ftr⊨fQW["f",hgnIO];
TOL⊢SMSReal[rdata$$["SubIterationTolerance"]];
iNR⊢SMSInteger[idata$$["Iteration"]];
staten⊢SMSReal[ed$$["hp",Ihg+nhg+1]];
SMSIf[(iNR==1 && staten==0)||(iNR>1 && ftr<TOL)];(*elastic*)
  hg⊣hgnIO;
  SMSExport[Join[hg,{0}],Table[ed$$["ht",Ihg+i],{i,nhg+1}]];
SMSElse[];(*plastic*)
  hgb⊣hgnIO;
  SMSDo[jNR,1,30,1,hgb];
    Qg⊨fQW["Q",hgb];
    Ag⊨SMSD[Qg,hgb];
    LU⊨SMSLUFactor[Ag];
    Δh⊨SMSLUSolve[LU,-Qg];
    SMSIf[Sqrt[Δh.Δh]<TOL,
      DhgDr⊣SMSLUSolve[LU,-SMSD[Qg,rg,"Constant"→hgb]];
      SMSExport[Join[hgb+Δh,{1}],
   Table[ed$$["ht",Ihg+i],{i,nhg+1}]];
```

```
        SMSBreak[];];
      𝕙gb⊣𝕙gb+Δ𝕙;
      SMSIf[jNR=="29",SMSExport[{1,2},
          {idata$$["SubDivergence"],idata$$["ErrorStatus"]}];
    SMSBreak[];];
    SMSEndDo[𝕙gb,D𝕙gDr];
    𝕙g⊣SMSFreeze[𝕙gb,"Dependency"→{𝕣g,D𝕙gDr}];
  SMSEndIf[𝕙g];
```

(5.86)

This input segment is an *AceGen* implementation of the abstract Algorithm 5.1 which describes the general idea of an operator split for integration of the constitutive model within a load step. First the flow rule is checked using the trial values for the plastic strains $\mathbf{h}_{gn}$ ≡ `𝕙gnIO`. The `rdata$$["SubIteration Tolerance"]` variable is a global variable and is specific for the chosen finite element environment. It specifies the tolerance with which the local problem has to be solved. Similarly, the `idata$$["Iteration"]` global variable specifies the index of the current global Newton iteration. The test $f < \text{TOL}$ is in the first global iteration (`iNR==1`) unreliable. Additional condition `iNR == 1 && staten == 0` enforces an elastic state in case that the Gauss point was in an elastic state at the end of previous time step. The state indicator `staten` is positioned in the $\mathbf{h}_g$ vector after the Gauss point unknowns, thus positioned at position $I_{hg} + n_{hg} + 1$.
In case that the flow rule is fulfilled, the trial values are exported as new plastic strains, the state indicator is set to "0" (elastic) and the computation of the constitutive equation is finished. In case that the flow rule is not fulfilled an iteration loop (max iteration = 30) is started to solve the nonlinear equation system (5.82) by the Newton algorithm described in Sect. 5.2.3. When convergence is achieved the plastic strains ε^p and $\hat{\alpha}$ are stored as new history values. Also the state indicator is set to "1" (plastic). The derivative of the plastic variables $\frac{\partial \mathbf{h}_{\hat{g}}}{\partial \mathbf{r}_g} = \widehat{D_{\mathbf{r}_g}\mathbf{h}_g}$ ≡ `D𝕙gDr` in (5.61) is computed as well and introduced by the `"Dependency"` directive. The latter is needed when the tangent matrix contribution at a Gauss point is derived. Note that the function `SMSLUFactor` performs full symbolic factorization of the system of linear equations and the function `SMSLUSolve` full symbolic back substitution.

Step 6: Evaluation of tangent matrix and residual

```
W⊨fQW["W",𝕙g];
SMSDo[m,1,np];
  Rgm⊨Jed SMSD[W,𝕡e,m,"Constant"->𝕙g];
  SMSExport[wgp Rgm,p$$[m],"AddIn"→True];
  SMSDo[n,m,np];
    Kgmn⊨SMSD[Rgm,𝕡e,n];
    SMSExport[wgp Kgmn,s$$[m,n],"AddIn"→True];
  SMSEndDo[];
SMSEndDo[];
```

Strain energy is here evaluated for the converged values of plastic variables. The internal load vector is calculated according to (5.59) and tangent matrix according to Algorithm 5.1 where the residual is computed for a constant "frozen" value of the plastic variables (enforced by `"Constant" ->hg` directive).

Step 7: Code generation

```
SMSEndDo[];
(*end of Gauss loop*)
SMSWrite[];
```

At the end of the session the `SMSWrite` function writes the generated formulas together with the code that provides interface to the chosen finite element environment to a file in the compiled language of the target finite element environment.

Other Plasticity Models. Inelastic material deformations occur for many engineering materials. However it is not possible to formulate one general model that covers all applications, Metals depict in most cases ductile behaviour which enables them to undergo large plastic strains. On the opposite ceramic materials are brittle and thus have only a small range of plastic deformations. In between are materials which behave, depending on the stress state, either brittle or ductile. Furthermore, some of the materials show associated or non-associated performance and different hardening behaviour can be observed in experiments. Thus it is necessary to design constitutive equations that model the major response characteristics of an inelastic material. In the next sections, some often applied constitutive equations are summarized.

Plasticity models can be based on different flow surfaces. One of them is the Tresca model that represents a multisurface model for plasticity. It results from the assumption that plastic yielding starts when the maximum shear stress reaches a critical level. Hardening is excluded in this model which additionally is pressure insensitive due to the chosen yield criterion. Since metals do not depict plastic flow under pure pressure loading the Tresca model can be applied to describe metal plasticity, see e.g. Khan and Huang (1995).

A further constitutive model is the Mohr–Coulomb model that can be applied to represent pressure sensitive materials such as soil and concrete. Since this model is based on micromechanical frictional sliding between particles it is well suited to describe the inelastic behaviour of certain soils, see e.g. Desai and Siriwardane (1984).

A criterion which is used often in soil mechanics is the Drucker–Prager flow condition, see Drucker and Prager (1952) or Khan and Huang (1995). It depends not only upon the second invariant II_s of the stress deviator $\boldsymbol{s}$ as in the classical von Mises theory of metal plasticity, see (5.70), but also on a hydrostatic stress term represented here by the first invariant I_σ of the stress tensor $\boldsymbol{\sigma}$.

Many more different constitutive models which stem from experimental observations and describe plastic flow can be formulated. These are related to different types of materials, see for an overview e.g. Desai and Siriwardane (1984), Lubliner (1990),

Hofstetter and Mang (1995) or Khan and Huang (1995). Furthermore different loading conditions like pulsating or dynamic loads can lead to so called ratcheting strains and hence special constitutive descriptions are needed to model such effects, see e.g. Ekh et al. (2000) and Johansson et al. (2005).

Such plasticity models models can be easily implemented only by changing the `fQW` auxiliary function defined in Box 5.3 and the input segment (5.84) where interface to the finite element environment is initialized. With small changes, the input can be adapted also for other element types (two-dimensional solids, beams, shells, etc.) and different kinematical equations (plane strain, plain stress, finite strain, etc.).

5.3 Elasto-Plastic Materials, Finite Deformations

Models for finite plasticity are essential for engineering analysis such as metal forming, cutting, crash simulations or pile driving in soils. The formulation of the underlying theoretical background has a long history. However the possibility to apply such models within numerical simulations has shed new light on some theoretical aspects.

Many authors start from a hypo-elastic constitutive equation for the elastic part of the deformation when finite elasto-plastic deformations have to be considered, see e.g. Khan and Huang (1995). Such material assumption does not represent elasticity in the strict sense, see e.g. Truesdell and Noll (1965) or Simo and Pister (1984). Besides this restriction which can result in unwanted effects[10] there exist also problems of numerical nature which will be discussed in the following.

A hypo-plastic constitutive law is presented in a rate form and relates a stress flux, see e.g. (1.79), to the symmetrical spatial velocity gradient $\boldsymbol{d}$: $\overset{\nabla}{\boldsymbol{\tau}} = \mathbb{C}\,[\boldsymbol{d}^e]$ with an incremental elasticity tensor $\mathbb{C}$. The idea behind this is an additive split of the spatial velocity gradient into an elastic and a plastic part $\boldsymbol{d} = \boldsymbol{d}^e + \boldsymbol{d}^p$ analogous to the assumptions made for the small strain case in the last section. For the correct choice of the stress rates, see e.g. Simo and Hughes (1998) and Khan and Huang (1995). The rate form requires a time integration which results in a costly algorithm. To overcome this disadvantage algorithms for plasticity were developed which base on hyperelastic constitutive equations defined in Sect. 5.1. These formulations use a so called operator-split technique, discussed already in the previous section, in which the elastic part follows directly via a function evaluation of the hyperelastic material law, see e.g. Simo and Ortiz (1985) or Simo (1988). This circumvents the time integration of the elastic constitutive equation.

[10]It is well known that by using the wrong rates for the description of the elastic response physically meaningless stress states can occur at large deformations, see e.g. Atluri (1984). However it should be noted that special stress rates can be defined that circumvent these problems, see Xiao et al. (1997).

5.3.1 General Formulation

Based on the single crystal model for metal plasticity a multiplicative split of the deformation gradient is introduced instead of the additive split of the strain rates into elastic and plastic parts. For a theoretical foundation see e.g. Lee and Liu (1967) or Lubliner (1990). This split is defined by

$$\boldsymbol{F} = \boldsymbol{F}^e\, \boldsymbol{F}^p, \tag{5.87}$$

where an intermediate configuration was introduced besides the initial- and spatial configuration, see Fig. 5.4.

Based on this split the right and left "elastic" Cauchy–Green tensor is introduced by using $\boldsymbol{F}^e$

$$\tilde{\boldsymbol{C}}^e = \boldsymbol{F}^{e^T}\, \boldsymbol{F}^e, \qquad \boldsymbol{b}^e = \boldsymbol{F}^e\, \boldsymbol{F}^{e^T}. \tag{5.88}$$

where the tilde indicates in $(5.88)_1$ that the right Cauchy–Green tensor is referred to the intermediate configuration.

Analogous to the definition of the spatial velocity gradient $\boldsymbol{l} = \dot{\boldsymbol{F}}\, \boldsymbol{F}^{-1}$, see (1.37), the corresponding elastic and plastic parts are given by

$$\boldsymbol{l}^e = \dot{\boldsymbol{F}}^e\, \boldsymbol{F}^{e^{-1}}, \qquad \tilde{\boldsymbol{L}}^p = \dot{\boldsymbol{F}}^p\, \boldsymbol{F}^{p^{-1}}. \tag{5.89}$$

Again tensor $\boldsymbol{l}^e$ is referred to the spatial configuration while $\tilde{\boldsymbol{L}}^p$ acts in the intermediate configuration. Since $\boldsymbol{F}^p$ can be written with (5.87) also as $\boldsymbol{F}^{e^{-1}}\,\boldsymbol{F}$ the following expression is derived with $\dot{\overline{\boldsymbol{F}^{-1}}} = -\boldsymbol{F}^{-1}\,\dot{\boldsymbol{F}}\,\boldsymbol{F}^{-1}$

$$\tilde{\boldsymbol{L}}^p = \frac{\partial}{\partial t}(\boldsymbol{F}^{e^{-1}}\,\boldsymbol{F})\,\boldsymbol{F}^{p^{-1}} = \boldsymbol{F}^{e^{-1}}\,[\,\dot{\boldsymbol{F}}\,\boldsymbol{F}^{-1} - \dot{\boldsymbol{F}}^e\,\boldsymbol{F}^{e^{-1}}\,]\,\boldsymbol{F}^e, \tag{5.90}$$

which yields with $(5.89)_1$ and (1.37)

$$\boldsymbol{F}^e\,\tilde{\boldsymbol{L}}^p\,\boldsymbol{F}^{e^{-1}} = \boldsymbol{l} - \boldsymbol{l}^e. \tag{5.91}$$

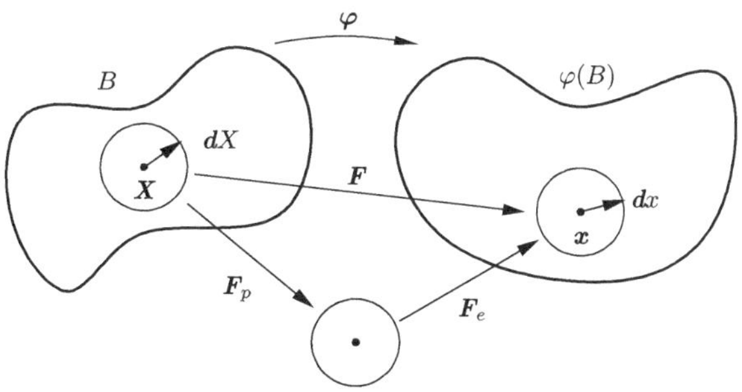

Fig. 5.4 Multiplicative decomposition of the deformation gradient $\boldsymbol{F}$

This equation motivates the following definition of the spatial plastic velocity gradient $\boldsymbol{l}^p = \boldsymbol{F}^e\,\tilde{\boldsymbol{L}}^p\,\boldsymbol{F}^{e^{-1}}$, which results as instead of (5.91) to

$$\boldsymbol{l}^p = \boldsymbol{l} - \boldsymbol{l}^e. \tag{5.92}$$

The decomposition of the spatial velocity gradient into its symmetric part $\boldsymbol{d}$ and its antisymmetric part $\boldsymbol{w}$ yields, with the previous definitions, the additive decomposition of the symmetric velocity gradient

$$\boldsymbol{d} = \boldsymbol{d}^e + \boldsymbol{d}^p. \tag{5.93}$$

This equation is often starting point of rate equations for finite elasto-plastic deformations. Note however that the following definitions are used in (5.91): $\boldsymbol{d}^p = \mathrm{sym}\,[\,\boldsymbol{F}^e\,\tilde{\boldsymbol{L}}^p\,\boldsymbol{F}^{e^{-1}}\,]$. The additional equation $\boldsymbol{w} = \boldsymbol{w}^e + \boldsymbol{w}^p$ for the antisymmetric part of the spatial velocity gradient (spin) results from (5.92) and needs further considerations, see e.g. the discussion in Besseling and van der Giessen (1994).

Like in the geometrical linear theory of elasto-plastic deformations the free energy Ψ is written as function of the elastic deformation and of m inner variables α_k $(k = 1, \ldots, m)$

$$\Psi(\tilde{\boldsymbol{C}}^e, \alpha_k) = \tilde{W}(\tilde{\boldsymbol{C}}^e) + \tilde{H}(\alpha_k). \tag{5.94}$$

The corresponding dependence of the strain energy function on $\tilde{\boldsymbol{C}}^e$ can be found e.g. in Mandel (1974). By considering the specific stress power $\boldsymbol{\tau} \cdot \boldsymbol{d}$ the local dissipation $\mathcal{D}$ is stated as

$$\mathcal{D} = \boldsymbol{\tau} \cdot \boldsymbol{d} - \dot{\Psi}(\tilde{\boldsymbol{C}}^e, \alpha_k) \geqslant 0. \tag{5.95}$$

The time derivative of the free energy function yields with $\dot{\tilde{\boldsymbol{C}}}^e = 2\,\boldsymbol{F}^{e^T}\,\boldsymbol{d}^e\,\boldsymbol{F}^e$

$$\mathcal{D} = \left(\boldsymbol{\tau} - 2\,\boldsymbol{F}^e\,\frac{\partial \tilde{W}}{\partial \tilde{\boldsymbol{C}}^e}\,\boldsymbol{F}^{e^T}\right) \cdot \boldsymbol{d}^e + \boldsymbol{\tau} \cdot \boldsymbol{d}^p - \sum_{k=1}^{m} \frac{\partial \tilde{H}}{\partial \alpha_k}\,\dot{\alpha}_k \geqslant 0. \tag{5.96}$$

From this inequality follow by using the standard arguments of material theory, see e.g. Truesdell and Noll (1965) or Lubliner (1990), the constitutive relations for stresses and hardening variables

$$\boldsymbol{\tau} = 2\,\boldsymbol{F}^e\,\frac{\partial \tilde{W}}{\partial \tilde{\boldsymbol{C}}^e}\,\boldsymbol{F}^{e^T} \quad \text{and} \quad q_k = -\frac{\partial \tilde{H}}{\partial \alpha_k}, \tag{5.97}$$

and further more the reduced form of the dissipation inequality

$$\mathcal{D} = \boldsymbol{\tau} \cdot \boldsymbol{d}^p + \sum_{k=1}^{m} q_k \, \dot{\alpha}_k \geqslant 0, \tag{5.98}$$

which represents a restriction for the evolution equations of plastic flow. The equations derived above can also be referred to the intermediate configuration which leads to the stress tensor in the intermediate configuration

$$\tilde{\boldsymbol{S}} = \boldsymbol{F}^p \, \boldsymbol{S} \, \boldsymbol{F}^{p\,T} = \boldsymbol{F}^{e\,-1} \, \boldsymbol{\tau} \, \boldsymbol{F}^{e\,-T} \tag{5.99}$$

and the constitutive equations

$$\tilde{\boldsymbol{S}} = 2 \, \frac{\partial \tilde{W}}{\partial \tilde{\boldsymbol{C}}^e} \quad \text{and} \quad q_k = - \frac{\partial \tilde{H}}{\partial \alpha_k} . \tag{5.100}$$

The first term in the reduced dissipation inequality (5.98) can be formulated with respect to the intermediate configuration

$$\boldsymbol{\tau} \cdot \boldsymbol{d}^p = \tilde{\boldsymbol{S}} \cdot (\tilde{\boldsymbol{C}}^e \, \tilde{\boldsymbol{L}}^p)^S = \boldsymbol{\Sigma} \cdot \boldsymbol{L}^p .$$

Here $\boldsymbol{\Sigma} = \tilde{\boldsymbol{C}}^e \, \tilde{\boldsymbol{S}}$ is the Mandel stress tensor which can be non-symmetric in the general case. Hence (5.98) can be rewritten as

$$\mathcal{D} = \boldsymbol{\Sigma} \cdot \tilde{\boldsymbol{L}}^p + \sum_{k=1}^{m} q_k \, \dot{\alpha}_k \geqslant 0, \tag{5.101}$$

see Mandel (1974). These relations are valid for general elasto-plastic material behaviour. They will be specified in the following for isotropic materials. For that the assumption of a free energy function is valid which does not depend upon any orientation in the initial configuration. Furthermore no orientation will enter the constitutive relations in the intermediate configuration which has the consequence of an undetermined plastic spin $\boldsymbol{w}^p$. In such cases, often the constitutive assumption $\boldsymbol{w}^p = \boldsymbol{0}$ is chosen. Further physical interpretations and considerations regarding the plastic spin can be found in Dafalias (1985) or Besseling and van der Giessen (1994).

The elastic domain of a given deformation state is described by a yield condition which is formulated in terms of the Kirchhoff stresses and the hardening variables

$$f\,(\boldsymbol{\tau}, \alpha_k) \leqslant 0. \tag{5.102}$$

The assumption of maximum plastic dissipation at a fixed configuration is valid in case of associative plasticity. This leads together with (5.98) to the evolution equations for the plastic variables

$$\boldsymbol{d}^p = \lambda \, \frac{\partial f}{\partial \boldsymbol{\tau}}, \qquad \dot{\alpha}_k = \lambda \, \frac{\partial f}{\partial \alpha_k}, \tag{5.103}$$

see e.g. Simo (1992) or Simo and Miehe (1992). These equations correspond to the formulation in the geometrically (5.103) can be obtained by computing the Lie derivative of the spatial strain measure $\boldsymbol{b}_e = \boldsymbol{F}_e \boldsymbol{F}_e^T$ using the multiplicative split (5.87). The *pull back* of $\boldsymbol{b}_e$ to the reference configuration leads with (1.44) to

$$\boldsymbol{F}^{-1} \boldsymbol{b}_e \boldsymbol{F}^{-T} = \boldsymbol{F}^{-1} (\boldsymbol{F} \boldsymbol{F}_p^{-1} \boldsymbol{F}_p^{-T} \boldsymbol{F}^T) \boldsymbol{F}^{-T} = \boldsymbol{F}_p^{-1} \boldsymbol{F}_p^{-T}.$$

The subsequent time derivative yields

$$\frac{\partial}{\partial t} \boldsymbol{F}_p^{-1} \boldsymbol{F}_p^{-T} = \dot{\boldsymbol{F}}_p^{-1} \boldsymbol{F}_p^{-T} + \boldsymbol{F}_p^{-1} \dot{\boldsymbol{F}}_p^{-T}.$$

This equation can be rewritten by using the identity $\boldsymbol{F}_p \boldsymbol{F}_p^{-1} = \mathbf{1}$, which results in $\dot{\boldsymbol{F}}_p^{-1} = -\boldsymbol{F}_p^{-1} \dot{\boldsymbol{F}}_p \boldsymbol{F}_p^{-1}$,

$$\frac{\partial}{\partial t} \boldsymbol{F}_p^{-1} \boldsymbol{F}_p^{-T} = -\boldsymbol{F}_p^{-1} \dot{\boldsymbol{F}}_p \boldsymbol{F}_p^{-1} \boldsymbol{F}_p^{-T} - \boldsymbol{F}_p^{-1} \boldsymbol{F}_p^{-T} \dot{\boldsymbol{F}}_p^{-T} \boldsymbol{F}_p^{-T}.$$

The final transformation to the spatial configuration by a *push forward* operation yields

$$\mathcal{L}_v \boldsymbol{b}_e = \boldsymbol{F} \left(\frac{\partial}{\partial t} \boldsymbol{F}_p^{-1} \boldsymbol{F}_p^{-T} \right) \boldsymbol{F}^T = -\boldsymbol{F}_e (\dot{\boldsymbol{F}}_p \boldsymbol{F}_p^{-1} + \boldsymbol{F}_p^{-T} \dot{\boldsymbol{F}}_p^{-T}) \boldsymbol{F}_e^T.$$

With the definition of a plastic velocity gradient $\tilde{\boldsymbol{L}}_p = \dot{\boldsymbol{F}}_p \boldsymbol{F}_p^{-1}$ in the intermediate plastic configuration analogous to (1.37), see Sect. 5.3, the Lie derivative of $\boldsymbol{b}_e$ is obtained as

$$\mathcal{L}_v \boldsymbol{b}_e = -2 \boldsymbol{F}_e \frac{1}{2} (\tilde{\boldsymbol{L}}_p + \tilde{\boldsymbol{L}}_p^T) \boldsymbol{F}_e^T. \tag{5.104}$$

This result can be interpreted as a *push forward* of the symmetric part of the plastic velocity gradient from the intermediate plastic configuration to the current configuration.

Equation (5.104) leads then with (5.91) and (5.92) to

$$\mathcal{L}_v \boldsymbol{b}^e = -2 \boldsymbol{F}^e \operatorname{sym} (\tilde{\boldsymbol{L}}^p) \boldsymbol{F}^{e^T} = -2 \operatorname{sym} (\boldsymbol{l}^p \boldsymbol{b}^e). \tag{5.105}$$

With the further assumption that the plastic spin is zero ($\boldsymbol{w}^p = \mathbf{0}$) the relation $\boldsymbol{d}^p = \boldsymbol{l}^p$ follows with (5.92) from (5.93). Thus Eq. (5.103)$_1$ can be rewritten as

$$-\frac{1}{2} \mathcal{L}_v \boldsymbol{b}^e = \operatorname{sym} (\lambda \frac{\partial f}{\partial \boldsymbol{\tau}} \boldsymbol{b}^e), \tag{5.106}$$

see Simo and Miehe (1992), where $\boldsymbol{b}^e$ denotes the Lie derivative defined by

$$\mathcal{L}_v\, \boldsymbol{b}^e = -2\, \boldsymbol{F}^e\, \frac{1}{2}\, (\tilde{\boldsymbol{L}}^p + \tilde{\boldsymbol{L}}^{p\,T})\, \boldsymbol{F}^{e\,T} . \tag{5.107}$$

For the isotropic case the Kirchhoff stresses can be expressed as function of the left Cauchy–Green tensor $\boldsymbol{b}^e$, see also (5.6),

$$\boldsymbol{\tau} = 2\, \frac{\partial \tilde{W}}{\partial \boldsymbol{b}^e}\, \boldsymbol{b}^e . \tag{5.108}$$

The left Cauchy–Green tensor follows from

$$\boldsymbol{b}^e = \boldsymbol{F}^e\, \boldsymbol{F}^{e\,T} = \boldsymbol{F}\, \boldsymbol{C}^{p-1}\, \boldsymbol{F}^T \qquad \text{with} \quad \boldsymbol{C}^p = \boldsymbol{F}^{p\,T}\, \boldsymbol{F}^p . \tag{5.109}$$

This means that $\boldsymbol{b}^e$ is related to the inverse of the right Cauchy–Green tensor $\boldsymbol{C}^{p-1}$ which is referred to the initial configuration.

Experiments substantiate that plastic flow does not depend upon the volume change in the body. This fact is synonymous with the assumption $\text{tr}(\boldsymbol{d}^p) = 0$ or $\det\, \boldsymbol{F}^p = 1$ which corresponds to incompressibility of rubber materials, see Sect. 5.1. There a split of the deformation in isochoric and volumetric parts lead to a decomposition of the strain energy function (5.14). Such split can be defined also for the above equation. It yields with (1.19) instead of (5.87) to

$$\boldsymbol{F} = J_F^{e\,\frac{1}{3}}\, \hat{\boldsymbol{F}}^e\, \boldsymbol{F}^p \qquad \text{with} \quad \det\, \boldsymbol{F}^p = 1 . \tag{5.110}$$

Since it is not obvious that the integration of evolution equation (5.106) preserves this constraint condition, special care is needed in the design of associated numerical algorithms, see Sect. 5.3.2.

5.3.2 Integration of Constitutive Equations for Finite Deformation Problems

Implicit Euler schemes were developed for the integration of inelastic constitutive equations of problems undergoing small strains. Such algorithms will now be developed for inelastic constitutive equations at finite strains. Constitutive equations for finite deformation problems can be formulated in different ways. This was subject - especially in view of numerical algorithms for finite element analysis - of many different research efforts, see e.g. Argyris and Kleiber (1977), Nagtegaal (1982), Argyris et al. (1982), Simo (1988), Nagtegaal et al. (1990), Peric et al. (1992), Simo and Hughes (1998) and Simo (1998). The choice of the mathematical description depends upon the material at hand but also upon the efficiency of specific solution methods. Two possible formulations and associated integration algorithms are discussed in this section for elasto-plastic material with isotropic hardening which was already described in Sect. 5.3. Interesting applications and examples of finite deformation plasticity can be found e.g. in Peric and Owen (1997).

These algorithms are developed for a time interval $[\, t_n , t \,]$ where it is assumed that the deformation φ and its gradient $\boldsymbol{F}$ are known at time t_n. This shall also be true for the internal variables $\{ \boldsymbol{F}^e , \boldsymbol{\xi}_\alpha \}$. Hence the set of initial values

$$\boldsymbol{F}_n = \operatorname{Grad} \varphi_n , \qquad \{ \boldsymbol{F}^e_n , \boldsymbol{\xi}_{\alpha_n} \} \tag{5.111}$$

is known. Now the algorithmic approximation of the evolution equations for plastic flow

$$\boldsymbol{l}^p = \sum_{g=1}^{m} \lambda_g \, \frac{\partial f_g(\boldsymbol{\tau}, q_\alpha)}{\partial \boldsymbol{\tau}} \tag{5.112}$$

$$\dot{\xi}_\alpha = \sum_{g=1}^{m} \lambda_g \, \frac{\partial f_g(\boldsymbol{\tau}, q_\alpha)}{\partial q_\alpha} \tag{5.113}$$

with the Kuhn–Tucker conditions

$$\lambda_g \geqslant 0, \quad f_g(\boldsymbol{\tau}, q_\alpha) \leqslant 0, \quad \lambda_g \, f_g(\boldsymbol{\tau}, q_\alpha) = 0. \tag{5.114}$$

has to be constructed. More details related to the derivation of the plastic evolution equations can be found in Sect. 5.3. Kinematic hardening is not considered.

The evolution equations in (5.113) include multi-surface plasticity. The sum sign can be neglected in case of only one yield surface.

In Sect. 5.3 several relations were derived and definitions introduced. These can be used to write the spatial plastic rate $\boldsymbol{l}^p$ as

$$\boldsymbol{l}^p = \boldsymbol{F}^e \, \boldsymbol{L}^p \boldsymbol{F}^{e\,-1} \qquad \text{with} \quad \dot{\boldsymbol{F}}^p = \boldsymbol{L}^p \, \boldsymbol{F}^p . \tag{5.115}$$

The structure of the last equations leads to an exponential approximation of the evolution of the plastic deformation gradient, see e.g. Simo (1992), Simo and Hughes (1998) and Miehe (1993),

$$\boldsymbol{F}^p = \exp[\, (t - t_n) \, \boldsymbol{L}^p \,] \, \boldsymbol{F}^p_n . \tag{5.116}$$

Some reformulations—considering the multiplicative split of the deformation gradient $\boldsymbol{F} = \boldsymbol{F}^e \, \boldsymbol{F}^p$ – yield with (5.116)

$$\begin{aligned} \boldsymbol{F} &= \boldsymbol{F}^e \, \exp[\, (t - t_n) \, \boldsymbol{L}^p \,] \, \boldsymbol{F}^{e\,-1} \boldsymbol{F}^e \boldsymbol{F}^p_n \\ &= \exp[\, (t - t_n) \, \boldsymbol{F}^e \, \boldsymbol{L}^p \, \boldsymbol{F}^{e\,-1} \,] \boldsymbol{F}^e \boldsymbol{F}^p_n , \end{aligned} \tag{5.117}$$

where the standard properties of an exponential map where utilized. Equation (5.117) can be resolved with the definition in $(5.115)_1$ and $\Delta t = t - t_n$ and with $\boldsymbol{F}^{e\,tr} = \boldsymbol{F}\,\boldsymbol{F}_n^{p\,-1}$ with respect to $\boldsymbol{F}^e$

$$\boldsymbol{F}^e = \exp[\,(\Delta t\,)\,\boldsymbol{l}^p]\,\boldsymbol{F}^{e\,tr}. \tag{5.118}$$

The definition of the trial value $\boldsymbol{F}^{e\,tr}$ is physically motivated since the computation of the elastic part of the deformation gradient is performed for frozen plastic variables $\boldsymbol{F}^p = \boldsymbol{F}_n^p$. By inserting $(5.113)_1$ into Eq. (5.118) relation

$$\boldsymbol{F}^e = \exp\left[-\sum_{g=1}^{m} \lambda_g \Delta t \frac{\partial f_g(\boldsymbol{\tau}, q_\alpha)}{\partial \boldsymbol{\tau}}\right] \boldsymbol{F}^{e\,tr} \tag{5.119}$$

can be determined. The definitions of the flow increment $\Delta\lambda_g = \Delta t\,\lambda_g$ and the implicit Euler approximation of $(5.113)_2$ lead to the algorithmic version of the flow rule (5.113)

$$\boldsymbol{F}^e = \exp\left[-\sum_{g=1}^{m} \Delta\lambda_g \frac{\partial f_g(\boldsymbol{\tau}, q_\alpha)}{\partial \boldsymbol{\tau}}\right] \boldsymbol{F}^{e\,tr} \tag{5.120}$$

$$\xi_\alpha = \xi_{\alpha_n} + \sum_{g=1}^{m} \Delta\lambda_g \frac{\partial f_g(\boldsymbol{\tau}, q_\alpha)}{\partial q_\alpha} \tag{5.121}$$

and the Kuhn–Tucker conditions

$$\Delta\lambda_g \geqslant 0, \quad f_g(\boldsymbol{\tau}, q_\alpha) \leqslant 0 \text{ and } \Delta\lambda_g\, f_g(\boldsymbol{\tau}, q_\alpha) = 0. \tag{5.122}$$

The stresses in these equations is given by

$$\boldsymbol{\tau} = \boldsymbol{F}^e \left[\frac{\partial W(\boldsymbol{C}^e)}{\partial \boldsymbol{C}^e}\right] \boldsymbol{F}^{e\,T} \tag{5.123}$$

and the hardening variables follow from

$$q_\alpha = -\frac{\partial H(\xi_\alpha)}{\partial \xi_\alpha}. \tag{5.124}$$

The solution of the nonlinear system (5.120) to (5.124) has to be performed locally at each Gauss point. It can be obtained by using Newton's method.

Note that the constraint condition of incompressibility of the plastic flow, $J^p = 1$, is exactly fulfilled by the algorithmic flow rule (5.120), see e.g. Simo (1992).

5.3.3 Automation of Finite Strain Plasticity

According to the classification of nonlinear computational problems finite strain elasto-plastic problems fall into the category of *time-dependent* Gauss *point coupled problems*. The idea of operator split for the integration of the constitutive model within a load step is employed in a same manner as for small strain elasto-plastic problems. In fact one can simply replace all occurrences of small strain symbol ε in Box 5.1 with a symbol for an appropriate strain measure on which strain energy function and plastic evolution equations Q_g depend explicitly. The result is an *ADB* formulation of the Gauss point contribution to the element residual and tangent matrix for finite strain elasto-plastic problems. For example, in the formulation presented in Sect. 5.3.2 evolution equations explicitly depend on the deformation gradient $\mathbf{F}$. Consequently, a set of intermediate variables $\mathbf{r}_g$ that is a part of the general algorithm Box 5.1 can be identified for the present formulation as a set of mutually independent variables that form the deformation gradient

$$\mathbf{r}_g = \mathrm{var}(\boldsymbol{F}). \tag{5.125}$$

The only true difference is related to the material model dependent auxiliary function fQW and to Step 4 in Sect. 5.2.6 where the strain measure is defined. The definition of the strain measure given in the input segment (5.85) has to be replaced by the following input segment

```
SMSFreeze[𝔽,IdentityMatrix[3]+ℍ];
𝕣g=SMSVariables[𝔽];
```
. (5.126)

The *AceGen* input presented in Sect. 5.2.6 is then directly applicable for finite strain problems.

5.3.4 Finite Strain Plasticity Example

In this section, an abstract symbolic formulation for the contribution of one finite element Ω_e to the internal force vector $\mathbf{R}_e$ and to the tangential stiffness matrix $\mathbf{K}_e$ is presented for a problem undergoing finite plastic strains.

The formulation employs the general implicit Euler scheme for the integration of the finite strain inelastic constitutive equations as stated in Sect. 5.3.2. The used Neo-Hooke strain energy W is defined in Sect. 5.1, see e.g. Eq. (5.8). The parts of the finite strain plasticity model, necessary for the abstract symbolic description, are summarized in Box 5.4 and the *AceGen* input with the corresponding definition of fQW function in Box 5.5.

Given: $\boldsymbol{F},\ \boldsymbol{F}_n^{p-1}-\mathbf{1},\ \hat{\alpha}_n$ **Find:** $\boldsymbol{F}^{p-1}-\mathbf{1},\ \hat{\alpha}$

1 $\boldsymbol{F}^e=\boldsymbol{F}\boldsymbol{F}^{p-1},\ \boldsymbol{C}^e=\boldsymbol{F}^{eT}\boldsymbol{F}^e,\ J_F^2=\det\boldsymbol{C}^e$

2 $W=\frac{\mu}{2}(\mathrm{tr}\boldsymbol{C}^e)-3-\ln J_F^2)+\frac{\lambda}{4}(J_F^2-1-\ln J_F^2)$

3 $\boldsymbol{\tau}=2\boldsymbol{F}^e\dfrac{\partial W}{\partial\boldsymbol{C}^e}\boldsymbol{F}^{eT}$ .... $\xrightarrow{\text{automation}}$ .... $\boldsymbol{\tau}=2\boldsymbol{F}^e\dfrac{\hat{\delta}W}{\hat{\delta}\boldsymbol{C}^e}\boldsymbol{F}^{eT}$

4 $\boldsymbol{s}=\boldsymbol{\tau}-\dfrac{\mathrm{tr}\boldsymbol{\tau}}{3}\mathbf{1}$

5 $f=\sqrt{\boldsymbol{s}\cdot\boldsymbol{s}}-\sqrt{\frac{2}{3}}\,(Y_0+H\hat{\alpha})$

6 $\boldsymbol{n}=\dfrac{\partial f}{\partial\boldsymbol{\tau}}$ .... $\xrightarrow{\text{automation}}$ .... $\boldsymbol{n}=\dfrac{\hat{\delta}f}{\hat{\delta}\boldsymbol{\tau}}$

7 $\mathbf{Q}_\varepsilon=\boldsymbol{F}^e-\exp(-(\hat{\alpha}-\hat{\alpha}_n)\sqrt{\frac{3}{2}}\,\boldsymbol{n})\boldsymbol{F}\boldsymbol{F}_n^{p-1}=0$

8 $\mathbf{Q}_g=\{Q_{\varepsilon 11},\ Q_{\varepsilon 22},\ Q_{\varepsilon 33},\ Q_{\varepsilon 12},\ Q_{\varepsilon 13},\ Q_{\varepsilon 23},\ Q_{\varepsilon 21},\ Q_{\varepsilon 31},\ Q_{\varepsilon 32},\ f\}^T$

9 $\mathbf{h}_g=\{\ F_{11}^{p-1}-1,\ F_{22}^{p-1}-1,\ F_{33}^{p-1}-1,\ F_{12}^p,\ F_{13}^p,\ F_{23}^p,\ F_{21}^p,\ F_{31}^p,\ F_{32}^p,\ \hat{\alpha}\ \}^T$

Box 5.4. Summary of finite strain elasto-plastic material model

```
fQW[task_,𝕙g_]:=Block[{},
                                    ┌─────────┬─────────┬─────────┐
                                    │ 𝕙g[[1]] │ 𝕙g[[4]] │ 𝕙g[[5]] │
  𝔽pinv=IdentityMatrix[3]+          │ 𝕙g[[7]] │ 𝕙g[[2]] │ 𝕙g[[6]] │;αh=𝕙g[[10]];
                                    │ 𝕙g[[8]] │ 𝕙g[[9]] │ 𝕙g[[3]] │
                                    └─────────┴─────────┴─────────┘
  𝔽e⊢𝔽.𝔽pinv;
  SMSFreeze[ℂe,𝔽eᵀ.𝔽e,"Symmetric"→True];JF2⊢Det[ℂe];
  {Em,ν,Y0,Hh}=𝕕B[[1;;4]];{λ,μ}⊢SMSHookeToLame[Em,ν];
  W⊢Simplify[μ/2 (Tr[ℂe]-3-Log[JF2])+λ/4 (JF2-1-Log[JF2])];
  If[task=="W",Return[W]];
  SMSFreeze[τ,2 𝔽e.SMSD[W,ℂe,"Symmetric"→True].𝔽eᵀ,"Symmetric"→True];
  s⊢τ-1/3 IdentityMatrix[3] Tr[τ];
  fg⊢SMSSqrt[Tr[s.s]]-Sqrt[2/3] (Y0+Hh αh);
  If[task=="f",Return[fg]];
  𝕟⊢SMSD[fg,τ,"Symmetric"→True];αhn=𝕙gnIO[[10]];
                                    ┌─────────────┬─────────────┬─────────────┐
                                    │ 𝕙gnIO[[1]]  │ 𝕙gnIO[[4]]  │ 𝕙gnIO[[5]]  │
  𝔽pinvn=IdentityMatrix[3]+         │ 𝕙gnIO[[7]]  │ 𝕙gnIO[[2]]  │ 𝕙gnIO[[6]]  │;
                                    │ 𝕙gnIO[[8]]  │ 𝕙gnIO[[9]]  │ 𝕙gnIO[[3]]  │
                                    └─────────────┴─────────────┴─────────────┘
  ℚF=Simplify[𝔽e-SMSMatrixExp[-(αh-αhn) Sqrt[3/2] 𝕟,1].𝔽.𝔽pinvn];
  If[task=="Q",
   Return[Append[{ℚF[[1,1]],ℚF[[2,2]],ℚF[[3,3]],ℚF[[1,2]],ℚF[[1,3]],ℚF[[2,3]],
      ℚF[[2,1]],ℚF[[3,1]],ℚF[[3,2]]},fg]]];
 ]
```

Box 5.5. *AceGen* input for elasto-plastic auxiliary function fQW for three-dimensional finite strain material model with linear isotropic hardening.

The derivation is here described for a three-dimensional case. The vector of the variables at Gauss point level contains the components of the plastic strains $\boldsymbol{F}^{p-1}-\mathbf{1}$

at time $t = t_{n+1}$ and the equivalent plastic strain $\hat{\alpha}$ at time $t = t_{n+1}$

$$\mathbf{h}_g = \{ F_{11}^{p-1} - 1,\ F_{22}^{p-1} - 1,\ F_{33}^{p-1} - 1,\ F_{12}^{p},\ F_{13}^{p},\ F_{23}^{p},\ F_{21}^{p},\ F_{31}^{p},\ F_{32}^{p},\ \hat{\alpha} \}^T \tag{5.127}$$

Traditionally all the unknowns are set automatically to zero at the start of the finite element analysis. The unity matrix is subtracted from $\boldsymbol{F}^{p-1}$ with purpose to start with all unknowns having zero value at time t_0. Otherwise additional initialization of the diagonal terms of $\boldsymbol{F}^{p-1}$ would be required.

It is worth noting that inaccurate integration of the plastic evolution equations, see e.g. (5.120), leads to a loss of volume in case of incompressible plasticity and furthermore to a non-symmetric global tangent matrix, even in case of isotropic plasticity, Hence objectivity of the resulting finite element is lost. These problems can be avoided by an exact exponential approximation of the evolution of the plastic deformation gradient, see e.g. Simo (1998). An exponential approximation requires a reliable evaluation of the matrix exponential and its derivatives which has proven to be a difficult task. Thus a numerical approximation is often used instead, see e.g. Itskov (2003) and Lu (2004). The *AceGen* function `SMSMatrixExp[M,1]` returns an exact, closed form solution of a matrix exponent and its first derivative obtained by automatic differentiation of an appropriate scalar function as presented in Korelc and Stupkiewicz (2014), Hudobivnik and Korelc (2016). The automatically generated closed form solution of a matrix exponential yields accurate results up to machine precision. This is also the case for multiple eigenvalues and hence significantly improves the reliability of the finite strain plasticity formulation.

The `fQW` function defined in Box 7.5 returns, depending on the value of the input parameter `task`, the yield condition `f`, the strain energy function `W` or the evolution equations `Qg` at a Gauss point. They are evaluated for the given values of the variables $\mathbb{h}$ at Gauss point level. Thus, the function returns, with respect to the supplied parameters, either trial, iterative or converged values.

References

Argyris, J.H., and M. Kleiber. 1977. Incremental formulation in nonlinear mechanics and large strain elasto-plasticity- Natural approach. I. *Computer Methods in Applied Mechanics and Engineering* 11: 215–247.

Argyris, J.H., J.S. Doltsinis, P.M. Pimenta, and H. Wuestenberg. 1982. Thermomechanical response of solids at high strains- natural approach. *Computer Methods in Applied Mechanics and Engineering* 32: 3–57.

Atluri, S.N. 1984. On constitutive relations at finite strain: hypo-elasticity and elasto-plasticity with isotropic and kinematic hardening. *Computer Methods in Applied Mechanics and Engineering* 43: 137–171.

Bertram, A. 1989. *Axiomatische Einführung in die Kontinuumsmechanik*. Mannheim, Wien, Zürich: BI-Wissenschaftsverlag.

Besseling, J.F., and E. van der Giessen. 1994. *Mathematical modelling of inelastic deformations*. London: Chapman & Hall.

Ciarlet, P.G. 1988. *Mathematical elasticity I: three-dimensional elasticity*. Amsterdam: North-Holland.

Dafalias, Y.F. 1985. The plastic spin. *Journal of Applied Mechanics* 52: 865–871.

de Souza Neto, E.A., D. Peric, and D.R.J. Owen. 2008. *Computational methods for plasticity, theory and applications*. Chichester: Wiley.

Desai, C.S., and H.J. Siriwardane. 1984. *Constitutive laws for engineering materials*. Englewood Cliffs: Prentice-Hall.

Doll, S., and K. Schweizerhof. 2000. On the Development of Volumetric Strain Energy Functions. *Transactions of the ASME - E - Journal of Applied Mechanics* 67: 17–21.

Drucker, D.C., and W. Prager. 1952. Soil mechanics and plastic analysis or limit design. *Quarterly Applied Mathametics* 10: 157–165.

Ekh, M., A. Johansson, H. Thorberntsson, and B. Josefson. 2000. Models for Cyclic Ratchetting PlasticityIntegration and Calibration. *Journal of Engineering Materials and Technology* 122: 49.

Gear, C.W. 1971. *Numerical initial value problems in ordinary differential equations*. Englewood Cliffs: Prentice-Hall.

Hill, R. 1950. *The mathematical theory of plasticity*. Oxford: Clarendon Press.

Hofstetter, G., and H.A. Mang. 1995. *Computational Mechanics of Reinforced Concrete Structures*. Berlin: Vieweg.

Hudobivnik, B., and J. Korelc. 2016. Closed-form representation of matrix functions in the formulation of nonlinear material models. *Finite Elements in Analysis and Design* 111: 19–32.

Itskov, M. 2003. Computation of the exponential and other isotropic tensor functions and their derivatives. *Computer Methods in Applied Mechanics and Engineering* 192: 3985–3999.

Johansson, G., M. Ekh, and K. Runesson. 2005. Computational modeling of inelastic large ratcheting strains. *International Journal of Plasticity* 21(5): 955–980.

Khan, A.S., and S. Huang. 1995. *Continuum theory of plasticity*. Chichester: Wiley.

Korelc, J. 2009. Automation of primal and sensitivity analysis of transient coupled problems. *Computational Mechanics* 44: 631–649.

Korelc, J., and S. Stupkiewicz. 2014. Closed-form matrix exponential and its application in finite-strain plasticity. *International Journal for Numerical Methods in Engineering* 98: 960–987.

Lee, E.H., and D.T. Liu. 1967. Finite Strain Elasto-Plastic Theory with Applications to Plane-Wave Analysis. *Journal of Applied Mechanics* 38: 19–27.

Lu, J. 2004. Exact expansions of arbitrary tensor functions and their derivatives. *International Journal of Solids & Structures* 41: 337–349.

Lubliner, J. 1985. A model of rubber viscoelasticity. *Mechancis Research Communications* 12: 93–99.

Lubliner, J. 1990. *Plasticity Theory*. London: MacMillan.

Luenberger, D.G. 1984. *Linear and nonlinear programming*, 2nd ed. Reading: Addison-Wesley.

Malvern, L.E. 1969. *Introduction to the mechanics of a continuous medium*. Englewood Cliffs: Prentice-Hall.

Mandel, J. 1974. Thermodynamics and plasticity. In *Foundations of continuum thermodynamics*, ed. J.J. Delgado Domingers, N.R. Nina, and J.H. Whitelaw, 283–304. London: MacMillan.

Marsden, J.E., and T.J.R. Hughes. 1983. *Mathematical foundations of elasticity*. Englewood Cliffs: Prentice-Hall.

Miehe C. 1993. Kanonische modelle multiplikativer elasto-plastizität. thermodynamische formulierung und numerische implementation. *Technical Report F 93/1*, Forschungs- und Seminarberichte aus dem Bereich der Mechanik der Universität Hannover.

Miehe, C. 1994. Aspects of the Formulation and Finite Element Implementation of Large Strain Isotropic Elasticity. *International Journal for Numerical Methods in Engineering* 37: 1981–2004.

Moreau, J.J. 1976. Application of convex analysis to the treatment of elastoplastic strucures. In *Applications of methods of functional analysis to problems in mechanics*, ed. P. Germain, and B. Nayroles. Berlin: Springer.

Nagtegaal, J.C. 1982. On the implementation of inelastic constitutive equations with special reference to large deformation problems. *Computer Methods in Applied Mechanics and Engineering* 33(1–3): 469–484.

Nagtegaal, J.C., D.M. Parks, and J.C. Rice. 1990. On numerically accurate finite element solutions in the fully plastic range. *Computer Methods in Applied Mechanics and Engineering* 4: 153–177.

Ogden, R.W. 1972. Large deformation isotropic elasticity: on the correlation of theory and experiment for incompressible rubberlike solids. *Proceedings of the Royal Society of London* 326: 565–584.

Ogden, R.W. 1984. *Non-linear elastic deformations*. Chichester: Ellis Horwood and John Wiley.

Oliver, J., A.E. Huespe, S. Blanco, and D.L. Linero. 2006. Stability and robustness issues in numerical modeling of material failure with the strong discontinuity approach. *Computer Methods in Applied Mechanics and Engineering* 195: 7093–7114.

Peric, D., D.R.J. Owen, and M.E. Honnor. 1992. A model for finite strain elasto-plasticity based on logarithmic strains: computational issues. *Computer Methods in Applied Mechanics and Engineering* 94(1): 35–61.

Peric, D., and D.R.J. Owen. 1997. Finite-element applications to the nonlinear mechanics of solids. *Reports on Progress in Physics* 61: 1495–1574.

Prager, W. 1955. *Probleme der Plastizitätstheorie*. Basel, Stuttgart: Birkäuser.

Schröder, J. 2010. Anisotropic polyconvex energies. In *Polyconvex analysis*, vol. 62, ed. J. Schröder, 1–53., CISM Wien: Springer.

Schröder, J., P. Neff, and V. Ebbing. 2008. Anisotropic polyconvex energies on the basis of crystallographic motivated structural tensors. *Journal of the Mechanics and Physics of Solids* 56: 3486–3506.

Sheng, D., and S.W. Sloan. 2001. Load stepping schemes for critical state models. *International Journal for Numerical Methods in Engineering* 50: 67–93.

Simo, J.C. 1988. A framework for finite strain elastoplasticity based on the multiplicative decomposition and hyperelastic relations. part ii: computational aspects. *Computer Methods in Applied Mechanics and Engineering* 67: 1–31.

Simo, J.C. 1992. Algorithms for static and dynamic multiplicative plasticity that preserve the classical return mapping schemes of the infinitesimal theory. *Computer Methods in Applied Mechanics and Engineering* 99: 61–112.

Simo, J.C. 1998. Numerical analysis and simulation of plasticity. In *Handbook of numerical analysis*, ed. P.G. Ciarlet, and J.L. Lions, 179–499. North-Holland: Elsevier.

Simo, J.C., and T.J.R. Hughes. 1998. *Computational inelasticity*. Berlin: Springer.

Simo, J.C., and C. Miehe. 1992. Associative coupled thermoplasticity at finite strains: formulation, numerical analysis and implementation. *Computer Methods in Applied Mechanics and Engineering* 98: 41–104.

Simo, J.C., and M. Ortiz. 1985. A Unified Approach to Finite Deformation Elastoplasticity Based on the Use of Hyperelastic Constitutive Equations. *Computer Methods in Applied Mechanics and Engineering* 49: 201–215.

Simo, J.C., and K.S. Pister. 1984. Remarks on rate constitutive equations for finite deformation problems. *Computer Methods in Applied Mechanics and Engineering* 46: 201–215.

Simo, J.C., and R.L. Taylor. 1985. Consistent tangent operators for rate-independent elastoplasticity. *Computer Methods in Applied Mechanics and Engineering* 48: 101–118.

Sloan, S.W. 1987. Substepping schemes for the numerical integration of elastoplastic stress-strain relations. *International Journal for Numerical Methods in Engineering* 24: 893–911.

Sloan, S.W., A.J. Abbo, and D. Sheng. 2001. Refined explicit integration of elastoplastic models with automatic error control. *Engineering Computations* 18: 121–154.

Stoer, J., and R. Bulirsch. 1990. *Numerische Mathematik 2*, dritte edition. Berlin: Springer.

Truesdell, C., and W. Noll. 1965. The nonlinear field theories of mechanics. In *Handbuch der Physik III/3*, ed. S. Flügge. Berlin: Springer.
Wriggers, P. 2008. *Nonlinear finite elements*. Berlin: Springer.
Xiao, H., O. Bruhns, and A. Meyers. 1997. Hypo-elasticity model based upon the logarithmic stress rate. *Journal of Elasticity* 47(1): 51–68.

Chapter 6
Continuum Elements

After a short introduction in the main targets of element development for continuum elements this chapter will consider different approaches ranging from standard up to special formulations. For all finite elements the associated *AceGen* input is provided such that the reader can see the effort needed to generate the different finite element formulations.

6.1 Requirements for Continuum Finite Elements

Ideally finite elements in solids mechanics should be applicable to many different problem classes. To achieve such goal elements have to fulfill the following requirements.

1. *Locking* free behaviour for incompressible materials,
2. good bending performance and *locking* free thin elements,
3. no sensitivity against mesh distortions,
4. good coarse mesh accuracy,
5. simple implementation of nonlinear constitutive equations and
6. efficiency (e.g. few necessary integration points).

The first point concerns special problem classes that include rubber like materials and elasto-plastic material equations in the framework of J_2-plasticity. Elements and related *AceGen* inputs for incompressibility will be described in Sects. 6.4 and 6.5.

The second point relates to three-dimensional elements that can be applied to solve beam- or shell problems. As it is long known from linear finite element analysis standard solid elements tend to lock. Thus the pure displacement element with bi- or tri-linear ansatz function has bad convergence behaviour in bending problems. Special elements were developed for the analysis of thin solids, see e.g. Hauptmann and Schweizerhof (1998).

The third point is essential when modern methods for mesh generation are employed. These methods lead for arbitrary geometries to so called unstructured

J. Korelc and P. Wriggers, *Automation of Finite Element Methods*,
DOI 10.1007/978-3-319-39005-5_6

meshes that consist of finite elements shapes with arbitrary geometry. Within this respect it is especially in three-dimensional application of concern to develop accurate and efficient tetrahedral elements since these are easier to generate for arbitrary geometries than hexahedral elements.

The fourth point is related to the fact that in real engineering applications often three-dimensional components have to be analyzed which size and complexity cannot be modeled using a converged mesh. Thus coarse mesh accuracy is important, especially when the simulation is nonlinear.[1]

The fifth point is directly related to the demand to use more accurate mathematical and physical models for constitutive equations in the simulation of nonlinear engineering structures and components. Within this process new complex nonlinear constitutive equations have to be implemented. Here a simple interface to the finite element, as provided in *AceGen*, should support the user in order to efficiently change existing finite elements and to be able to implement new complex constitutive equations.

The sixth point concerns efficiency of the element formulation. With increasing speed of the equation solvers the computing time for an element plays a essential role in the overall efficiency of a finite element simulations. This is especially true for simulation of dynamical or of fatigue behavior. Here *AceGen* provides with its code optimizing tools an ideal platform for the generation of efficient elements.

Different formulations have been developed in order to construct finite elements that fulfill requirements stated above. These are

- techniques which base on an reduced integration of the integrals leading to the element matrices,
- stabilization methods, see e.g. Belytschko et al. (1984),
- hybrid or mixed variational principles that base on complimentary energy written in terms of the stress field,
- mixed variational principle of Hu-Washizu type,
- mixed variational principle for rotational fields,
- nodally based elements, see e.g. Puso and Solberg (2006),
- composite or macro formulations, see e.g. Guo et al. (2000),
- higher order displacement elements of Lagrange type, see e.g. Düster et al. (2003),
- NURBS based isogeometric elements, see e.g. Hughes et al. (2005) and
- formulations based on the Cosserat point theory, see Nadler and Rubin (2003).

Some of the formulations are considered in the next chapters. However the discussion of all these methodologies is beyond the scope of this book. An in depth discussion of these different element formulations can be found in the cited papers or books on the finite element method, e.g. Hughes (1987) for small strain elements and in Zienkiewicz and Taylor (2000), Bathe (1996), Belytschko et al. (2000) and Wriggers (2008) for large strain applications.

[1] The importance of this point will diminish with the increasing computing power, but at the moment it is still of concern.

6.2 Two-Dimensional Elements

A higher order displacement element with a cubic ansatz function is developed in this section for plane strain hyper elastic deformations. Then it is shown how to extend the formulation for axisymmetric elements and finally a special two dimensional element concerning deformation dependent load is formulated. All formulations are implemented by using *AceGen*.

6.2.1 *Hyperelastic Triangular Element*

Higher order elements can be produced with *AceGen* in a similar manner like lower order elements. They actually can be generated using the concept of additional nodes that starts from the geometry of the linear triangular element and then adds as many additional nodes as needed to obtain the correct interpolation within the element. Here this concept is applied to a cubic triangular element that has in total 10 nodes.

Starting from the linear triangle with three nodes the additional 6 nodes at the element edges and their coordinates are generated as follows, see Fig. 6.1

$$
\begin{aligned}
&\mathbf{X}_4 = \mathbf{X}_1 + \frac{1}{3}\left(\mathbf{X}_2 - \mathbf{X}_1\right), \quad \mathbf{X}_5 = \mathbf{X}_1 + \frac{2}{3}\left(\mathbf{X}_2 - \mathbf{X}_1\right), \\
&\mathbf{X}_6 = \mathbf{X}_2 + \frac{1}{3}\left(\mathbf{X}_3 - \mathbf{X}_2\right) \quad \text{etc.}
\end{aligned}
$$

The mid node is then obtained from

$$
\mathbf{X}_{10} = \frac{1}{3}\left(\mathbf{X}_1 + \mathbf{X}_2 + \mathbf{X}_3\right) \tag{6.1}
$$

The shape function of the element are given with $\kappa = 1 - \xi - \eta$ by

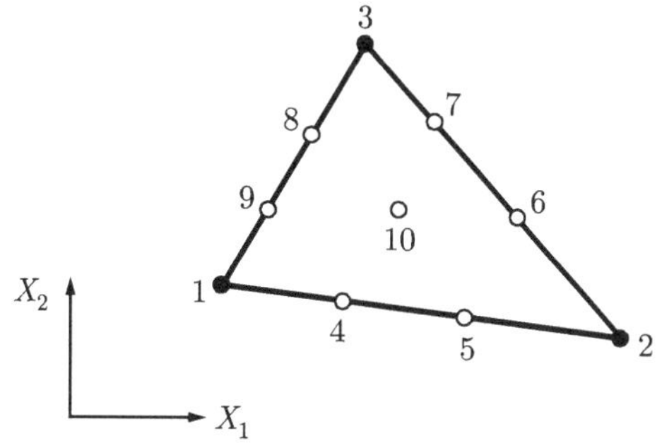

Fig. 6.1 Vertices and mid nodes at the edges and element for the triangular element with cubic shape functions

$$\begin{array}{ll} N_1 = \frac{1}{2}\,\kappa\,(3\,\kappa-1)(3\,\kappa-2) & N_2 = \frac{1}{2}\,\xi\,(3\,\xi-1)(3\,\xi-2) \\ N_3 = \frac{1}{2}\,\eta\,(3\,\eta-1)(3\,\eta-2) & N_4 = \frac{9}{2}\,\kappa\,\xi\,(3\,\kappa-1) \\ N_5 = \frac{9}{2}\,\kappa\,\xi\,(3\,\xi-1) & N_6 = \frac{9}{2}\,\eta\,\xi\,(3\,\xi-1) \\ N_7 = \frac{9}{2}\,\eta\,\xi\,(3\,\eta-1) & N_8 = \frac{9}{2}\,\kappa\,\eta\,(3\,\eta-1) \\ N_9 = \frac{9}{2}\,\kappa\,\eta\,(3\,\kappa-1) & N_{10} = 27\,\eta\,\xi\,\kappa \end{array} \tag{6.2}$$

Thus the components of the displacement field within an element Ω_e are approximated by

$$u = \sum_{I=1}^{10} N_I(\xi\,,\eta)\,u_I \quad \text{and} \quad v = \sum_{I=1}^{10} N_I(\xi\,,\eta)\,v_I \tag{6.3}$$

The deformation gradient and its determinant is then given by

$$\boldsymbol{F} = \mathbf{1} + \text{Grad}\,\boldsymbol{u} \qquad J_F = \det \boldsymbol{F} \tag{6.4}$$

where the gradient with respect to the coordinates **X** is computed by using the isoparametric map

$$X = \sum_{I=1}^{10} N_I(\xi\,,\eta)\,X_I \quad \text{and} \quad Y = \sum_{I=1}^{10} N_I(\xi\,,\eta)\,Y_I. \tag{6.5}$$

```
<<"AceGen`"
SMSInitialize["T3T1-Int","Environment"→"AceFEM"];
SMSTemplate["SMSTopology"→"T1","SMSNoNodes" → 10,
  "SMSAdditionalNodes"→
   Function[{X1,X2,X3},{X1+1/3 (X2-X1),X1+2/3 (X2-X1),
     X2+1/3 (X3-X2),X2+2/3 (X3-X2),X3+1/3 (X1-X3),X3+2/3 (X1-X3),
     (X1+X2+X3)/3}],
  "SMSSegments"→{{1,4,5,2,6,7,3,8,9}},
  "SMSSegmentsTriangulation"→
   {{{1,4,9},{4,5,10},{5,2,6},{4,10,9},{5,6,10},{9,10,8},
     {10,7,8},{10,6,7},{8,7,3}}},"SMSSymmetricTangent"→True,
  "SMSDomainDataNames"->{"λ","μ"},"SMSDefaultData"->{1,0}
 ];
nen=SMSNoNodes;ndim=SMSNoDimensions;np=SMSNoDOFGlobal;
(* Constants for 12 point integration (6^th order)*)
{a,b,c,d,e,f,g,pw,qw,rw}=
  SetPrecision[{0.873821971016996,0.063089014491502,
    0.501426509658179,0.249286745170910,0.636502499121399,
    0.310352451033785,0.053145049844816,0.050844906370207,
    0.116786275726379,0.082851075618374},SMSEvaluatePrecision];
ξall ={a,b,b,c,d,d,e,e,f,g,f,g};
ηall ={b,a,b,d,c,d,f,g,e,e,g,f};
wgpall={pw,pw,pw,qw,qw,qw,rw,rw,rw,rw,rw,rw};
```

Box 6.1. Initialization part of *AceGen* input for the cubic triangular plain strain element

For a strain energy that uses the split into volumetric and deviatoric strains the multiplicative split (1.19)

$$\hat{\boldsymbol{F}} = J_F^{-\frac{1}{3}} \boldsymbol{F} \tag{6.6}$$

has to be used. With this split the right Cauchy Green tensor can be written as

$$\hat{\boldsymbol{C}} = \hat{\boldsymbol{F}}^T \hat{\boldsymbol{F}} = J_F^{-\frac{2}{3}} \boldsymbol{C} \tag{6.7}$$

Now the strain energy is formulated in terms of the volumetric and deviatoric kinematical variables as

$$W(\hat{\boldsymbol{C}}\,, J_F) = \frac{\Lambda}{2}\left[\frac{1}{2}(J_F^2 - 1) - \ln J_F\right] + \mu \frac{1}{2}\,(I_{\hat{C}} - 3). \tag{6.8}$$

The internal residual of the finite element is now obtained using *AceGen*. All quantities in (6.8) depend upon the displacement field leading to the residual

$$\mathbf{R}_e = \int_{\Omega_e} \frac{\partial W[\hat{\boldsymbol{C}}(\mathbf{p}_e), J_F(\mathbf{p}_e)]}{\partial \mathbf{p}_e}\, d\Omega. \tag{6.9}$$

where $\mathbf{p}_e$ is the unknown vector related to the element which contains all nodal displacement vectors: $\mathbf{p}_e^T = \{u_1\,, v_1\,, u_2\,, v_2\,, \ldots\,, u_{10}\,, v_{10}\}$. This yields the element residual, as shown in Chap. 4. The total size of $\mathbf{R}_e$ is related to the interpolation order, and in total has for the 10 node element 20 entries. Note that the integration has to be performed with respect to the initial coordinates, thus the Jacobi matrix $\mathbf{J}_e$ has to be used to transform from the initial state to the reference state, see Fig. 4.6. This leads to the numerical integration (Weighting w_{gp}, integration points $\xi_g\,, \eta_g$)

$$\mathbf{R}_e \approx \sum_{g=1}^{n_g} w_{gp} \frac{\delta W[\hat{\boldsymbol{C}}(\xi_g\,, \eta_g)(\mathbf{p}_e), J_F(\xi_g\,, \eta_g)(\mathbf{p}_e)]}{\delta \mathbf{p}_e} J_e(\xi_g\,, \eta_g) = \sum_{g=1}^{n_g} w_{gp}\, \mathbf{R}_g \tag{6.10}$$

Since the element has a cubical interpolation a 12 point integration rule has to be applied that is of 6th order. Since the 12 point integration is not implemented among the standard integration rules in *AceGen* the coordinates and weighting factors have to be explicitly stated in the code as can be seen in the lower part of Box 6.1. Box 6.1 also contains initialization of *AceGen* and template constants where additionally to the standard set of constants (described in Sect. 2.7) we need to specify:

`"SMSNoNodes"` constants has to be given when the element has more nodes than the number of nodes of the element with the choosen topology (in our case `"T1"` topology defines triangular element with three nodes)

`"SMSAdditionalNodes"` constant specifies pure function that returns additional nodes. Arguments of the function are the coordinates of topological nodes given as mesh input data.

(for example `Function[{X1,X2,X3},{(X1+X2)/2}]` adds one additional node in the middle of nodes 1 and 2).

`"SMSSegments"` constant is used for visualization and defines the sequence of the element node indices that define the edge of the segment.

`"SMSSegmentsTriangulation"` constant is also used for visualization and specifies how to split an arbitrary shaped elements into a set of triangular or quadrilateral segments.

```
SMSStandardModule["Tangent and residual"];
{λ,μ}⊢SMSReal[Table[es$$["Data",i],{i,2}]];
XIO⊢Table[SMSReal[nd$$[i,"X",j]],{i,nen},{j,ndim}];
uIO⊢SMSReal[Table[nd$$[i,"at",j],{i,nen},{j,ndim}]];pe=Flatten[uIO];
SMSDo[Ig,1,12];
Ξ={ξ,η}⊢SMSReal[{SMSPart[ξall,Ig],SMSPart[ηall,Ig]}];
wgp⊢SMSPart[wgpall,Ig];κ⊨1-ξ-η;
Nh⊨{κ (3 κ-1) (3 κ-2)/2,ξ (3 ξ-1) (3 ξ-2)/2,η (3 η-1) (3 η-2)/2,
   9 κ ξ (3 κ-1)/2, 9 κ ξ (3 ξ-1)/2, 9 η ξ (3 ξ-1)/2, 9 η ξ (3 η-1)/2,
   9 κ η (3 η-1)/2, 9 κ η (3 κ-1)/2,27 η ξ κ};
X⊢SMSFreeze[Nh.XIO];u⊨Nh.uIO;Je⊨SMSD[X,Ξ];
Jed⊨Det[Je];H⊨SMSD[u,X,"Dependency"→{Ξ,X,SMSInverse[Je]}];
   ⎛1+H⟦1,1⟧   H⟦1,2⟧   0⎞
F⊨ ⎜ H⟦2,1⟧  1+H⟦2,2⟧  0⎟;JF⊨Det[F];
   ⎝   0         0     1⎠
FISO⊨JF^(-1/3) F;CISO⊨FISOᵀ.FISO;
W⊨λ/2 (1/2 (JF^2-1)-Log[JF])+μ/2 (Tr[CISO]-3);
SMSDo[m,1,np];
  Rgm⊨Jed SMSD[W,pe,m];
  SMSExport[wgp Rgm,p$$[m],"AddIn"→True];
  SMSDo[n,m,np];
    Kgmn⊨SMSD[Rgm,pe,n];
    SMSExport[wgp Kgmn,s$$[m,n],"AddIn"→True];
  SMSEndDo[];
SMSEndDo[];
SMSEndDo[];
SMSWrite[];
```

Box 6.2. `"Tangent and residual"` subroutine part of *AceGen* input for the cubic triangular plain strain element

Differentiation of the residual in (6.10) with respect to the element displacements $\mathbf{p}_e$ yields the tangent stiffness matrix

$$\mathbf{K}_e \approx \sum_{g=1}^{n_g} w_{gp} \frac{\hat{\delta}\mathbf{R}_g}{\hat{\delta}\mathbf{p}_e}. \tag{6.11}$$

The second part of the *AceGen* input that concerns the numerical integration loop with the definition of the shape functions, kinematics, strain energy function and tangent matrix and residual can be found for the above element formulation in Box 6.2.

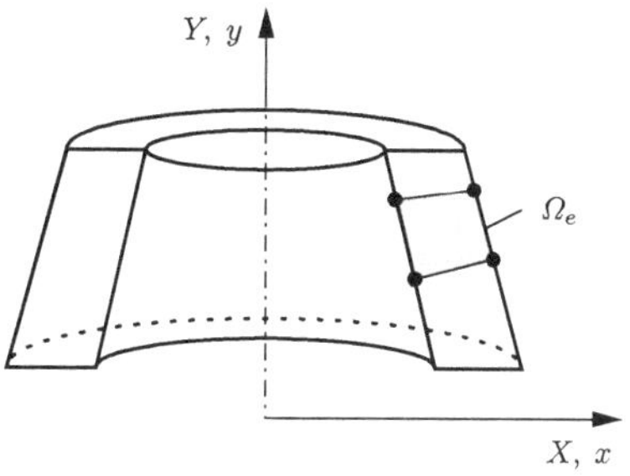

Fig. 6.2 Axisymmetric finite 4-node element

6.2.2 *Axisymmetric Element*

In this section the matrix formulation for an axisymmetric finite element undergoing finite elastic deformations is derived with respect to the initial configuration. The constitutive behaviour will be described by the compressible Neo-Hooke material as given in the previous sections. For the approximation of geometry and deformation either bi-linear or bi-quadratic shape functions are selected.

For an axisymmetric problem the geometry of a structure under consideration as well as the loading has to be axisymmetric, see Fig. 6.2. It is assumed that the axis of symmetry coincides with coordinate Y.

The strain energy will be the starting point for the development of the finite element. For the computation of the residuum the strain energy in (1.98) is applied. Its internal energy part can be written for an element Ω_e in the initial configuration (Ω_e) as

$$\int_{B_e} W(\boldsymbol{C})\, dV = 2\pi \int_{\Omega_e} W(\boldsymbol{C})\, R\, d\Omega. \tag{6.12}$$

where the volume element is now described by $dV = 2\pi\, R\, dA$ with R being the radius of the circumferential shape of the axisymmetric volume.

The two ingredients needed in this formulation are W and the right Cauchy–Green tensor $\boldsymbol{C}$. The latter has to be computed from the displacement field using the selected finite element ansatz. Here the Neo-Hooke strain energy (5.10) with the parameters related to (5.11) is used, leading to

$$W(\boldsymbol{C}) = \frac{\lambda}{4}(J_F^2 - 1) - \frac{\lambda}{2} \ln J_F - \mu \ln J_F + \frac{\mu}{2}(\text{tr}\, \boldsymbol{C} - 3). \tag{6.13}$$

Since the Jacobian J_F can be written as $J_F = \det \boldsymbol{F} = \sqrt{\det \boldsymbol{C}}$ and the first invariant as $I_C = \text{tr}\, \boldsymbol{C}$, all terms in (6.13) are known as functions of $\boldsymbol{C}$ in Ω_e.

Now the right Cauchy–Green tensor has to be computed from the displacement interpolation within the element Ω_e. Here isoparametric interpolation is selected for a quadrilateral and higher order elements for displacements and coordinates, see Sect. 4.1.2. The components of the right Cauchy Green strain tensor, needed in (6.13), follow in the axisymmetric case for a finite element Ω_e as

$$
\boldsymbol{C} = \boldsymbol{F}^T \boldsymbol{F} \quad \text{with} \quad \boldsymbol{F} = \begin{bmatrix} \frac{\partial x}{\partial X} & \frac{\partial x}{\partial Y} & 0 \\ \frac{\partial y}{\partial X} & \frac{\partial y}{\partial Y} & 0 \\ 0 & 0 & \frac{x}{X} \end{bmatrix} . \tag{6.14}
$$

where the deformation gradient $\boldsymbol{F}$ is computed using the classical coordinate transformations related to the isoparametric formulations. The components of the nodal coordinates ($x_I = X_I + u_I$ and $y_I = Y_I + v_I$) are related to the current configuration $\bar{\varphi}_e$. Furthermore the radius is given by the initial coordinate in X-direction, see Fig. 6.2. Thus it is interpolated as

$$
R = X = \sum_{I=1}^{n} N_I(\xi\,,\eta)\, X_I .
$$

The internal residual of the finite element is now obtained using *AceGen* based on (6.13). Since all quantities depend upon the displacement field the residual is given, as shown in the previous sections, by

$$
\mathbf{R}_e = 2\pi \int_{\Omega_e} \frac{\hat{\delta} W(\boldsymbol{C}(\mathbf{p}_e))}{\hat{\delta}\mathbf{p}_e} \, R \, d\Omega \tag{6.15}
$$

where $\mathbf{p}_e$ is the unknown vector related to the element which contains all nodal displacement vectors: $\mathbf{p}_e^T = \{ u_1\,, v_1\,, u_2\,, v_2\,, \dots\,, u_{n_{en}}\,, v_{n_{en}} \}$.

The total size of $\mathbf{R}_e$ is related to the interpolation order, and in total has $2\,n_{en}$ entries. Note that the integration has to be performed with respect to the initial coordinates, thus the Jacobi matrix $\mathbf{J}_e$ has to be applied to transform from the initial state to the reference state, see Fig. 6.2. This leads by using the Gauss integration (weighting w_{gp}, integration points $\xi_g\,, \eta_g$ and $J_e = \frac{dV}{d\Box} = 2\pi\, R(\xi_g\,, \eta_g)\; \det\, \mathbf{J}_e(\xi_g\,, \eta_g))$ to

$$
\begin{aligned}
\mathbf{R}_e &\approx \sum_{g=1}^{n_g} w_{gp}\, J_e(\xi_g, \eta_g) \frac{\hat{\delta} W(\boldsymbol{C}(\mathbf{p}_e(\xi_g, \eta_g)))}{\hat{\delta}\mathbf{p}_e} \\
&= \sum_{g=1}^{n_g} w_{gp}\, \mathbf{R}_g(\mathbf{p}_e, \xi_g, \eta_g)
\end{aligned} \tag{6.16}
$$

Differentiation of this term with respect to the element displacements $\mathbf{p}_e$ yields the Gauss point tangent stiffness matrix

$$
\mathbf{K}_g = \frac{\hat{\delta}\mathbf{R}_g(\mathbf{p}_e, \xi_g, \eta_g)}{\hat{\delta}\mathbf{p}_g} . \tag{6.17}
$$

The *AceGen* input for this element formulation can be found in Box 6.3.

6.2.3 Deformation Dependent Loads

Applied loads can depend upon the deformation in some technical problems. On one hand the load direction can change (in such case also the term *follower loads* is used) on the other hand the magnitude of the load can de- or increase. As an example loading of structures due to fluids or wind can be mentioned. Interaction of structural systems with fluids and gas are special engineering application of high relevance. A related treatment can be found in Hassler and Schweizerhof (2008). An in depth discussion of the algorithmic treatment of deformation dependent loads and their discretization can be found in Schweizerhof (1982), Schweizerhof and Ramm (1984) for the general case and in Simo et al. (1991) and Yosibash et al. (2007) for axisymmetric deformations, while the latter paper discusses discretization using high order finite element interpolations.

These follower loads act at the surface of a solid. They possess a residual vector and a tangent matrix due to their deformation dependency. Thus they can be discretized and assembled like standard finite elements.

```
SMSInitialize["Q1-rot-symm","Environment"→"AceFEM"];
SMSTemplate["SMSTopology"→"Q1","SMSSymmetricTangent"→True,
  "SMSDomainDataNames"->{"E","ν"},"SMSDefaultData"->{21000,0.3}];
nen=SMSNoNodes;ndim=SMSNoDimensions;np=SMSNoDOFGlobal;
SMSStandardModule["Tangent and residual"];
{Em,ν}⊢SMSReal[Table[es$$["Data",i],{i,2}]];{λ,μ}⊨SMSHookeToLame[Em,ν];
𝕏IO⊢Table[SMSReal[nd$$[i,"X",j]],{i,nen},{j,ndim}];
𝕦IO⊢SMSReal[Table[nd$$[i,"at",j],{i,nen},{j,ndim}]];𝕡e=Flatten[𝕦IO];
SMSDo[Ig,1,SMSInteger[es$$["id","NoIntPoints"]]];
Ξ={ξ,η}⊢Table[SMSReal[es$$["IntPoints",i,Ig]],{i,2}];
wgp⊢SMSReal[es$$["IntPoints",4,Ig]];
ℕh⊨1/4 {(1-ξ) (1-η),(1+ξ) (1-η),(1+ξ) (1+η),(1-ξ) (1+η)};
𝕏⊢SMSFreeze[ℕh.𝕏IO];𝕦⊨ℕh.𝕦IO;Je⊨SMSD[𝕏,Ξ];Jed⊨2 π 𝕏⟦1⟧ Det[Je];
ℍ⊨SMSD[𝕦,𝕏,"Dependency"→{Ξ,𝕏,SMSInverse[Je]}];
    ⎛1+ℍ⟦1,1⟧   ℍ⟦1,2⟧      0      ⎞
𝔽⊨  ⎜ ℍ⟦2,1⟧   1+ℍ⟦2,2⟧     0      ⎟ ;JF⊨Det[𝔽];ℂt⊨𝔽ᵀ.𝔽;
    ⎝   0         0      1+𝕦⟦1⟧/𝕏⟦1⟧ ⎠
W⊨λ/4 (JF^2-1)-λ/2 Log[JF]-μ Log[JF]+1/2 μ (Tr[ℂt]-3);
SMSDo[m,1,np];
  Rgm⊨Jed SMSD[W,𝕡e,m];
  SMSExport[wgp Rgm,p$$[m],"AddIn"→True];
  SMSDo[n,m,np];
    Kgmn⊨SMSD[Rgm,𝕡e,n];
    SMSExport[wgp Kgmn,s$$[m,n],"AddIn"→True];
  SMSEndDo[];
SMSEndDo[];
SMSEndDo[];
SMSWrite[];
```

Box 6.3. *AceGen* input for axisymmetric element

In this section only loads which are direction depending are considered. In more detail loads which are always normal to the deformed surface of a solid or structure will be discussed. Thus the term describing the surface loads $\int_{\Gamma_\sigma} \boldsymbol{\eta} \cdot \bar{\boldsymbol{t}}\, dA$ will be considered where $\bar{\boldsymbol{t}} = \hat{p}\, \boldsymbol{n}$ is introduced. This leads to the surface load term

$$g_p(\boldsymbol{\varphi}\,,\boldsymbol{\eta}) = - \int\limits_{\varphi(\Gamma_\sigma)} \boldsymbol{\eta} \cdot \hat{p}\, \boldsymbol{n}\, da. \tag{6.18}$$

Here the integration has to be performed with respect to the current configuration. The normal vector $\boldsymbol{n}$ depends on the deformation. Note that this form is associated to the weak form for the description of solids. A potential form for this loading term is in general not available. However a pseudo potential $\Pi^P(\boldsymbol{\varphi})$ can be introduced as

$$\Pi^P(\boldsymbol{\varphi}) = - \int\limits_{\varphi(\Gamma_\sigma)} (\boldsymbol{\varphi} \cdot \hat{p}\, \boldsymbol{n})\, da. \tag{6.19}$$

Its variation yields—under the condition that $p\boldsymbol{n}$ is kept constant during the variation—the correct weak form (6.18).

The discretization of (6.19) can be obtained by introducing convective coordinates with base vectors $\boldsymbol{g}_\alpha$, see Appendix B.1.2. For a discretization of the surface of a three-dimensional finite element with quadratic shape functions the coordinates are depicted in Fig. 6.3.

The normal vector of the discretized surface is given in the spatial configuration by

$$\boldsymbol{n} = \frac{\boldsymbol{g}_1 \times \boldsymbol{g}_2}{\| \boldsymbol{g}_1 \times \boldsymbol{g}_2 \|}. \tag{6.20}$$

The base vectors are computed from the deformation field by a partial derivative with respect to the convective coordinates ξ and η, $\boldsymbol{g}_\alpha = \boldsymbol{\varphi}_{,\alpha}$, where $\alpha = 1$ stands for ξ and $\alpha = 2$ for η. Hence the normal vector can be described by

$$\boldsymbol{n} = \frac{\boldsymbol{\varphi},_\xi \times \boldsymbol{\varphi},_\eta}{\| \boldsymbol{\varphi},_\xi \times \boldsymbol{\varphi},_\eta \|}. \tag{6.21}$$

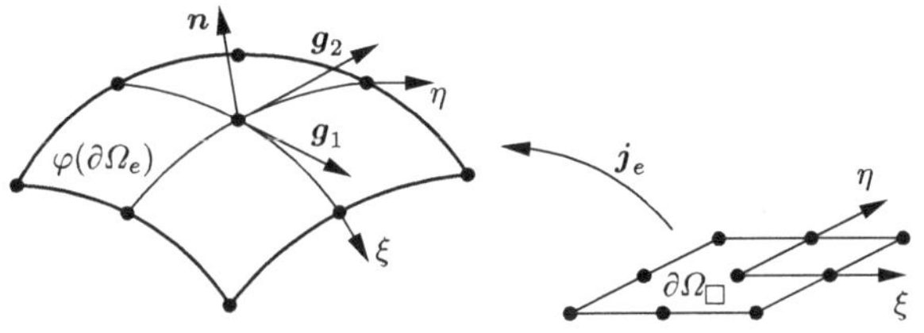

Fig. 6.3 Coordinate systems for deformation dependent loads

Since the area element da in (6.19) can be computed with respect to the reference configuration by $da = \| \varphi,_\xi \times \varphi,_\eta \| \, d\xi \, d\eta$ the pseudo potential Π^P can be transformed to the reference configuration of the loaded element surface, see right part of Fig. 6.3,

$$\Pi^P(\varphi) = -\int\limits_{\varphi(\Gamma_\sigma)} \varphi \cdot \hat{\boldsymbol{t}} \, da = -\mathop{\mathsf{A}}_{r=1}^{n_r} \int\limits_{\partial\Omega_{\square_r}} \varphi \cdot \hat{\boldsymbol{t}} \, d\xi \, d\eta. \tag{6.22}$$

with $\hat{\boldsymbol{t}} = \hat{p} \, \hat{\boldsymbol{n}} = \hat{p} \, (\boldsymbol{\varphi}_{,\xi} \times \boldsymbol{\varphi}_{,\eta})$ and where $\mathop{\mathsf{A}}_{r=1}^{n_r}$ denotes the assembly of the n_r loaded surfaces and $\partial\Omega_{\square_r}$ the surface of the reference element, see right part of Fig. 6.3. Based on this form the finite element discretization is straight forward. Using the isoparametric shape functions yields with $\varphi = \boldsymbol{x} = \{x\,, y\,, z\}^T$ to the components

$$x = \sum_{I=1}^{m} N_I(\xi, \eta)\, x_I \,, \; y = \sum_{I=1}^{m} N_I(\xi, \eta)\, y_I \quad \text{and} \quad z = \sum_{I=1}^{m} N_I(\xi, \eta)\, z_I \tag{6.23}$$

where $x_I = X_I + u_I$, $y_I = Y_I + v_I$ and $z_I = Z_I + w_I$ depend on the nodal displacements. The derivatives of the components of the position vector $\boldsymbol{x}$ are then given by

$$x_{,\alpha} = \sum_{I=1}^{m} N_I(\xi, \eta)_{,\alpha}\, x_I \,, \; y_{,\alpha} = \cdots \quad \text{and} \quad z_{,\alpha} = \sum_{I=1}^{m} N_I(\xi, \eta)_{,\alpha}\, z_I \tag{6.24}$$

for the loaded element surface. Now the cross product in (6.22) can be computed $\boldsymbol{\varphi}_{,\xi} \times \boldsymbol{\varphi}_{,\eta}$ with $\boldsymbol{\varphi}_{,\xi} = \{x_{,\xi}\,, y_{,\xi}\,, z_{,\xi}\}^T$ and $\boldsymbol{\varphi}_{,\eta} = \{x_{,\eta}\,, y_{,\eta}\,, z_{,\eta}\}^T$. Thus we can also write $\boldsymbol{\varphi}_{,\xi} \times \boldsymbol{\varphi}_{,\eta} = \boldsymbol{x}_{,\xi} \times \boldsymbol{x}_{,\eta} = \boldsymbol{g}_1 \times \boldsymbol{g}_2$. The discretization of (6.22) follows with these definitions and can be numerically integrated. This leads for one surface element with $\hat{\boldsymbol{t}}(\xi_g\,, \eta_g) = \hat{p}\,[\boldsymbol{x}_{,\xi}(\xi_g\,, \eta_g) \times \boldsymbol{x}_{,\eta}(\xi_g\,, \eta_g)]$ to

$$\int\limits_{\partial\Omega_{\square_r}} \varphi \cdot \hat{p}(\boldsymbol{x}_{,\xi} \times \boldsymbol{x}_{,\eta})\, d\xi\, d\eta = \sum_{g=1}^{n_g} \boldsymbol{x}(\xi_g\,, \eta_g) \cdot \hat{\boldsymbol{t}}(\xi_g\,, \eta_g)\, w_g \tag{6.25}$$

where $\boldsymbol{x}$ depends through (6.23) on the nodal displacement variables $\mathbf{p}_e$. Now the residual load vector at a Gauss point is given by

$$\mathbf{R}_g = \left. \frac{\hat{\delta} W^P}{\hat{\delta} \mathbf{p}_e} \right|_{\hat{t}=const.} \tag{6.26}$$

where $W^P = \boldsymbol{x}(\xi_g\,, \eta_g) \cdot \hat{\boldsymbol{t}}(\xi_g\,, \eta_g)$ represents an appropriate pseudo potential.

```
SMSInitialize["UL-P2-2d-def","Environment"→"AceFEM"];
SMSTemplate["SMSTopology"→"P2","SMSSymmetricTangent"→False,
  "SMSDomainDataNames"->{"ph -pressure"},"SMSDefaultData"->{1}];
nen=SMSNoNodes;ndim=SMSNoDimensions;np=SMSNoDOFGlobal;
SMSStandardModule["Tangent and residual"];
XIO⊢Table[SMSReal[nd$$[i,"X",j]],{i,nen},{j,ndim}];
uIO⊢SMSReal[Table[nd$$[i,"at",j],{i,nen},{j,ndim}]];
pe=Flatten[uIO];ph⊢SMSReal[es$$["Data",1]];
mult⊨SMSReal[rdata$$["Multiplier"]];
SMSDo[Ig,1,SMSInteger[es$$["id","NoIntPoints"]]];
Ξ={ξ,η}⊢Table[SMSReal[es$$["IntPoints",i,Ig]],{i,2}];
wgp⊢SMSReal[es$$["IntPoints",4,Ig]];
κ⊨1-ξ-η;Nh⊨{(2 ξ-1) ξ,(2 η-1) η,(2 κ-1) κ,4 ξ η,4 κ η,4 κ ξ};
X⊢SMSFreeze[Nh.XIO];u⊨Nh.uIO;x=X+u;
{g1,g2}⊨{SMSD[x,ξ],SMSD[x,η]};th⊢SMSFreeze[ph Cross[g1,g2]];
WP⊨-mult th.x;
SMSDo[m,1,np];
  Rgm⊨SMSD[WP,pe,m,"Constant"→th];
  SMSExport[wgp Rgm,p$$[m],"AddIn"→True];
  SMSDo[n,1,np];
    Kgmn⊨SMSD[Rgm,pe,n];
    SMSExport[wgp Kgmn,s$$[m,n],"AddIn"→True];
  SMSEndDo[];
SMSEndDo[];
SMSEndDo[];
SMSWrite[];
```

Box 6.4. *AceGen* input for quadratic triangular deformation dependent load element

Linearization of (6.26) yields a non-symmetric tangent matrix for pressure loading[2] since now the derivative is computed without holding $\hat{\boldsymbol{t}}$ constant

$$\mathbf{K}_g = \frac{\hat{\delta}\mathbf{R}_g(\mathbf{p}_e)}{\hat{\delta}\mathbf{p}_e}. \tag{6.27}$$

The *AceGen* input for a triangular quadratic surface load element with constant pressure can be found in Box. 6.4. Here the shape function have to be used as presented in (4.32). Due to the fact that the element will be attached to a three-dimensional solid the coordinates and degree of freedoms have three components. All other definitions and equations were derived above. Note that in the *AceGen* input `g1` stands for the base vectors $\boldsymbol{x}_{,\xi}$, `g2` for $\boldsymbol{x}_{,\eta}$, `ph` for the pressure $\hat{p}$ and `th` represents the surface traction $\hat{\boldsymbol{t}}$. The value `mult` denotes the loading parameter λ used to

[2] In general, no potential is associated with the pressure load. However under certain boundary conditions the total assembled tangent matrix can be symmetrical, e.g. for internal pressure in a closed solid. An in depth discussion can be found in e.g. Sewell (1967) and Schweizerhof (1982).

increment the load within a load stepping procedure, see Eq. (3.23) in Sect. 3.2. The nonsymmetry of the resulting tangent matrix is reflected in the *AceGen* input by setting `"SMSSymmetricTangent"->False`.

6.3 Three-Dimensional Elements

Finite elements for solids that undergo finite deformations will be described in this section. The procedures and formulations within *AceGen* do not change much when compared to two-dimensional elements as described above. The standard isoparametric three-dimensional, hexahedral element was already presented as a part of introductory example in Chap. 4. The only differences are the introduction of three-dimensional vectors, the change of shape functions, the three-dimensional formulation of the constitutive equations and the Gauss integration. However all these changes are mostly reflected in the *AceGen* input by changing some numbers according to the dimension of the problem and by using the templates for three-dimensional elements, e.g. the definition of a linear or quadratic triangle by `"T1"` or `"T2"` is now replaced by a linear or quadratic tetrahedral `"O1"` or `"O2"` and analogously a linear quadrilateral `"Q1"` now becomes a linear hexahedral element `"H1"`. Additionally more complex strain energy functions are discussed in this section than the ones used in Sect. 4.2. It will be observed that, despite the increase in complexity, the length of the input for *AceGen* does not change dramatically.

6.3.1 Hyperelastic Solid Elements

Finite elastic deformations occur in many structures like foams or rubber parts. Elements to tackle these problems have to be selected according to the needs of the formulations. Especially rubber elasticity requires an element that can reproduce incompressible behaviour. Such elements are described in Sect. 6.4.

In order to emphasize that the *AceGen* input for standard finite strain three-dimensional elements is very short a simple hexahedral Neo-Hookian element with linear shape function was generated in Sect. 4.2.1. The strain energy was given in (4.50) with the compressible part $\lambda\,/\,2(J_F - 1)^2$ where J_F is the determinant of the deformation gradient $\boldsymbol{F}$. The *AceGen* input is provided in Box 4.3. The input follows the derivations of the element in Sect. 6.2.1, however here the volumetric deviatoric split is not used. Thus the right Cauchy–Green tensor is simply given by $\boldsymbol{C} = \boldsymbol{F}^T\boldsymbol{F}$.

Next we will develop a finite element based on a more complex strain energy function. This specific strain energy function includes all invariants of the right Cauchy Green tensor, $I_C = \operatorname{tr}\boldsymbol{C}$, $II_C = \operatorname{tr}[\operatorname{cof}\boldsymbol{C}]$ and $III_C = \det[\boldsymbol{C}]$. For details regarding the strain energy function, see e.g. Schröder (2010) and Schröder et al. (2011). The strain energy function is selected as

$$W(\boldsymbol{C}) = \frac{\alpha}{2}\,(\mathrm{tr}\,\boldsymbol{C})^2 + \frac{\beta}{2}\,(\mathrm{tr}\,[\mathrm{cof}\,\boldsymbol{C}])^2 - \gamma\ln(\sqrt{\det\boldsymbol{C}})\,. \tag{6.28}$$

Here the material parameters have to be selected according to the relations $\alpha > 0, \beta > 0$ and $\gamma > 0$. Furthermore the condition of a stress-free reference configuration has to be met. This yields the additional restriction $\gamma = 6\,\alpha + 12\,\beta$. Note that the cofactor of the right Cauchy Green tensor is given by cof $\boldsymbol{C} = \det\,\boldsymbol{C}\,\boldsymbol{C}^{-1} = (\det\,\boldsymbol{F})^2\,\boldsymbol{C}^{-1}$. This strain energy is relatively complex and needs a lot of effort in order to compute the 2nd Piola–Kirchhoff stresses for the formulation of the weak form and residual. Last not least the derivation of the tangent matrix for such element is quite time consuming, additionally the implementation needs often a thourough debugging. By using *AceGen* all this effort is not noticed. We can use the same scheme as before and obtain for a three-dimensional quadratic tetrahedral element as easy as an element with isotropic Neo-Hooke material. The input file is shown in Box 6.5.

```
SMSInitialize["O2_cofC","Environment"→"AceFEM"];
SMSTemplate["SMSTopology"→"O2","SMSSymmetricTangent"→True,
  "SMSDomainDataNames"->{"α -constant 1","β -constant 2"},
  "SMSDefaultData"->{21000,0.3}];
nen=SMSNoNodes;ndim=SMSNoDimensions;np=SMSNoDOFGlobal;
SMSStandardModule["Tangent and residual"];
{α,β}⊢SMSReal[Table[es$$["Data",i],{i,2}]];
XIO⊢Table[SMSReal[nd$$[i,"X",j]],{i,nen},{j,ndim}];
uIO⊢SMSReal[Table[nd$$[i,"at",j],{i,nen},{j,ndim}]];
pe=Flatten[uIO];
SMSDo[Ig,1,SMSInteger[es$$["id","NoIntPoints"]]];
Ξ={ξ,η,ζ}⊢Table[SMSReal[es$$["IntPoints",i,Ig]],{i,3}];
wgp⊢SMSReal[es$$["IntPoints",4,Ig]];
κ=1-ξ-η-ζ;Nh⊨{(2 ξ-1) ξ,(2 η-1) η,(2 ζ-1) ζ,
  (2 κ-1) κ,4 ξ η,4 η ζ,4 ζ ξ,4 ξ κ,4 η κ,4 ζ κ};
X⊢SMSFreeze[Nh.XIO];u⊨Nh.uIO;Je⊨SMSD[X,Ξ];Jed⊨Det[Je];
H⊨SMSD[u,X,"Dependency"→{Ξ,X,SMSInverse[Je]}];
F⊨IdentityMatrix[3]+H;Ct⊨Fᵀ.F;JC⊨Det[Ct];cofC⊨JC Inverse[Ct];
W⊨1/2 α Tr[Ct]^2+β/2 Tr[cofC]^2-(6 α+12 β) Log[SMSSqrt[JC]];
SMSDo[m,1,np];
 Rgm⊨Jed SMSD[W,pe,m];
 SMSExport[wgp Rgm,p$$[m],"AddIn"→True];
 SMSDo[n,m,np];
  Kgmn⊨SMSD[Rgm,pe,n];
  SMSExport[wgp Kgmn,s$$[m,n],"AddIn"→True];
 SMSEndDo[];
SMSEndDo[];
SMSEndDo[];
SMSWrite[];
```

Box 6.5. *AceGen* input for quadratic solid tetrahedral element

It is interesting to note that the formulation of such displacement element in *AceGen* is very general. In order to obtain e.g. a tri-linear hexahedral element only the `"SMSTopology"` constant has to be changed from `"O2"` (quadratic tet) to `"H1"` (linear hex) and the shape functions $\mathbb{N}\mathtt{h}$ have to be exchanged from the quadratic shape functions for the tetrahedral element (4.44) to the linear ones of the hexahedral element, see (4.41). The other variables, like number of nodes or total number of degrees of freedom are automatically adjusted by selecting `"SMSTopology"`. Hence it is rather simple to change to a different element formulation as long as the topology is provided by the system, for a more complex formulation using topologies that are not included in *AceGen*, see e.g. Sect. 6.2.1. The same is true when a different strain energy is selected. In the latter case only the potential W has to be replaced by a different formulation.

Element based on an anisotropic strain energy function. Based on the same formulation now a finite element is developed that accounts for transversely isotropic material behaviour at finite strains by using the transversely isotropic strain energy function (5.21) together with the isotropic strain energy function from line four in Table 5.1. In summary the isotropic strain energy is given by

$$W^{iso} = \frac{\alpha}{2}(\text{tr}\ \boldsymbol{C})^2 + \frac{\beta}{2}(\text{tr}\ [\text{Cof}\ \boldsymbol{C}])^2 - \gamma \log J_F + \varepsilon_1 \,(\det \boldsymbol{C}^{\varepsilon_2} + \det \boldsymbol{C}^{-\varepsilon_2} - 2) \tag{6.29}$$

where the latter part is a penalty term that is related to the enforcement of the incompressibility constraint. The transversely isotropic strain energy function has the form

$$\begin{aligned} W^{ti} = g_0 \Big[& \frac{1}{g_C + 1}(\text{tr}\ [\boldsymbol{C}\,\boldsymbol{M}])^{g_C+1} + \frac{1}{g_H + 1}(\text{tr}\ [\text{Cof}\ \boldsymbol{C}\,\boldsymbol{M}])^{g_H+1} \\ & + \frac{1}{g_J}\,(\det \boldsymbol{C})^{-g_J} \Big] . \end{aligned}$$

Again a tetrahedral element with quadratic shape functions will be generated. Note that derivation of the residual and the consistent tangent for the above strain energy is very complicated and includes a lot of complex derivations. Its explicit form can be found e.g. in Schröder et al. (2011). When using *AceGen* only the input in Box 6.6 has to be created, the rest is done by the system. The time for the automatic generation of the element when using *AceGen* is then only a few seconds on a modern notebook.

The material parameters in these strain energy functions are now selected to analyze the so-called Cook's membrane problem shown in Fig. 6.4a. The tapered beam is clamped at $X = 0$ and loaded at the $X = 48$ in vertical direction by a constant load. For the isotropic strain energy the following values are used: $\alpha = 42\,\text{kPa}$,

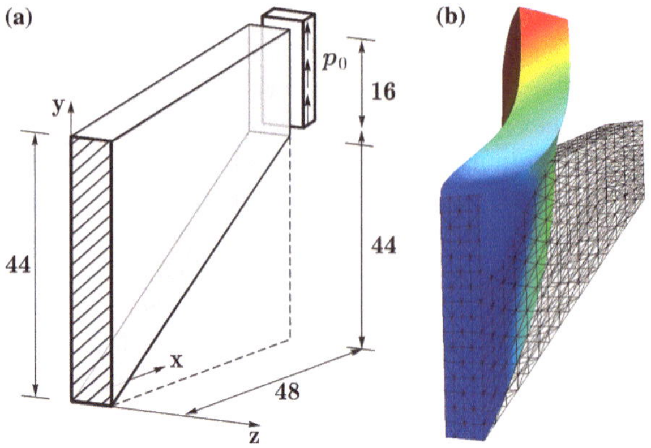

Fig. 6.4 **a** Cook's membrane problem and **b** Undeformed and deformed mesh

$\beta = 84\,\text{kPa}, \gamma = 6\alpha + 12\beta, \varepsilon_1 = 100\,\text{kPa}$ and $\varepsilon_2 = 10$. For the transversely isotropic function the parameters $g_0 = 3000\,\text{kPa}$, $g_C = 4$, $g_H = 8$, $g_J = 1$ are introduced, see also Schröder et al. (2011). Furthermore the direction vector $\boldsymbol{a}$ in $\boldsymbol{M} = \boldsymbol{a} \otimes \boldsymbol{a}$ was selected as $\boldsymbol{a} = [\,1, 1, 1\,]/\sqrt{3}$. Due to the anisotropic strain energy function it is expected that the tapered beam will depict an out of plane deformation even if loaded only in planar direction.

The applied load is incremented into 10 steps with a maximum load of 1000 kPa. The deformed state of the membrane is computed using a mesh with $16 \times 16 \times 4$ finite elements leading to 5248 elements with 25,440 unknown nodal values. The mesh is depicted in Fig. 6.4b which shows also the deformed configuration. The out of plane deformation due to the anisotropy is clearly visible by the twist of the structure that would not appear for a purely isotropic material equation.

Within the analysis Newton's method is applied. The convergence within each load step is—as expected—quadratically, see e.g. Table 6.1 for the second load step. Thus the automation of this complex finite element has the same features as a correctly handwritten code. However several advantages when using *AceGen* will be emphasized.

Table 6.1 Cook's membrane problem: Newton convergence in load step 2

Iteration	Residual Norm
1	3.431×10^{2}
2	1.044×10^{1}
3	1.018×10^{0}
4	7.088×10^{-2}
5	8.967×10^{-6}
6	3.591×10^{-10}

- The derivation of the tangent matrix is very complicated and thus time consuming for strain energy functions like the transversely isotropic function (5.21). This derivation is done by *AceGen* in a fast and reliable way.
- Due to the internal structure of *AceGen* the automatically generated code is often faster—when executed within a finite element code—and thus more efficient than handwritten code.

```
SMSInitialize["O2_TIso","Environment"→"AceFEM"];
SMSTemplate["SMSTopology"→"O2","SMSSymmetricTangent"→True,
  "SMSDomainDataNames"->{"α","β","ϵ1","ϵ2","g0","gC","gH","gJ","ax",
    "ay","az"},"SMSDefaultData"->{42,84,100,10,3000,4,8,1,1,0,0}];
nen=SMSNoNodes;ndim=SMSNoDimensions;np=SMSNoDOFGlobal;
SMSStandardModule["Tangent and residual"];
{α,β,ϵ1,ϵ2,g0,gC,gH,gJ,ax,ay,az}⊢
  SMSReal[Table[es$$["Data",i],{i,11}]];
XIO⊢Table[SMSReal[nd$$[i,"X",j]],{i,nen},{j,ndim}];
uIO⊢SMSReal[Table[nd$$[i,"at",j],{i,nen},{j,ndim}]];
pe=Flatten[uIO];
SMSDo[Ig,1,SMSInteger[es$$["id","NoIntPoints"]]];
Ξ={ξ,η,ζ}⊢Table[SMSReal[es$$["IntPoints",i,Ig]],{i,3}];
wgp⊢SMSReal[es$$["IntPoints",4,Ig]];
κ=1-ξ-η-ζ;Nh⊨{(2 ξ-1) ξ,(2 η-1) η,(2 ζ-1) ζ,
  (2 κ-1) κ,4 ξ η,4 η ζ,4 ζ ξ,4 ξ κ,4 η κ,4 ζ κ};
X⊢SMSFreeze[Nh.XIO];u⊨Nh.uIO;Je⊨SMSD[X,Ξ];Jed⊨Det[Je];
H⊨SMSD[u,X,"Dependency"→{Ξ,X,SMSInverse[Je]}];
F⊨IdentityMatrix[3]+H;Ct⊨Fᵀ.F;JC⊨Det[Ct];
a⊨{ax,ay,az};M⊨Outer[Times,a ,a];
cofC⊨JC Inverse[Ct];cofCM⊨cofC.M;CM⊨Ct.M;
W⊨α/2 ( Tr[Ct])^2+β/2 (Tr[cofC])^2-(6 α+12 β) Log[SMSSqrt[JC]]+
   ϵ1 (JC^ϵ2+JC^(-ϵ2)-2)+ g0 (1/(gC+1) (Tr[CM])^(gC+1)+
      1/(gH+1) (Tr[cofCM])^(gH+1)+1/gJ JC^(-gJ));
SMSDo[m,1,np];
  Rgm⊨Jed SMSD[W,pe,m];
  SMSExport[wgp Rgm,p$$[m],"AddIn"→True];
  SMSDo[n,m,np];
    Kgmn⊨SMSD[Rgm,pe,n];
    SMSExport[wgp Kgmn,s$$[m,n],"AddIn"→True];
  SMSEndDo[];
SMSEndDo[];
SMSEndDo[];
SMSWrite[];
```

Box 6.6. *AceGen* input for quadratic transversely isotropic tetrahedral solid element

The latter is especially valuable in explicit dynamics computations or iterative solutions of implicit equations when the computing time is governed by the speed of the loop over the elements.

6.4 Mixed Elements for Incompressibility

Pure displacement elements are not suitable for problems in which the constitutive behaviour exhibits incompressibility since they tend to lock, see e.g. Braess (2007). The constraint conditions due to incompressibility which are related to the pure volumetric mode[3] can only be fulfilled with an considerable stiffening of the bending modes, see e.g. Hueck et al. (1994). More specifically, this behaviour is called volume locking.

Mixed finite element methods can help to avoid locking, see e.g. Zienkiewicz and Taylor (1989) and Brezzi and Fortin (1991). There exist different possibilities to construct mixed elements:

- **Method of Lagrangian multipliers**. Here the constraint condition of incompressibility will be directly introduced via a Lagrangian multiplier. Hence a strain energy

$$W = W^{\text{inc}} + p\,G(J_F) \quad \text{with} \quad G(J_F) = 0 \tag{6.30}$$

is formulated. For finite deformations the constraint condition is given by $G(J_F) = J_F - 1$ with $J_F = \det \boldsymbol{F}$.
Finite elements based on this methodology have the disadvantage that contrary to the pure displacement elements additional unknowns occur. These are the Lagrangian multipliers that can be shown to be equivalent to the pressure p. Furthermore special techniques are needed to solve the associated incremental equation system for displacements and Lagrangian multipliers

$$\begin{bmatrix} \mathbf{K}_{uu} & \mathbf{B}_{up} \\ \mathbf{B}_{pu}^T & \mathbf{0} \end{bmatrix} \begin{Bmatrix} \Delta\mathbf{u} \\ \Delta\mathbf{p} \end{Bmatrix} = - \begin{Bmatrix} \mathbf{R}_u \\ \mathbf{R}_p \end{Bmatrix} \tag{6.31}$$

which has zero entries in the diagonal. The sub-matrix $\mathbf{K}_{uu}$ follows from W^{inc} while $\mathbf{B}_{up}$ is related to the discretization of the term $p\,G(J_F)$. Associated finite

---

[3] In the elastic case the condition is $J_F = \det \boldsymbol{F} = 1$ and for plastic flow the condition $J_p = \det \boldsymbol{F}_p = 1$ holds.

element formulations can be found in Oden and Key (1970) and Duffet and Reddy (1983) and will be developed in the next section using *AceGen*.

- **Perturbed Lagrangian method**. To have a greater variability for the formulation of ansatz functions the following strain energy function

$$W = W^{\text{inc}} + p\, G(J_F) - \frac{1}{2\,\epsilon}\, p^2 \tag{6.32}$$

can be introduced. The constraint condition is again given by $G(J_F) = J_F - 1$. $\epsilon > 0$ is a perturbation parameter. Choosing now a continuous ansatz function for displacements and pressure the following incremental equation system can be derived

$$\begin{bmatrix} \mathbf{K}_{uu} & \mathbf{B}_{up} \\ \mathbf{B}_{pu}^T & -\frac{1}{\epsilon}\,\mathbf{K}_{pp} \end{bmatrix} \begin{Bmatrix} \Delta\mathbf{u} \\ \Delta\mathbf{p} \end{Bmatrix} = - \begin{Bmatrix} \mathbf{R}_u \\ \mathbf{R}_{p\,\epsilon} \end{Bmatrix}. \tag{6.33}$$

Here contrary to (6.31) the incremental displacements and pressures can be computed using standard equation solvers.

- **Penalty method.** Finally the strain energy function

$$W = W^{\text{inc}} + \frac{\epsilon}{2} G(J_F)^2 \tag{6.34}$$

can be introduced. It leads to an incremental system of equations that depends purely upon the displacement field. The solution of (6.34) depends on the penalty parameter ϵ. For small values of ϵ the influence of the constraint condition disappears. For large values of ϵ the constraint is fulfilled more and more exactly but the condition number of the linear equation system will be very large and the solution algorithms do not converge with standard solvers. Papers regarding penalty formulations were published for the linear case by Malkus and Hughes (1978) and for the large strain case of rubber elasticity by e.g. Häggblad and Sundberg (1983) and Sussman and Bathe (1987).

- **Selective Hu-Washizu functional**. In this functional the incompressibility constraint is introduced as in the penalty method but is formulated via a constitutive equations for the pressure. In that case the functional

$$H(\varphi\,, p\,, \theta\,) = W(\hat{\mathbf{C}}) + K\,[\,G(\theta)]^2 + p\,(J_F - \theta) \tag{6.35}$$

is formulated, see also Sect. 1.1.1. Within the finite element discretization ansatz functions are selected for the deformation φ, the pressure p and the volumetric strain θ. $G(\theta)$ defines the constitutive equation for the pressure term, here K is the modulus of compression. The formulation of $W(\hat{\mathbf{C}})$ is provided by (5.14). The associated discretization within the finite element method was firstly presented in Simo et al. (1985).

Finite elements that are derived for mixed methods have to fulfill additional mathematical conditions which guarantee the stability of the element formulation. This condition is known as BB-condition, named after its inventors Babuska and Brezzi. Its fulfillment is related to the condition that matrix $\mathbf{B}_{pu}$ in (6.33) is not rank deficient.[4] The BB-condition has the disadvantage that it cannot be formulated for e single element. Always a patch of elements has to be considered, see e.g. Brezzi and Fortin (1991) or Braess (2007). A numerical method to show fulfillment of the BB-condition was derived in Chapelle and Bathe (1993).

6.4.1 Mixed T2-P1 Element

A classical formulation that is know to fulfill the BB-condition in linear applications is a two-dimensional triangle element with quadratic interpolations for the deformation field and linear interpolation for the pressure. The principle of stationary elastic potential (1.98) is written with respect to the initial configuration in this case as

$$\begin{aligned} \Pi(\varphi\,,p) &= \int\limits_B \left[\, W[\boldsymbol{C}(\varphi)] + p\,(\,J_F - 1\,) - \bar{\boldsymbol{b}}\cdot\varphi\,\right] dV \\ &= \int\limits_B \Pi^* dV \Rightarrow STAT \end{aligned} \tag{6.36}$$

where W is the elastic strain energy and $\bar{\boldsymbol{b}}$ the volume load. In this example the strain energy is chosen as

$$W = \mu\left[\frac{1}{2}(\,\mathrm{tr}\,\boldsymbol{C} - I_{tr}) - \ln J_F\right] \tag{6.37}$$

with the Lame constant μ.[5] The number I_{tr} is equal to "2" for two-dimensional and equal to "3" for three-dimensional problems.

The quadratic shape functions N_I^{φ} for the interpolation of the deformation φ can be found in (4.32). They are related to the six nodes defining the triangular

---

[4]For general nonlinear applications there exists no formulation of the BB-condition. However it can be applied the to the tangent spaces that belong to a given state of deformation and pressure, as e.g. provided in (6.31).

[5]Here the Lame constant λ has to be neglected since incompressibility is directly enforced by the Lagrange multiplier term $p\,(\,J_F - 1\,)$ in (6.36).

element. The linear pressure interpolation N_I^p is only based on the vertices of the triangle, the shape functions are given in (4.31). Thus the element has different numbers of local unknowns at the nodes as depicted in Fig. 6.5. The vertices have three unknowns $\mathbf{p}_I^T = \{ u_I , v_I , p_I \}$, $(I = 1 , 2 , 3)$ while the mid nodes at the edges have two unknowns $\mathbf{p}_I^T = \{ u_I , v_I \}$, $(I = 4 , 5 , 6)$.

```
<<"AceGen`"
SMSInitialize["T2-P1","Environment"→"AceFEM"];
SMSTemplate["SMSTopology"→"T2","SMSNoNodes" → 6,
  "SMSDOFGlobal"→{3,3,3,2,2,2},
  "SMSNodeID"→{"MIX -L","MIX -L","MIX -L","D","D","D"},
  "SMSDefaultData"->{1},"SMSSymmetricTangent"→True,
  "SMSDomainDataNames"->{"μ"}];
nen=SMSNoNodes;ndim=SMSNoDimensions;np=SMSNoDOFGlobal;
SMSStandardModule["Tangent and residual"];
XIO⊢Table[SMSReal[nd$$[i,"X",j]],{i,nen},{j,ndim}];
peIO⊢SMSReal[Table[nd$$[i,"at",j],{i,nen},{j,SMSDOFGlobal[[i]]}]];
μ⊢SMSReal[es$$["Data",1]];pe=Flatten[peIO];
uIO=peIO[[All,{1,2}]];pIO=peIO[[{1,2,3},3]];
SMSDo[Ig,1,SMSInteger[es$$["id","NoIntPoints"]]];
Ξ={ξ,η}⊢Table[SMSReal[es$$["IntPoints",i,Ig]],{i,2}];
wgp⊢SMSReal[es$$["IntPoints",4,Ig]];κ=1-ξ-η;Np⊨{ξ,η,κ};
Nu⊨{(2 ξ-1) ξ,(2 η-1) η,(2 κ-1) κ,4 ξ η,4 η κ,4 ξ κ};
X⊢SMSFreeze[Nu.XIO];Je⊨SMSD[X,Ξ];Jed⊨Det[Je];
u⊨Nu.uIO;p⊨Np.pIO;
H⊨SMSD[u,X,"Dependency"→{Ξ,X,SMSInverse[Je]}];
F⊨IdentityMatrix[2]+H;Ct⊨F^T.F;JF=Det[F];
W=μ/2 (Tr[Ct]-ndim)-μ Log[JF];Πs⊨W+p (JF-1);
SMSDo[m,1,np];
  Rgm⊨Jed SMSD[Πs,pe,m];
  SMSExport[wgp Rgm,p$$[m],"AddIn"→True];
  SMSDo[n,m,np];
    Kgmn⊨SMSD[Rgm,pe,n];
    SMSExport[wgp Kgmn,s$$[m,n],"AddIn"→True];
  SMSEndDo[];
SMSEndDo[];
SMSEndDo[];
SMSWrite[];
```

Box 6.7. *AceGen* input for the mixed T2-P1 element

Modern finite elements environments based on C or C++ such as *AceFEM* do not require that all nodes have the same number of unknowns as it is the case with older Fortran based finite element environments. This special arrangement of nodal unknowns can be handled naturally just by adding additional constants to the

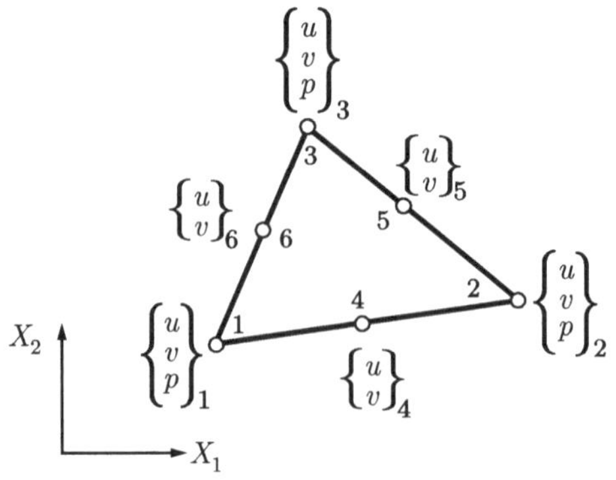

Fig. 6.5 Nodal values at vertices and mid nodes of the mixed T2-P1 triangular element

`SMSTemplate` function at the element initialization phase. This is reflected in the input for *AceGen* in Box 6.7. Additionally to the standard set of constants (described in Sect. 2.7) we need to specify:

`"SMSDOFGlobal"` contains the number of degree's of freedom per node for all nodes,

`"SMSNodeID"` contains for all nodes the node identification or *NodeID*. The *NodeID* is an arbitrary keyword that is used for identification of the nodes according to the physical meaning of the nodal degree's of freedom. In order to have a consistent set of elements one has to use the same *NodeID* for the nodes with the same physical meaning. Node identification can have additional switches. For example `"-L"` switch indicates that formulation is leading to a zero diagonal terms (e.g. method of Lagrangian multipliers).

The *NodeID* of the first three nodes (vertices) is `"MIX -L"` where `-L` switch denotes the presence of the Lagrangian multiplier while the *NodeID* of the mid nodes is `"D"` standing for deformation or displacement. The nested set `peIO` contains all degree's of freedom of the element which are in this case 12 dof's for the deformation and 3 dof's for the pressure totaling 15 dof's: $\mathbf{p}_e^T = \{\mathbf{p}_1^T, \ldots, \mathbf{p}_6^T\} \equiv$ `peIO`. By this definition the interpolations functions for the deformation φ and the pressure p are given by

$$\varphi_e = \sum_{I=1}^{6} N_I^{\varphi}(\xi, \eta)\, \mathbf{u}_I \qquad p_e = \sum_{I=1}^{3} N_I^{p}(\xi, \eta)\, p_I \tag{6.38}$$

The element formulation is not different from the formulations in Sect. 6.2. The deformation gradient $\boldsymbol{F}$ and the right Cauchy–Green tensor $\boldsymbol{C} = \boldsymbol{F}^T\boldsymbol{F}$ can be computed as well as the Jacobian $J_F = \det \boldsymbol{F}$. Once these kinematical quantities are known the strain energy function and with this the variational functional can be formulated at a Gauss point and evaluated using *AceGen*.[6] Note that the differentiation of the variational form (6.36) will be automatically performed with respect to the displacement and pressure degrees of freedom since both are present in the element unknown vector $\mathbf{p}_e$.

The complete *AceGen* input for this mixed two-dimensional element formulation can be found in Box 6.7.

6.4.2 Mixed T2-P0 Element

A different mixed interpolation can be achieved by using a quadratic shape function for the deformation and a constant pressure approximation within a triangular element. This element is not BB-stable but discussed here in order to show the possibilities of *AceGen* for different types of mixed interpolations. Again the potential formulation (6.36) is basis of the formulation together with the strain energy (6.37). The interpolation is now given by

$$\varphi_e = \sum_{I=1}^{6} N_I^{\varphi}(\xi\,,\eta)\,\mathbf{u}_I \qquad p_e = p_0 = const \tag{6.39}$$

where the quadratic shape functions N_I^{φ} can be found in (4.32). The right Cauchy–Green tensor $\boldsymbol{C} = \boldsymbol{F}^T\mathbf{F}$ is computed based for the shape functions via the deformation gradient $\boldsymbol{F}$. The functional Π (6.36) depends via $J_F = \det \boldsymbol{F}$ and $\boldsymbol{C}$ on the displacement field and is constant in the pressure p.

The input for *AceGen* can now be generated. The constant pressure term is not continuous thus it can be attached to a local auxiliary node. The element combines the concept of additional nodes (see Sect. 6.2.1) with the concept of *NodeID*-s (see Sect. 6.4.1). Thus the total number of nodes is 7 but the seventh node has an additional switches attached to its *NodeID* that specify a local auxiliary node (`"-LP"`) with an attached mixed variable leading to a zero diagonal (`"-L"`). Since the 7th node does not belong to a specific position in space a `Null` symbol is given instead of actual coordinates. The mixed element has 12 displacement unknowns and one constant pressure. Thus in total 13 unknowns are present.

The *AceGen* input for this mixed element formulation can be found in Box 6.8.

[6] The three-dimensional version of this element can be developed exactly along the lines of the two-dimensional formulation. Only the number of shape quadratic functions for the displacement field changes to 10 for the tetrahedral element and thee mixed interpolations for the pressure are linear with values at the vertices of the tetrahedra.

```
<<"AceGen`"
SMSInitialize["T2-P0","Environment"→"AceFEM"];
SMSTemplate[
  "SMSTopology"→"T2","SMSNoNodes"→7,"SMSDOFGlobal"→{2,2,2,2,2,2,1},
  "SMSNodeID"→{"D","D","D","D","D","D","MIX -LP -L"},
  "SMSAdditionalNodes"→Function[{},{Null}],
  "SMSDomainDataNames"->{"μ"},"SMSSymmetricTangent"→True,
  "SMSDefaultData"->{1}];
nen=SMSNoNodes;ndim=SMSNoDimensions;np=SMSNoDOFGlobal;
SMSStandardModule["Tangent and residual"];
μ⊢SMSReal[es$$["Data",1]];
XIO⊢Table[SMSReal[nd$$[i,"X",j]],{i,nen-1},{j,ndim}];
peIO⊢SMSReal[Table[nd$$[i,"at",j],{i,nen},{j,SMSDOFGlobal⟦i⟧}]];
pe=Flatten[peIO];uIO=peIO⟦1;;6⟧;pIO=peIO⟦7,1⟧;
SMSDo[Ig,1,SMSInteger[es$$["id","NoIntPoints"]]];
Ξ={ξ,η}⊢Table[SMSReal[es$$["IntPoints",i,Ig]],{i,2}];
wgp⊢SMSReal[es$$["IntPoints",4,Ig]];
κ=1-ξ-η;
Nu⊨{(2 ξ-1) ξ,(2 η-1) η,(2 κ-1) κ,4 ξ η,4 η κ,4 ξ κ};
X⊢SMSFreeze[Nu.XIO];Je⊨SMSD[X,Ξ];Jed⊨Det[Je];
u⊨Nu.uIO;p⊨pIO;
H⊨SMSD[u,X,"Dependency"→{Ξ,X,SMSInverse[Je]}];
F⊨IdentityMatrix[2]+H;Ct⊨Fᵀ.F;JF=Det[F];
W=μ/2 (Tr[Ct]-ndim)-μ Log[JF];Πs⊨W+p (JF-1);
SMSDo[m,1,np];
  Rgm⊨Jed SMSD[Πs,pe,m];
  SMSExport[wgp Rgm,p$$[m],"AddIn"→True];
  SMSDo[n,m,np];
    Kgmn⊨SMSD[Rgm,pe,n];
    SMSExport[wgp Kgmn,s$$[m,n],"AddIn"→True];
  SMSEndDo[];
SMSEndDo[];
SMSEndDo[];
SMSWrite[];
```

Box 6.8. *AceGen* input for mixed T2-P0 element

6.4.3 Mixed Q1-P0 Element

Finally a large deformation finite element is derived that is based on the selective Hu-Washizu variational formulation

$$\Pi(\varphi, p, \theta) = \int_B [\, W(\hat{\mathbf{C}}) + W(\theta) + p(J_F - \theta)\,]\, dV - \Pi^{ext} \Rightarrow STAT. \qquad (6.40)$$

The continuum mechanical basis for the mixed Q1-P0 element was already discussed in Sect. 1.3.3. Equation (1.104) describes the potential form with respect to the initial configuration. A simple choice for the strain energies $W(\hat{\boldsymbol{C}})$ and $W(\theta)$ in (6.40) is

$$W(\hat{\boldsymbol{C}}) = \frac{\mu}{2}(\text{tr}\ \hat{\boldsymbol{C}} - 3) \quad \text{and} \quad W(\theta) = \frac{\kappa}{2}\ (\theta - 1)^2 \tag{6.41}$$

with κ being the bulk modulus and μ the shear modulus.

```
SMSInitialize["H1-P0","Environment"→"AceFEM"];
SMSTemplate["SMSTopology"→"H1","SMSSymmetricTangent"→True,
  "SMSNoDOFCondense"→2,"SMSDomainDataNames"->{"κ","μ"},
  "SMSDefaultData"->{10 000.,5000.}];
nen=SMSNoNodes;ndim=SMSNoDimensions;
np=SMSNoDOFGlobal;nhe=SMSNoDOFCondense;
SMSStandardModule["Tangent and residual"];
{κ,μ}⊢SMSReal[Table[es$$["Data",i],{i,2}]];
XIO⊢Table[SMSReal[nd$$[i,"X",j]],{i,nen},{j,ndim}];
uIO⊢SMSReal[Table[nd$$[i,"at",j],{i,nen},{j,ndim}]];
heIO⊢SMSReal[Array[ed$$["ht",#]&,nhe]];
pae=Flatten[{uIO,heIO}];
SMSDo[Ig,1,SMSInteger[es$$["id","NoIntPoints"]]];
Ξ={ξ,η,ζ}⊢Table[SMSReal[es$$["IntPoints",i,Ig]],{i,3}];
wgp⊢SMSReal[es$$["IntPoints",4,Ig]];
Ξn={{-1,-1,-1},{1,-1,-1},{1,1,-1},{-1,1,-1},
    {-1,-1,1},{1,-1,1},{1,1,1},{-1,1,1}};
Nh⊨Table[1/8 (1+ξ Ξn⟦i,1⟧) (1+η Ξn⟦i,2⟧) (1+ζ Ξn⟦i,3⟧),{i,1,nen}];
X⊢SMSFreeze[Nh.XIO];u⊨Nh.uIO;Je⊨SMSD[X,Ξ];Jed⊨Det[Je];
H⊨SMSD[u,X,"Dependency"→{Ξ,X,SMSInverse[Je]}];
F⊨IdentityMatrix[3]+H;Ct⊨Fᵀ.F;JF⊨Det[F];
θ⊨heIO⟦1⟧+1;p⊨heIO⟦2⟧;
Πs⊨p (JF-θ)+μ/2 (JF^-2/3 Tr[Ct]-3)+κ/2 (θ-1)^2;
SMSDo[m,1,np+nhe];
  Ragm⊨Jed SMSD[Πs,pae,m];
  SMSExport[wgp Ragm,p$$[m],"AddIn"→True];
  SMSDo[n,m,np+nhe];
    Kagmn⊨SMSD[Ragm,pae,n];
    SMSExport[wgp Kagmn,s$$[m,n],"AddIn"→True];
  SMSEndDo[];
SMSEndDo[];
SMSEndDo[];
SMSWrite[];
```

Box 6.9. *AceGen* input for mixed H1-P0 element

Due to its robustness and good performance in incompressible solid problems the Q1-P0 element can be found in many existing finite element codes. Linear shape functions are used for the deformation field related to the deviatoric kinematical variables. Additionally a constant ansatz function is applied to discretize the pressure and volumetric strain.

The *AceGen* input, see Box 6.9, is written for a three-dimensional brick element H1/P0 using the tri-linear shape functions provided in (4.41). It is sufficient to define the selective Hu-Washizu potential function to describe the internal energy. Accordingly to the classification of computational problems given in Sect. 3.1 the presented mixed problem can be classified as time-independent locally coupled problem. Due to the fact that the pressure p and the dilatation θ are constant they can be eliminated at the element level. This is denoted in the *AceGen* input by setting `"SMSNoDOFCondense"->2` to condense the variables $\mathbf{h}_e$ ($\theta = h_{e_1}$ and $p = h_{e_2}$) at element level.[7] A detailed description of the solution and automation of locally coupled problems using static elimination of local unknowns is given in Sect. 3.3.3. Note that the differentiation of the terms in the potential $\Pi^* \equiv$ `Πs` is performed with respect to all unknown variables. These are the 24 nodal displacements $\mathbf{u} \equiv$ `uIO` and the two mixed variables $\mathbf{h}_e \equiv$ `he`. The produced code has a final size of 20 kBytes and thus a very fast execution time at element level.

6.5 Enhanced Strain Element

In the last twenty years many different finite elements were developed for finite deformation problems and successfully applied to special problem classes. One example is the Q1-P0 element which is well suited for incompressible materials. Good element performance for bending dominated structural problems is also necessary when arbitrary structural parts have to be discretized and simulated using three-dimensional solids. Last not least elements should be robust when large mesh distortion occur due to large deformations. Here enhanced strain elements (sometimes also called incompatible mode elements) can have some advantages.

A variational formulation of the discretization using incompatible modes was derived in Simo and Rifai (1990) based on the incompatible mode or non-conforming elements derived in Taylor et al. (1976). This concept has the advantage that it can also be applied to nonlinear problems like finite elastic or inelastic deformations. The concept followed in the work by Simo and Armero (1992) and Simo et al. (1993) is based on the principle of Hu-Washizu. The finite elements derived by this formulation are called *enhanced strain* or *enhanced assumed strain* (EAS) elements.

While very well suited for linear elastic problem, the enhanced strain elements do not provide a solution for all problem classes mentioned above in nonlinear applications. The elements become unstable under compression which was shown for the first time in Wriggers and Reese (1994), see also Wriggers and Reese (1996). Stabilized versions of the enhanced strain elements have been formulated to overcome

[7] The order of α and p is here not arbitrary since the condensation is based on the solution of a 2×2 equation system with a zero on the diagonal related to p. A standard solver is used for the condensation, hence the dilatation variable α has to be first.

this disadvantage. However, until lately, these stabilizations could not solve all defects found in Wriggers and Reese (1996) in a satisfactory way. Refined ansatz functions for the enhanced modes solved the instabilities for two-dimensional problems in the compression range, see Korelc and Wriggers (1996) and Glaser and Armero (1997), But they lead to instabilities in tension states. An in depth discussion of these phenomena and possible solutions can be found in Reese (2003, 2005), Wriggers (2008), Mueller-Hoeppe et al. (2009) and Korelc et al. (2010). The paper of Korelc et al. (2010) includes the *AceGen* code for the stabilized element formulation.

In the following section elements based on the enhanced strain concept will be constructed using *AceGen*.

6.5.1 General Concept and Formulation

The development of the nonlinear version of the *enhanced strain* elements is generally based on a mixed variational principle. Following Simo and Armero (1992) Hu-Washizu's principle is applied, see Sect. 1.3.3. Here the Hu-Washizu principle is formulated in terms of the deformation φ, the deformation gradient $\boldsymbol{F}$ and the first Piola–Kirchhoff stress tensor $\boldsymbol{P}$ which act as independent variables

$$\Pi(\varphi, \boldsymbol{F}, \boldsymbol{P}) = \int_B [\, W(\boldsymbol{F}) + \boldsymbol{P} \cdot (\, \mathrm{Grad}\, \varphi - \boldsymbol{F}\,)\,]\, dV - \int_B \varphi \cdot \rho_0\, \bar{\boldsymbol{b}}\, dV - \int_{\partial B_\sigma} \varphi \cdot \bar{\boldsymbol{t}}\, dA. \tag{6.42}$$

$W(\boldsymbol{F})$ denotes the strain energy function of the elastic material under consideration. This formulation is equivalent to the principle provided in (1.101). To simplify notation the last two terms in (6.42) which describe external forces will be combined and denoted by Π^{ext}.[8]

The variational principle of Hu-Washizu was formulated in this way in order to be able to additively decompose the deformation gradient, see Simo and Armero (1992). In this decomposition the local deformation gradient Grad φ is complemented by the independent gradient $\bar{\boldsymbol{F}}$

$$\boldsymbol{F} = \mathrm{Grad}\, \varphi + \bar{\boldsymbol{F}}. \tag{6.43}$$

[8] It is also possible to formulate the Hu-Washizu principle in other work conjugate variables. Examples are the 2nd Piola–Kirchhoff stress tensor and the Green–Lagrangian strain tensor $\boldsymbol{E}$ or the application of the Biot stress tensor $\boldsymbol{T}_B$ together with the right stretch tensor $\boldsymbol{U}$. From the viewpoint of continuum mechanics these formulations are equivalent. However due to the fact that the strain measures $\boldsymbol{F}$, $\boldsymbol{E}$ and $\boldsymbol{U}$ are different, their enhancement will lead to different finite element approximations and discretizations.

Thus the deformation gradient $\boldsymbol{F}$ is enriched by the *enhanced* gradient $\bar{\boldsymbol{F}}$ which can be incompatible with the deformation. With Eq. (6.43) relation

$$\Pi(\varphi, \bar{\boldsymbol{F}}, \boldsymbol{P}) = \int_B [\, W(\boldsymbol{F}) - \boldsymbol{P} \cdot \boldsymbol{F} \,] \, dV - \Pi^{ext}. \tag{6.44}$$

is obtained from (6.42). Equations (6.43) and (6.44) provide a variational basis which can be employed to incorporate the incompatible (enhanced) modes in a consistent way into the finite element formulation. Due to the fact that the interpolation for the enhanced gradient $\bar{\boldsymbol{F}}$ is constructed in such a way that the enhanced strains are orthogonal to the stress field at element level the $\boldsymbol{P} \cdot \boldsymbol{F}$ term in (6.44) can be neglected. Thus only the strain energy $W(\boldsymbol{F})$ enters the formulation that is used to derive the enhanced element leading to

$$\Pi(\varphi, \bar{\boldsymbol{F}}) = \int_B W(\boldsymbol{F}) \, dV - \Pi^{ext}. \tag{6.45}$$

To complete the model, a strain energy function W is needed. Different variants can be found for hyperelastic materials in Sect. 5.1.

6.5.2 *Discretization of the Enhanced Strain Element*

An isoparametric ansatz, see (4.7), is introduced to discretize the displacement field and the geometry of the current configuration in (6.43)

$$\boldsymbol{x}_e = \boldsymbol{X}_e + \boldsymbol{u}_e = \sum_{I=1}^{n} N_I(\boldsymbol{\Xi}) \, \mathbf{x}_I \qquad \text{with} \quad x_I = X_I + u_I. \tag{6.46}$$

In the two-dimensional case the bi-linear shape functions (4.33) are applied. In case of three-dimensional discretizations the shape functions (4.41) are used.

The conforming part of the deformation gradient can now be determined from (6.46). With (4.16), (4.17) and (4.55) the deformation gradient follows

$$\text{Grad}\, \varphi_e = \sum_{I=1}^{n} \mathbf{x}_I \otimes \nabla_X N_I(\boldsymbol{\Xi}) = \mathbf{1} + \left. \frac{\hat{\delta} \boldsymbol{u}}{\hat{\delta} \boldsymbol{X}} \right|_{\frac{D\boldsymbol{\Xi}}{D\boldsymbol{X}} = \mathbf{J}_e^{-1}} . \tag{6.47}$$

For the enhanced part of the deformation gradient an interpolation has to be selected which even can be incompatible. Following Glaser and Armero (1997) a product form is defined for the enriched part $\bar{\boldsymbol{F}}$

$$\bar{\boldsymbol{F}} = \boldsymbol{F}_0 \, \bar{\boldsymbol{H}}(\mathbf{h}_e). \tag{6.48}$$

$\mathbf{h}_e$ denote the *enhanced* parameters, $\bar{\boldsymbol{H}}$ contains the interpolation of enriched part. $\boldsymbol{F}_0$ is the constant part of the deformation gradient (6.47), which is evaluated at the element midpoint

$$\boldsymbol{F}_0 = \boldsymbol{F}|_{\boldsymbol{\Xi}=\mathbf{0}} = \sum_{I=1}^{n} \mathbf{x}_I \otimes \nabla_X N_I(\mathbf{0}). \tag{6.49}$$

The ansatz (6.48) fulfills the requirements for objectivity of the enhanced element formulation for arbitrary interpolations $\bar{\boldsymbol{H}}$, see Glaser and Armero (1997).[9]

The interpolations of the enriched part $\bar{\boldsymbol{H}}$ are related to the initial configuration of a finite element Ω_e. Since the incompatible interpolations have to be formulated with respect to the reference configuration $\Omega_\square$, like the isoparametric interpolations, $\bar{\boldsymbol{H}}$ has to be transformed to $\Omega_\square$ (for the relevant notation, see Fig. 2.3). This is performed by using the vector transformation when the enhanced gradient in reference coordinates can be expressed as a gradient of the incompatible displacements as follows

$$\bar{\boldsymbol{H}} = \frac{J_0}{J_e}\, \bar{\boldsymbol{H}}_{ref}(\boldsymbol{\Xi}, \mathbf{h}_e)\, \mathbf{J}_0^{-1} \tag{6.50}$$

or a full tensor transformation otherwise

$$\bar{\boldsymbol{H}} = \frac{J_0}{J_e}\, \mathbf{J}_0^{-T}\, \bar{\boldsymbol{H}}_{ref}(\boldsymbol{\Xi}, \mathbf{h}_e)\, \mathbf{J}_0^{-1}. \tag{6.51}$$

Here $\mathbf{J}_0$ defines the mapping between Ω_e and $\Omega_\square$ which is evaluated at the element midpoint ($\boldsymbol{\Xi} = \mathbf{0}$). The determinant of the transformation is denoted by $J_e = \det \mathbf{J}_e$. Its evaluation at the element midpoint is denoted by $J_0 = \det \mathbf{J}_0$.

Now the interpolation for the enhanced modes have to be selected. These can be incompatible since no derivatives of the enriched deformation gradient appear in (6.44). The interpolation can be written for two-dimensional elements in the compact form

$$\bar{\boldsymbol{H}}_{ref}(\boldsymbol{\Xi}, \mathbf{h}_e) = \begin{bmatrix} H_1\,(\xi,\eta)\; h_{e_1} & H_2\,(\xi,\eta)\; h_{e_2} \\ H_3\,(\xi,\eta)\; h_{e_3} & H_4\,(\xi,\eta)\; h_{e_4} \end{bmatrix}. \tag{6.52}$$

The interpolations H_I have to obey the orthogonality condition that follow from the first variation of (6.42). For further details see e.g. Wriggers (2008). Related to this the simplest interpolation with four enhanced or incompatible modes is given by

$$\bar{\boldsymbol{H}}_{ref}^{\mathrm{Q1/E4}}(\boldsymbol{\Xi}, \mathbf{h}_e) = \begin{bmatrix} \xi\, h_{e_1} & \eta\, h_{e_2} \\ \xi\, h_{e_3} & \eta\, h_{e_4} \end{bmatrix}. \tag{6.53}$$

The finite element based on this ansatz is called Q1/E4 element, see Simo and Armero (1992).

[9] This representation deviates from the form advocated in Simo and Armero (1992) in such way that $\bar{\boldsymbol{H}}$ was introduced as a gradient and hence could be interpolated without using $\boldsymbol{F}_0$.

The corresponding interpolation for the three-dimensional case leads to an ansatz for the enhanced deformation gradient with nine modes

$$\bar{\boldsymbol{H}}_{ref}^{\text{H1/E9}}(\boldsymbol{\Xi}, \mathbf{h}_e) = \begin{bmatrix} \xi\, h_{e_1} & \eta\, h_{e_2} & \zeta\, h_{e_3} \\ \xi\, h_{e_4} & \eta\, h_{e_5} & \zeta\, h_{e_6} \\ \xi\, h_{e_7} & \eta\, h_{e_8} & \zeta\, h_{e_9} \end{bmatrix}. \tag{6.54}$$

It is simply the extension of the two-dimensional interpolation and yields the so called H1/E9 element. As already shown in Simo et al. (1993) this ansatz is not sufficient to prevent locking. Thus additional enhanced modes have to be introduced in order to prevent volume locking. The related element has 12 incompatible modes and hence is called H1/E12 element. It is characterized by the following interpolation of the enhanced deformation gradient

$$\bar{\boldsymbol{H}}_{ref}^{\text{H1/E12}}(\boldsymbol{\Xi}, \mathbf{h}_e) = \bar{\boldsymbol{H}}_{ref}^{\text{H1/E9}} + (\xi\,\eta\, h_{e_{10}} + \xi\,\zeta\, h_{e_{11}} + \eta\,\zeta\, h_{e_{12}})\,\mathbf{1} \tag{6.55}$$

and a full tensor transformation (6.51) from reference to the actual frame.

The three-dimensional H1/E9 element is developed based on the mixed form (6.45) with respect to the initial configuration and enhanced strain interpolation (6.54). Thus the quantities above have to be described within an *AceGen* input file, see Box 6.10. The enhanced variables can be eliminated at element level using a block elimination scheme. It provides an efficient implementation since the part of the tangent matrix related to the enhanced variables can be inverted directly at element level due to the choice of incompatible interpolation functions. This task can be automated in *AceGen* by setting the initialization constant `SMSNoDOFCondense` that invokes a Schur complement elimination (static condensation) of the enhanced variables as described in Sects. 6.4.3 and 3.3.3, see Box 6.10

```
SMSInitialize["H1-E9","Environment"→"AceFEM"];
SMSTemplate["SMSTopology"→"H1","SMSSymmetricTangent"→True,
  "SMSNoDOFCondense"→9,"SMSDomainDataNames"->{"λ","μ"},
  "SMSDefaultData"->{10 000.,5000.}];
nen=SMSNoNodes;ndim=SMSNoDimensions;
np=SMSNoDOFGlobal;nhe=SMSNoDOFCondense;
SMSStandardModule["Tangent and residual"];
{λ,μ}⊢SMSReal[Table[es$$["Data",i],{i,2}]];
XIO⊢Table[SMSReal[nd$$[i,"X",j]],{i,nen},{j,ndim}];
uIO⊢SMSReal[Table[nd$$[i,"at",j],{i,nen},{j,ndim}]];
heIO⊢SMSReal[Array[ed$$["ht",#]&,nhe]];pae=Flatten[{uIO,heIO}];
SMSDo[Ig,1,SMSInteger[es$$["id","NoIntPoints"]]];
Ξ={ξ,η,ζ}⊢Table[SMSReal[es$$["IntPoints",i,Ig]],{i,3}];
wgp⊢SMSReal[es$$["IntPoints",4,Ig]];
Ξn={{-1,-1,-1},{1,-1,-1},{1,1,-1},{-1,1,-1},
    {-1,-1,1},{1,-1,1},{1,1,1},{-1,1,1}};
Nh⊨Table[1/8 (1+ξ Ξn⟦i,1⟧) (1+η Ξn⟦i,2⟧) (1+ζ Ξn⟦i,3⟧),{i,1,nen}];
X⊢SMSFreeze[Nh.XIO];u⊨Nh.uIO;Je⊨SMSD[X,Ξ];Jed⊨Det[Je];
```

```
J0⊨SMSReplaceAll[Je,{ξ→0,η→0,ζ→0}];J0d⊨Det[J0];
ℍ⊨SMSD[𝕦,𝐗,"Dependency"→{Ξ,𝐗,SMSInverse[Je]}];
ℍbref={{ξ 𝕙eIO⟦1⟧,η 𝕙eIO⟦2⟧,ζ 𝕙eIO⟦3⟧},
    {ξ 𝕙eIO⟦4⟧,η 𝕙eIO⟦5⟧,ζ 𝕙eIO⟦6⟧},
    {ξ 𝕙eIO⟦7⟧,η 𝕙eIO⟦8⟧,ζ 𝕙eIO⟦9⟧}};
ℍb⊨J0d/Jed ℍbref.SMSInverse[J0];
𝔽⊨IdentityMatrix[3]+ℍ+ℍb;ℂt⊨𝔽ᵀ.𝔽;JF⊨Det[𝔽];
W⊨λ/2 (JF-1)^2+μ ((Tr[ℂt]-3)/2-Log[JF]);
SMSDo[m,1,np+nhe];
  Ragm⊨Jed SMSD[W,𝕡ae,m];
  SMSExport[wgp Ragm,p$$[m],"AddIn"→True];
  SMSDo[n,m,np+nhe];
    Kagmn⊨SMSD[Ragm,𝕡ae,n];
    SMSExport[wgp Kagmn,s$$[m,n],"AddIn"→True];
  SMSEndDo[];
SMSEndDo[];
SMSEndDo[];
SMSWrite[];
```

Box 6.10. *AceGen* input for the three-dimensional enhanced element with nine modes

6.6 Example

The performance of different solid elements is shown by means of an example suitable to point out important properties of the different elements such as high coarse mesh accuracy, locking free response for incompressibility dominated problems.

The element formulation which are compared in this section are standard isoparametric as well as special elements for good bending performance and for incompressible problems. The following elements were selected:

- two standard elements H1 and H2 that use tri-linear and tri-quadratic interpolations, respectively, see Sect. 4.2,
- the mixed H1/P0 element as proposed by Simo et al. (1985) for finite deformations, see also Sect. 6.4.3,
- the mixed tetrahedral element O2/P1, its two-dimensional counterpart T2/P1 was developed in Sect. 6.4.1,
- the classical enhanced element H1/E9, developed in Sect. 6.5.2, and
- the H1/EI9 element described in Mueller-Hoeppe et al. (2009).

All elements use a hyperelastic material model, see (5.8). The nearly incompressible elements are based on the compressible part $g(J_F) = \frac{\Lambda}{2}\,(J_F - 1)^2$.

A nearly incompressible block of length l, width w and height h is loaded by an equally distributed surface load q at its top center, as shown in Fig. 6.6. Furthermore, all nodes on the top of the block are fixed in the x_1- and x_2-direction. For symmetry reasons, only a quarter of the block is discretized. The bottom face of the block is fixed in the x_3-direction. The symmetry boundary conditions are set such that nodes at $x_1 = 0.5\,w$ are fixed in x_1-direction and nodes at $x_2 = 0.5\,l$ are fixed in x_2-direction.

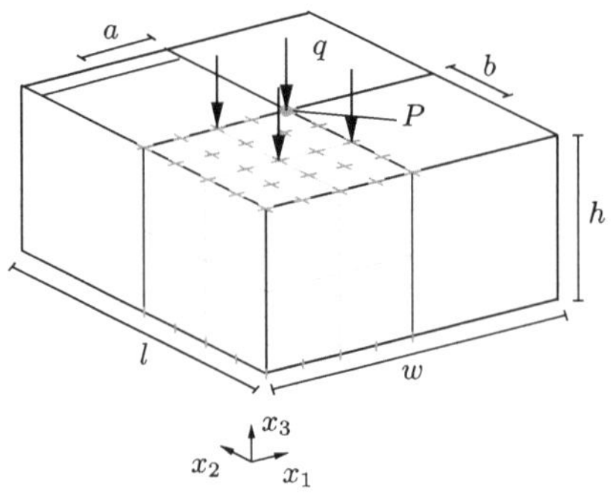

Geometry
$h = 50\,\text{mm}$
$w = 100\,\text{mm}$
$l = 100\,\text{mm}$
$a = 25\,\text{mm}$
$b = 25\,\text{mm}$

Load
$q = 3\,\text{MPa}$

Material
$\Lambda = 499.92568\,\text{MPa}$
$\mu = 1.61148\,\text{MPa}$

Fig. 6.6 Nearly incompressible block: system, load and material data

These boundary conditions are chosen according to a similar test presented in Reese et al. (2000).

The geometry and the material as well as the applied load and the boundary conditions are provided in Fig. 6.6. The convergence of the vertical displacement w_P in x_3-direction at the point P, in Fig. 6.6, is investigated for the H1/EI9, H1, H2, H1/P0, O2/P1 and the H1/E9 element for regular meshes with $4 \times 4 \times 4$, $8 \times 8 \times 8$, $16 \times 16 \times 16$ and $32 \times 32 \times 32$ elements. Since all elements are developed using *AceGen* they exhibit quadratic convergence within the Newton–Raphson solution algorithm.

In Table 6.2 the vertical displacement w_P in x_3-direction of point P is shown as a function of the number of degrees of freedom for all elements. It can be observed that the Q1 element locks, as can be expected for this nearly incompressible problem. Both enhanced strain elements and the mixed O2/P1[10] and the H1/P0 elements are softer than the H2 element. Thus still mild locking occurs for the higher order quadratic displacement element.

Again, all elements except the H1/E9 element converge to the same solution. For the H1/E9 element, solutions can only be obtained for the two coarsest meshes. For finer mesh resolutions the H1/E9 element depicts nonphysical hourglass instabilities.

It should be noted that the convergence behavior is different for the elements. The O2/P1 element is very robust and needs only three load steps to reach the final load level. Almost equally robust are the H1/P0, H1, H2 and H1/EI9 elements that however need for finer meshes sizes more load steps (increase from 4 to 6 load steps). The less robust element for this application is the classical enhanced element that needs for the coarse meshes more load steps to reach the final load level.

The deformed configuration at final load level is plotted in Fig. 6.7 to visualize the results for the elements that are known to perform well for this test.

[10] Due to the fact that the data are related to a nearly incompressible material with a Poisson ratio of $\nu = 0.4983$ the mixed O2/P1 element converges to a slightly different solution since the incompressibility is enforced exactly in O2/P1.

Table 6.2 Nearly incompressible block: displacement w_P (mm) for the H1/EI9, H1, H2, H1/P0, O2/P1 and H1/E9 element

Degrees of freedom	H1/EI9	H1	H2	H1/P0	O2/P1	H1/E9
260	19.09	7.78	18.32	19.87	19.80	20.00
1800	19.98	13.17	19.54	20.02	19.93	20.05
13328	20.01	17.54	19.98	20.01	19.97	–
102432	20.00	19.52	20.01	20.00	19.98	–

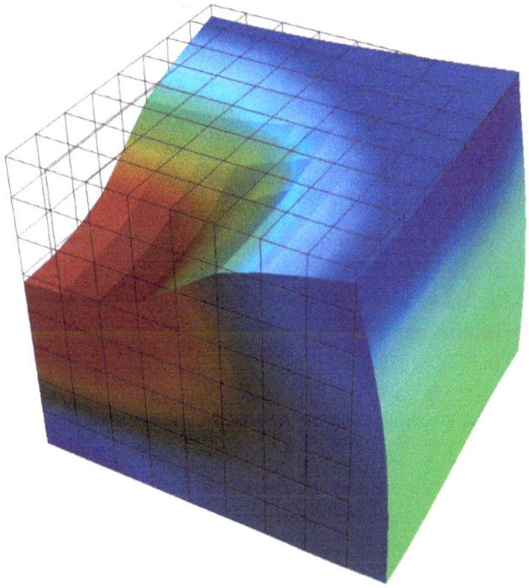

Fig. 6.7 Nearly incompressible block: deformed configuration with $8 \times 8 \times 8$ mesh

References

Bathe, K.J. 1996. *Finite Element Procedures*. Englewood Cliffs: Prentice-Hall.

Belytschko, T., J.S.J. Ong, W.K. Liu, and J.M. Kennedy. 1984. Hourglass control in linear and nonlinear problems. *Computer Methods in Applied Mechanics and Engineering* 43: 251–276.

Belytschko, T., W.K. Liu, and B. Moran. 2000. *Nonlinear Finite Elements for Continua and Structures*. Chichester: Wiley.

Braess, D. 2007. *Finite Elements: Theory, Fast Solvers, and Applications in Solid Mechanics*. Cambridge: Cambridge University Press.

Brezzi, F., and M. Fortin. 1991. *Mixed and Hybrid Finite Element Methods*. Berlin: Springer.

Chapelle, D., and K.J. Bathe. 1993. The inf-sup test. *Computers and Structures* 47: 537–545.

Duffet, G., and B.D. Reddy. 1983. The analysis of incompressible hyperelastic bodies by the finite element method. *Computer Methods in Applied Mechanics and Engineering* 41: 105–120.

Düster, A., S. Hartmann, and E. Rank. 2003. p-FEM applied to finite isotropic hyperelastic bodies. *Computer Methods in Applied Mechanics and Engineering* 192: 5147–5166.

Glaser, S., and F. Armero. 1997. On the formulation of enhanced strain finite elements in finite deformations. *Engineering Computations* 14: 759–791.

Guo, Y., M. Ortiz, T. Belytschko, and E.A. Repetto. 2000. Triangular composite finite elements. *International Journal for Numerical Methods in Engineering* 47: 287–316.

Häggblad, B., and J.A. Sundberg. 1983. Large strain solutions of rubber components. *Computers and Structures* 17: 835–843.
Hassler, M., and K. Schweizerhof. 2008. On the static interaction of fluid and gas loaded multi-chamber systems in large deformation finite element analysis. *Computer Methods in Applied Mechanics and Engineering* 197: 1725–1749.
Hauptmann, R., and K. Schweizerhof. 1998. A systematic development of 'solid-shell' element formulation for linear and nonlinear analyses employing only displacement degree of freedom. *International Journal for Numerical Methods in Engineering* 42: 49–69.
Hueck, U., B. Reddy, and P. Wriggers. 1994. On the stabilization of the rectangular four-node quadrilateral element. *Communications in Applied Numerical Methods* 10: 555–563.
Hughes, T.R.J. 1987. *The Finite Element Method*. Englewood Cliffs: Prentice Hall.
Hughes, T.J.R., J.A. Cottrell, and Y. Bazilevs. 2005. Isogeometric analysis: CAD, finite elements, NURBS, exact geometry, and mesh refinement. *Computer Methods in Applied Mechanics and Engineering* 194(39–41): 4135–4195.
Korelc, J., and P. Wriggers. 1996. Consistent gradient formulation for a stable enhanced strain method for large deformations. *Engineering Computations* 13: 103–123.
Korelc, J., U. Solinc, and P. Wriggers. 2010. An improved EAS brick element for finite deformation. *Computational Mechanics* 46: 641–659.
Malkus, D.S., and T.J.R. Hughes. 1978. Mixed finite element methods - reduced and selective integration techniques: a unification of concepts. *Computer Methods in Applied Mechanics and Engineering* 15: 63–81.
Mueller-Hoeppe, D.S., S. Loehnert, and P. Wriggers. 2009. A finite deformation brick element with inhomogeneous mode enhancement. *International Journal for Numerical Methods in Engineering* 78: 1164–1187.
Nadler, B., and M. Rubin. 2003. A new 3-d finite element for nonlinear elasticity using the theory of a Cosserat point. *International Journal of Solids and Structures* 40: 4585–4614.
Oden, J.T., and J.E. Key. 1970. Numerical analysis of finite axisymmetrical deformations of incompressible elastic solids of revolution. *International Journal of Solids and Structures* 6: 497–518.
Puso, M.A., and J. Solberg. 2006. A stabilized nodally integrated tetrahedral. *International Journal for Numerical Methods in Engineering* 67: 841–867.
Reese, S. 2003. On a consistent hourglass stabilization technique to treat large inelastic deformations and thermo-mechanical coupling in plane strain problems. *International Journal for Numerical Methods in Engineering* 57: 1095–1127.
Reese, S. 2005. On a physically stabilized one point finite element formulation for three-dimensional finite elasto-plasticity. *Computer Methods in Applied Mechanics and Engineering* 194: 4685–4715.
Reese, S., P. Wriggers, and B.D. Reddy. 2000. A new locking-free brick element technique for large deformation problems in elasticity. *Computers and Structures* 75: 291–304.
Schröder, J. 2010. Anisotropic polyconvex energies. In *Polyconvex Analysis*, ed. J. Schröder, 1–53., CISM Wien: Springer. (62).
Schröder, J., P. Wriggers, and D. Balzani. 2011. A new mixed finite element based on different approximations of the minors of deformation tensors. *CMAME* 200: 3583–3600.
Schweizerhof, K. 1982. Nichtlineare Berechnung von Tragwerken unter verformungsabhängiger Belastung mit finiten Elementen. Technical Report 82–2, Institut für Baustatik, Stuttgart.
Schweizerhof, K., and E. Ramm. 1984. Displacement dependent pressure loads in nonlinear finite element analysis. *Computers and Structures* 18: 1099–1114.
Sewell, M.J. 1967. On configuration-dependent loading. *Archives of Rational Mechanics* 23: 321–351.
Simo, J., R. Taylor, and P. Wriggers. 1991. A note on finite element implementation of pressure boundary loading. *Communications in Applied Numerical Methods* 7: 513–525.
Simo, J.C., and F. Armero. 1992. Geometrically non-linear enhanced strain mixed methods and the method of incompatible modes. *International Journal for Numerical Methods in Engineering* 33: 1413–1449.

Simo, J.C., and M.S. Rifai. 1990. A class of assumed strain methods and the method of incompatible modes. *International Journal for Numerical Methods in Engineering* 29: 1595–1638.

Simo, J.C., R.L. Taylor, and K.S. Pister. 1985. Variational and projection methods for the volume constraint in finite deformation elasto-plasticity. *Computer Methods in Applied Mechanics and Engineering* 51: 177–208.

Simo, J.C., F. Armero, and R.L. Taylor. 1993. Improved versions of assumed enhanced strain tri-linear elements for 3D finite deformation problems. *Computer Methods in Applied Mechanics and Engineering* 110: 359–386.

Sussman, T., and K.J. Bathe. 1987. A finite element formulation for nonlinear incompressible elastic and inelastic analysis. *Computers and Structures* 26: 357–409.

Taylor, R.L., P.J. Beresford, and E.L. Wilson. 1976. A non-conforming element for stress analysis. *International Journal for Numerical Methods in Engineering* 10: 1211–1219.

Wriggers, P. 2008. *Nonlinear Finite Elements*. Berlin: Springer.

Wriggers, P., and S. Reese. 1994. A note on enhanced strain methods for large deformations. Technical Report 3/94, Bericht des Instituts für Mechanik.

Wriggers, P., and S. Reese. 1996. A note on enhanced strain methods for large deformations. *Computer Methods in Applied Mechanics and Engineering* 135: 201–209.

Yosibash, Z., S. Hartmann, U. Heisserer, A. Düster, E. Rank, and M. Szanto. 2007. Axisymmetric pressure boundary loading for finite deformation analysis using p-FEM. *Computer Methods in Applied Mechanics and Engineering* 196: 1261–1277.

Zienkiewicz, O.C., and R.L. Taylor. 1989. *The Finite Element Method*, vol. 1, 4th ed. London: McGraw Hill.

Zienkiewicz, O.C., and R.L. Taylor. 2000. *The Finite Element Method*, vol. 2, 5th ed. Oxford: Butterworth-Heinemann.

Chapter 7
Structural Elements

Trusses, beams and shells belong to the most important structural elements in engineering practice. Many structures in civil engineering—like masts, domes, frames or cooling towers—consist of such structural elements. But also in mechanical engineering—car bodies, robots or general machines—can be modeled by beams and shells. The reliable mechanical and mathematical description of trusses, beams and shells is of great significance. Hence it is since a long time under investigation and associated with great names like Galileo, Leibniz, Mariotte, Bernoulli, Euler and Kirchhoff. Linear and approximate nonlinear theories are known for a long time and have been introduced to the engineering codes. Especially stability problems were solved by different approximate theories and associated numerical methods. However due to the development of inexpensive computer hardware it is today possible to perform numerical simulations based on completely nonlinear theories. Thus the general description of finite deformation states of such structural members has found its way into modern numerical simulation tools like the finite element method. Due to this development it is not necessary to discuss the validity of approximate theories since no restrictions with respect to deflections and rotations are made in this approach.

The structural elements are modeled in case of trusses and beams by one-dimensional models which however are imbedded in three-dimensional space. The same holds for two-dimensional shell models. All models are characterized by a description of the geometry as a curve or surface in space. Formulations which can be applied to describe the spatial curves or surfaces are provided by the introduction of an arc-length of a curve, convective coordinates for surfaces or simply reference to a cartesian coordinate system using an approximation of the initial geometry by polynomial patches. The last is often chosen within finite element approximations since it naturally fits into the isoparametric concept, see Sect. 4.1.

The latter leads often to a discretization of curved spatial beams by a number of straight finite elements with linear interpolation. Such approximation of the geometry however will create additional errors besides the discretization error of the deformation field. The error, due to the approximation of the geometry, vanishes with

J. Korelc and P. Wriggers, *Automation of Finite Element Methods*,
DOI 10.1007/978-3-319-39005-5_7

increasing number of finite elements. In that case it is essential that the additional element coordinates are always related to the exact geometry. These considerations also hold for shells. However they react more sensitive to errors in geometry since the surface properties can easily changed locally. A bi-linear isoparametric element, for example, approximates in general hyperbolic surfaces for unstructured meshes, even if the global surface is a sphere or a cylinder. These errors diminish for higher order finite element approximations. Additionally, in the last few years different approaches have been proposed which link the geometry directly to the finite element approximation by using the same ansatz, like *NURBS*, for geometry and finite element interpolation. This approach was firstly introduced by Cirak et al. (2000) for shells and later generalized and named isogeometric formulation, see Hughes et al. (2005).

In the following first truss and beam elements will be discussed then general shell elements for arbitrary three-dimensional surfaces will be derived.

7.1 Nonlinear Truss Element

This section is concerned with the formulation of a three-dimensional truss element. The underlying assumptions for truss structures are: trusses are straight members, they are connected by hinges and loaded only at their connection points. Thus trusses have to sustain only tension and compression forces but no bending and torsional moments. A geometrically exact formulation will be used thus the kinematical relations are not restricted by the magnitude of deformation. The constitutive equations are first formulated for purely elastic behaviour. Especially the St. Venant material relation for small strains, see (5.12), and the hyperelastic Ogden material for finite strain, see (5.7), are applied.

7.1.1 Kinematics and Strains

For trusses the three-dimensional kinematical relations have to be specialized for the one-dimensional case, see Sect. 1.1. The most simple way is to use the strain measures based on the spectral decomposition since the truss has only one principal direction along its axis. Thus the Green–Lagrange strain can be written in terms of the longitudinal stretch λ_X as

$$E_X = \frac{1}{2}(\lambda_X^2 - 1). \tag{7.1}$$

Here the stretch in the direction of the truss axis can be obtained from the change in length of the beam axis during deformation. From Fig. 7.1 it can be observed that the undeformed length of the beam is given by

Fig. 7.1 Truss: undeformed and deformed configuration

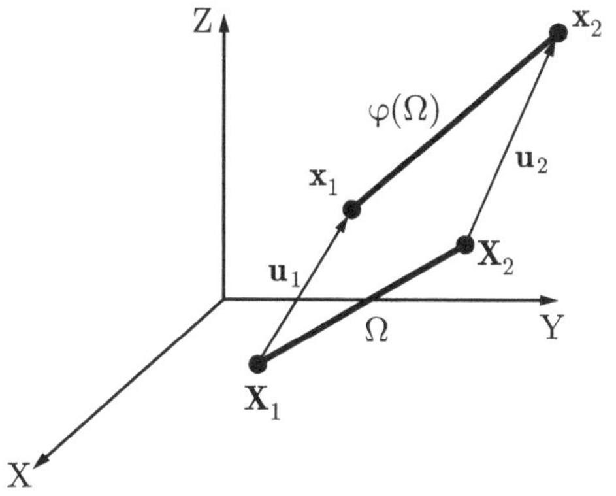

$$L = \|\mathbf{X}_2 - \mathbf{X}_1\| = \sqrt{(\mathbf{X}_2 - \mathbf{X}_1) \cdot (\mathbf{X}_2 - \mathbf{X}_1)} \tag{7.2}$$

where e.g. $\mathbf{X}_1$ and $\mathbf{X}_2$ denote the coordinates of the initial configuration at beginning and end of the truss, respectively. The deformed length of the truss can be computed from

$$\begin{aligned} l = \|\mathbf{x}_2 - \mathbf{x}_1\| &= \sqrt{(\mathbf{x}_2 - \mathbf{x}_1) \cdot (\mathbf{x}_2 - \mathbf{x}_1)} \quad \text{with} \\ \mathbf{x}_\alpha &= \mathbf{X}_\alpha + \mathbf{u}_\alpha \qquad (\alpha = 1, 2) \end{aligned} \tag{7.3}$$

where $\mathbf{x}_2$ and $\mathbf{x}_1$ are the current coordinates that can be obtained with the components of the displacement vector $\mathbf{u}_2$ and $\mathbf{u}_1$ at the beginning and end of the truss.[1] Based on these definition the stretch in direction of the truss axis is given by

$$\lambda_X = \frac{l}{L} \tag{7.4}$$

7.1.2 Constitutive Equations for the Truss

Two elastic constitutive equations are presented here for the truss. These are the St. Venant material for small strains and the Ogden material for finite strains.

St. Venant material. If a truss undergoes large displacements but only endures small strains then St. Venant material (5.12) is adequate for the constitutive description. It provides a linear relation between the Green–Lagrange strains $\mathbf{E}$ and the 2nd

[1] Since in the global configuration of the truss is already taken into account in (7.2) and (7.3) it is not necessary to introduce a transformation between local and global truss coordinates.

Piola–Kirchhoff stresses **S**, see Sect. 5.1.2. Since the truss element is loaded only along its local axis it is sufficient to consider the first component of the three-dimensional constitutive equation. This leads with Young's modulus E to

$$S_X = E\,E_X. \tag{7.5}$$

Relation (7.5) is only valid for small strains, see the remark related to Eq. (5.12). By using (7.5) and (7.1) the elastic St. Venant constitutive equation for the truss

$$S_X = E\,\frac{1}{2}\,(\lambda_X^2 - 1) \tag{7.6}$$

is obtained. The related strain energy can be written as

$$W^{SV} = \frac{1}{2}\,E\,E_X^2. \tag{7.7}$$

Ogden material. Finite elastic deformations can be described by the material equation of Ogden (5.7), see Sect. 5.1. The strain energy is given in terms of the principal stretches

$$W(\lambda_i) = \sum_r \frac{\mu_r}{\alpha_r}\left[\,\lambda_1^{\alpha_r} + \lambda_2^{\alpha_r} + \lambda_3^{\alpha_r} - 3\,\right]. \tag{7.8}$$

The constants μ_r and α_r are material parameters. λ_i denote the principal stretches which follow in general from the spectral decomposition of the strain tensor. Due to the one-dimensional loading of a truss element only the principal stretch λ_1 is relevant. It can be directly computed from (7.4): $\lambda_1 = \lambda_X$.

The constitutive relations of Ogden are often applied to rubber-like materials. In that case incompressible material behaviour has to be considered additionally which leads to the constraint equation $J_F = \lambda_1\,\lambda_2\,\lambda_3 = 1$. With the assumption that $\lambda_2 = \lambda_3$ the latter relation yields $\lambda_2 = \lambda_1^{-\frac{1}{2}}$. Inserting these relations into (7.8) yields the strain energy of Ogden for an incompressible one-dimensional case

$$W^O = \sum_r \frac{\mu_r}{\alpha_r}\,[\,\lambda_X^{\alpha_r} + 2\lambda_X^{-\frac{1}{2}\alpha_r} - 3]. \tag{7.9}$$

Since both strain energy functions (7.7) and (7.9) of the St. Venant and the Ogden material are related to the initial configuration it is possible to derive a uniform variational formulation for both materials.[2]

[2] In case that the stresses are needed for post-processing purposes they simply can be computed from (5.1) by $S_X = \partial W^{SV,O} \,/\, \partial E_X = 1\,/\,\lambda_X\,\partial W^{SV,O} \,/\, \partial \lambda_X$. Cauchy stresses then follow from (1.71) with $J_F = 1$: $\sigma_x = S_X\,\lambda_X^2$.

7.1.3 Variational Formulation

The finite element formulation of the nonlinear truss element using *AceGen* is based on the strain energy function W. The following one-dimensional version of the elastic potential, see Sect. 1.3.3, can be stated for a truss element

$$\Pi(\boldsymbol{u}) = \int\limits_{(X)} W^{SV,O} A\, dX - \sum_{k=1}^{n} \boldsymbol{u}_k \cdot \boldsymbol{P}_k \Longrightarrow STAT, \tag{7.10}$$

where the strain energy $W^{SV,O}$ is provided by one of the Eq. (7.7) or (7.9). $\boldsymbol{u}_k$ is the displacement at a node (hinge) k and $\boldsymbol{P}_k$ is the associated force acting at node k. A is the cross sectional area in the initial configuration. The strain energy functions W^{SV} in (7.7) and W^O in (7.9) are highly nonlinear. This yields a strongly nonlinear elastic potential (7.10). Hence Newton's method is usually applied for the solution of (7.10).

7.1.4 Finite-Element-Model

The discretization of Eq. (7.10) is obtained using finite elements. The displacements $u,\ v,\ w$ are approximated by linear shape functions.[3] Thus the interpolation

$$u_e = \sum_{I=1}^{2} N_I(\xi)\, u_{cI}, \quad v_e = \sum_{I=1}^{2} N_I(\xi)\, v_{cI} \ \text{ and } \ w_e = \sum_{I=1}^{2} N_I(\xi)\, w_{cI} \tag{7.11}$$

is used. Here $N_I(\xi)$ are given in terms of the linear shape functions (4.27). Note that the nodal values at the beginning and end point of the element define the interpolation within the truss element. Hence the computation of the stretch in (7.4) can be computed directly from (7.2) and (7.3) without further use of the shape functions for one element. In that case (7.2) describes the initial length L_e of the element while (7.3) is the length of the deformed truss element l_e. Thus the element stretch is given as

$$\lambda_X^e = \frac{l_e}{L_e}, \tag{7.12}$$

see Fig. 7.2.

By inserting (7.12) into the variational formulation (7.10) the potential for one element is obtained

[3] Quadratic or higher order interpolations could be introduced as well, but for most applications linear elements are sufficient since the linear shape functions are solution of the homogeneous differential equations of the truss in the geometrically linear case when linear elastic constitutive behaviour is assumed.

Fig. 7.2 Finite truss element

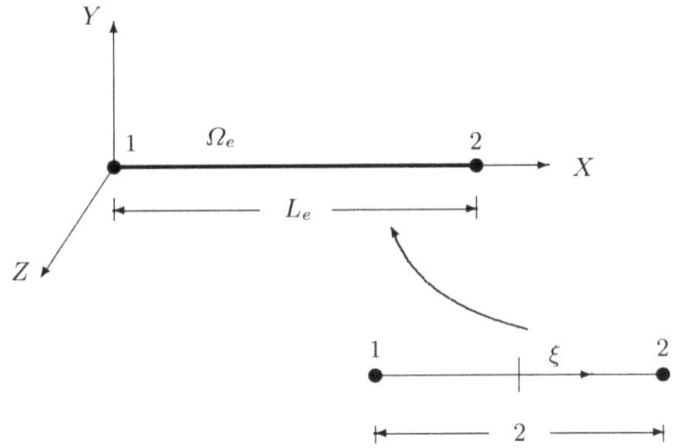

$$\Pi(\mathbf{p}_e) = \int_{(L_e)} W(\lambda_X^e)\, A\, dX - \mathbf{p}_e^T \mathbf{P}_e \Longrightarrow STAT \tag{7.13}$$

where $\mathbf{p}_e$ is the vector of the unknown nodal values related to the element L_e

$$\mathbf{p}_e^T = \{\, u_1, v_1, w_1, u_2, v_2, w_2\}. \tag{7.14}$$

and $\mathbf{P}_e^T = \{P_{11}, P_{21}, P_{31}, P_{12}, P_{22}, P_{32}\}$ contains the components of the forces acting at the element nodes.[4] Note that the first term in (7.13) can be integrated analytically since λ_X^e is constant within the element. This leads for the first term to

$$\int_{(L_e)} W^{SV,O}(\lambda_X^e)\, A\, dX = W^{SV,O}(\lambda_X^e)\, A\, L_e. \tag{7.15}$$

The residual $\mathbf{R}_e$, that describes the stress divergence term, follows from the derivative of (7.15)

$$\mathbf{R}_e(\mathbf{p}_e) = A\, L_e\, \frac{\partial W^{SV,O}[\lambda_X^e(\mathbf{p}_e)]}{\partial \mathbf{p}_e} = A\, L_e\, \frac{\hat{\delta} W^{SV,O}[\lambda_X^e(\mathbf{p}_e)]}{\hat{\delta} \mathbf{p}_e} \tag{7.16}$$

This formulations is valid as well for the material model of St. Venant as for the model of Ogden, only the associated strain energy expression (7.7) or (7.9) has to be inserted. Again the element tangent matrix is given by

$$\mathbf{K}_e = \frac{\partial \mathbf{R}_e(\mathbf{p}_e)}{\partial \mathbf{p}_e} = \frac{\hat{\delta} \mathbf{R}_e(\mathbf{p}_e)}{\hat{\delta} \mathbf{p}_e}. \tag{7.17}$$

[4]The second term in (7.13) can be considered at global level after assembly of the element residuals, as is well known from the linear finite element method.

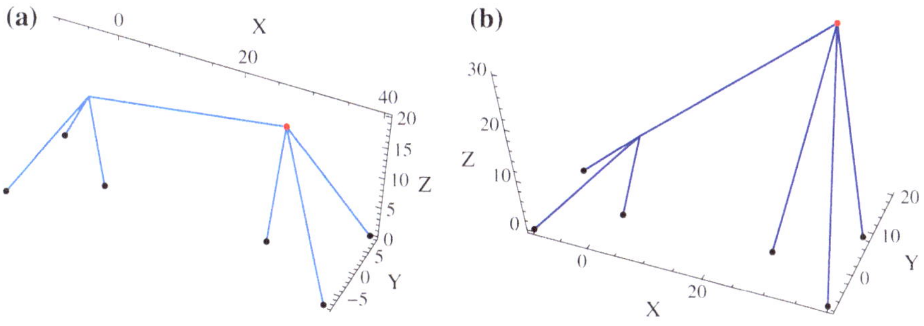

Fig. 7.3 **a** Truss system: undeformed state, **b** deformed state

The finite element code for a truss element can now be obtained by using *AceGen*. The input for an Ogden material with a two term strain energy ($r = 2$) is presented in Box 7.1. Note that the input follows directly the derivations presented above.

As an example, consider the system depicted in Fig. 7.3a consisting of 7 trusses undergoing finite deformations. All trusses are fixed at ground level. A two term strain energy function for an Ogden material is used with the data $\mu_1 = 100$, $\mu_2 = -10$, $\alpha_1 = 2$ and $\alpha_2 = -2$ for all trusses. The cross section of all trusses is set to $A = 10$.

```
SMSInitialize["Truss-3D-Ogden","Environment"→"AceFEM"];
SMSTemplate["SMSTopology"→"C1","SMSSymmetricTangent"→True,
  "SMSDOFGlobal"→3,"SMSDefaultIntegrationCode"→0,
  "SMSDomainDataNames"->
   {"μ1 -shear modulus 1","μ2 -shear modulus 2","α1 -factor 1",
    "α2 -factor 2","A -cross area"},
  "SMSDefaultData"->{1,1,1,1,1}];
nen=SMSNoNodes;ndim=SMSNoDimensions;np=SMSNoDOFGlobal;
SMSStandardModule["Tangent and residual"];
{μ1,μ2,α1,α2,A}⊢SMSReal[Table[es$$["Data",i],{i,5}]];
𝕏IO⊢Table[SMSReal[nd$$[i,"X",j]],{i,nen},{j,ndim}];
𝕦IO⊢SMSReal[Table[nd$$[i,"at",j],{i,nen},{j,ndim}]];
𝕩⊨𝕏IO+𝕦IO;𝕡e=Flatten[𝕦IO];
Le⊨SMSSqrt[(𝕏IO⟦1⟧-𝕏IO⟦2⟧).(𝕏IO⟦1⟧-𝕏IO⟦2⟧)];
le⊨SMSSqrt[(𝕩⟦1⟧-𝕩⟦2⟧).(𝕩⟦1⟧-𝕩⟦2⟧)];λX⊨le/Le;
WO⊨μ1/α1 (λX^α1+2 λX^(-1/2 α1)-3)+μ2/α2 (λX^α2+2 λX^(-1/2 α2)-3);
ℝe⊨A Le SMSD[WO,𝕡e];
SMSExport[ℝe,p$$,"AddIn"→True];
𝕂e⊨SMSD[ℝe,𝕡e];
SMSExport[𝕂e,s$$,"AddIn"→True];
SMSWrite[];
```

Box 7.1. *AceGen* input for the truss element with a two-parameter Ogden material

The truss system is loaded at the right top by prescribed displacements in Y and Z direction: $\bar{u}_1 = 20$ and $\bar{u}_3 = 10$. This load is prescribed in a single load

Table 7.1 Truss problem: Newton convergence for single load step

Iteration	Residual norm
1	5.09×10^{2}
2	3.93×10^{2}
3	3.34×10^{-1}
4	2.20×10^{-2}
5	2.74×10^{-5}
6	2.44×10^{-10}
7	2.30×10^{-15}

step. The convergence behaviour is documented in Table 7.1 and depicts clearly quadratic convergence up to computer precision. The deformed configuration of the truss system is shown in Fig. 7.3b.

7.2 Two-Dimensional Geometrically Exact Beam Elements

Nonlinear theories for beams were developed in the last four decades. They all can be applied for finite element discretization. Generally three approaches have to be distinguished.

- The first introduces a frame undergoing finite rigid rotations and formulates the strains and stresses relative to the rotations of the frame and is known as *co-rotational* formulation. Here the strains are assumed to be small but large deflections and rotations can be computed. Finite beam elements that base on such formulations can be found in z.B. in Oran and Kassimali (1976), Wempner (1969), Rankin and Brogan (1984), Lumpe (1982), Crisfield (1991) and Crisfield (1997).
- The second approach uses the three-dimensional continuum equations and introduces the beam kinematics by special isoparametric finite element interpolations. This approach is known as degenerated continuum approach, see e.g. Bathe and Bolourchi (1979), Dvorkin et al. (1988) or the textbooks by Bathe (1996) and Crisfield (1997).
- The third approach starts directly with beam kinematics but does not introduce any kinematical restrictions besides the classical assumption of "plane cross sections remain plane". No additional approximations are made and strains, deflections and rotations can be finite. These theories are called *geometrically exact*. Their development goes back to the work by Reissner (1972). A generalization for the three-dimensional case can be found in Simo (1985). Several other authors developed associated finite element formulations, see Simo and Vu-Quoc (1986), Pimenta and Yojo (1993), Jelenic and Saje (1995), Gruttmann et al. (1998) and Mäkinen (2007). In Gruttmann et al. (2000) elasto-plastic material and in Romero and Armero (2002) dynamics is considered within the geometrically exact framework.

Such beam theories include arbitrary loading and can be used for arbitrary beam and cable structures and thus can be applied in a general way to model one-dimensional construction elements.

Since there exists no general analytical solution of these nonlinear beam theories the framework of finite element methods has to be applied which enables the solution of many complex engineering problems.

In this chapter different beam theories are considered for the two-dimensional and three-dimensional case. Again *AceGen* is applied to obtain the associated finite element implementations.

In the first part of this chapter the theoretical basis for two-dimensional geometrically exact beam elements is derived and the related *AceGen* input is developed.

7.2.1 Two-Dimensional Beam Kinematics

The two-dimensional theory for the geometrically exact beam element is based on the assumptions of plane cross sections. It was derived in Reissner (1972) and is valid for finite deflections, rotations and strains. The nonlinear strain measures for the shear elastic beam can be found in Reissner (1972). They are based on the kinematical assumption for the beam deformation

$$\boldsymbol{\varphi} = \begin{Bmatrix} X_1 + u(X_1) \\ w(X_1) \end{Bmatrix} + X_2 \begin{Bmatrix} -\sin\psi(X_1) \\ \cos\psi(X_1) \end{Bmatrix} = \boldsymbol{\varphi}|_{X_2=0} + X_2\,\boldsymbol{t}, \tag{7.18}$$

see also Fig. 7.4. Here the initial configuration of the beam is straight and the local axis coincides with the global axis. When the beam is arbitrarily located in space then an additional transformation has to be applied.

The associated strain-deflection relations were derived in Reissner (1972) by using the principle of virtual work. This leads to strain measures for the axial strain ϵ, the shear strain γ and the curvature κ:

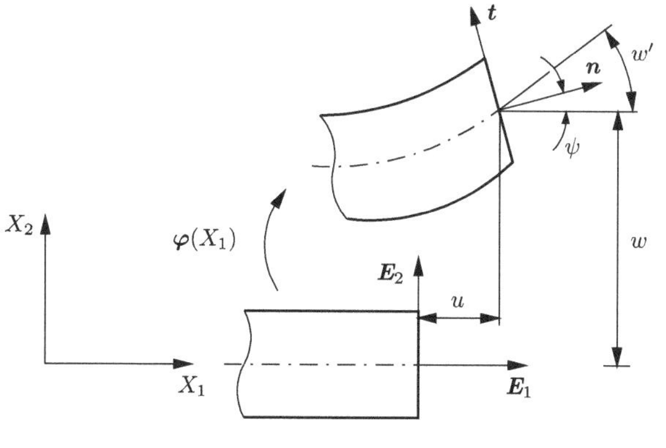

Fig. 7.4 Beam kinematics

$$\begin{aligned} \epsilon &= (1+u')\,\cos\psi + w'\,\sin\psi - 1, \\ \gamma &= w'\,\cos\psi - (1+u')\,\sin\psi, \\ \kappa &= \psi', \end{aligned} \tag{7.19}$$

where u is the displacement in axial direction, w the deflections and ψ the rotation, see Fig. 7.4. By $()'$ the derivative with respect to the coordinate X_1 are denoted. In this formulation the strain $(7.19)_3$ for the curvature is linear in ψ.[5]

The components in (7.19) can be summarized in the strain vector

$$\boldsymbol{\varepsilon} = \begin{Bmatrix} \epsilon \\ \gamma \\ \kappa \end{Bmatrix} = \mathbf{T}(\psi)\,\boldsymbol{u}' - \hat{\mathbf{T}}(\psi) \tag{7.20}$$

with

$$\mathbf{T}(\psi) = \begin{bmatrix} \cos\psi & \sin\psi & 0 \\ -\sin\psi & \cos\psi & 0 \\ 0 & 0 & 1 \end{bmatrix}, \; \boldsymbol{u}' = \begin{Bmatrix} u' \\ w' \\ \psi' \end{Bmatrix} \text{ and } \hat{\mathbf{T}}(\psi) = \begin{Bmatrix} 1-\cos\psi \\ \sin\psi \\ 0 \end{Bmatrix}.$$

7.2.2 Constitutive Equations

Constitutive equations describing finite elastic strains and elasto-plastic deformations can be included in different ways in the beam formulation.

For most engineering applications of beams it can be assumed that the strain are small, even for large deflections and rotations. Hence it is possible to describe elastic material behaviour by the classical Hooke law of the linear theory. It relates within the geometrically exact theory the 1st Piola-Kirchhoff stresses that are back-rotated using matrix $\boldsymbol{T}_B = \mathbf{T}^T \boldsymbol{P}$ with the strains in (7.19). These stresses can be interpreted as Biot stresses.[6] With the engineering strains $\boldsymbol{E}_B$ the stresses

$$\begin{Bmatrix} T_{11} \\ T_{12} \end{Bmatrix} = \begin{bmatrix} E & 0 \\ 0 & G \end{bmatrix} \begin{Bmatrix} \epsilon + X_2\,\kappa \\ \gamma \end{Bmatrix} \quad \text{or} \quad \boldsymbol{T}_B = \boldsymbol{C}\,\boldsymbol{E}_B. \tag{7.21}$$

are obtained, with Young's modulus E and the shear modulus G.

The integration of the stresses over the cross sectional area (width b and height h) yields the stress resultants

[5] This unexpected feature leads to a linear solution for special cases, like rolling up a beam under an end moment where the normal and shear forces are zero.

[6] Note that the stress $\boldsymbol{T}_B$ does not follow from the polar decomposition of the deformation gradient like the Biot stress, see Table 1.2.

$$N = \int_{(h)} T_{11}\, b\, dX_2, \quad Q = \int_{(h)} T_{12}\, b\, dX_2 \quad \text{and} \quad M = \int_{(h)} T_{11}\, X_2\, b\, dX_2 \tag{7.22}$$

leading with $\boldsymbol{S} = \{N, Q, M\}^T$ to the compact form

$$\boldsymbol{S} = \boldsymbol{D}\,\boldsymbol{\varepsilon}, \qquad \text{with } \boldsymbol{D} = \begin{bmatrix} EA & 0 & 0 \\ 0 & G\hat{A} & 0 \\ 0 & 0 & EI \end{bmatrix}, \tag{7.23}$$

with the cross sectional area A and the moment of inertia I. The introduction of the shear area $\hat{A}$ is related to a shear correction term that is needed within to correct the violation of the boundary condition for the shear stresses ($T_{12} = 0$) at $X_2 = \pm h/2$ due to the beam model assumption of "plane sections remain plane".

Other classical concepts for the formulation of constitutive equations for beams are based on the introduction of plastic hinges in order to avoid cross sectional integration. The related equations were developed in e.g. Kahn (1987) and Ehrlich and Armero (2005) within the framework of geometrically exact beams.

7.2.3 *Strain Energy Function*

As has been shown in the previous chapters, the most efficient approach for the derivation of the element vectors and matrices starts from the potential energy function. This can be formulation for the small strain case of a beam undergoing finite deformations as

$$\Pi = \frac{1}{2} \int_{l} \boldsymbol{\varepsilon} \cdot \boldsymbol{D}\boldsymbol{\varepsilon}\, dX - \int_{l} \boldsymbol{n} \cdot \boldsymbol{u}\, dX \tag{7.24}$$

with $\boldsymbol{n} = \{n_1, n_2, 0\}^T$ being the continuous loading in X_1 and X_2 and $\boldsymbol{u}$ being the generalized displacements $\boldsymbol{u} = \{u, w, \psi\}^T$. The strain $\boldsymbol{\varepsilon}$ is given explicitly in (7.20) and the elasticity matrix $\boldsymbol{D}$ for the stress resultants is provided in (7.23).

By formulation of the potential energy the nonlinear beam model is complete. This model and its approximations are now basis for the finite element discretization procedures discussed in the next section.

7.2.4 *Finite Element Formulation for the Two-Dimensional Beam*

For the finite element discretization of the shear elastic geometrically exact beam model linear shape functions can be selected since the weak form needs only ansatz

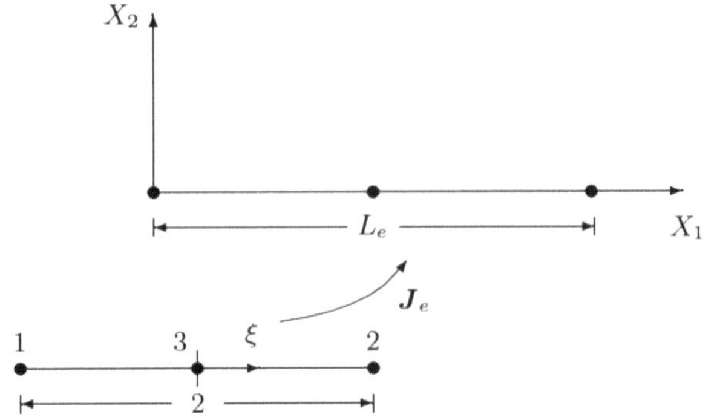

Fig. 7.5 Finite beam element

functions which are C^0-continuous. This is equivalent to the linear case, see e.g. Hughes (1987).

Hence polynomials (4.27) or (4.28) can be applied as finite element interpolations for the axial displacements u, the deflection w and the rotation ψ. Mathematically it can be shown for the geometrically linear case that quadratic shape functions depict a higher convergence order than linear shape functions. Computations show that this holds also for the geometrically nonlinear beam elements. Quadratic interpolations are based on elements with three nodes, see Fig. 7.5. The associated interpolation functions can be found in (4.28). In general the finite element approximation for a finite element is given by

$$u_e = \sum_{I=1}^{n} N_I(\xi)\, u_{cI}, \quad w_e = \sum_{I=1}^{n} N_I(\xi)\, w_{cI} \quad \text{and} \quad \psi_e = \sum_{I=1}^{n} N_I(\xi)\, \psi_{cI}, \tag{7.25}$$

for u, w and ψ. n is the number of nodes defining an element (linear ansatz: $n = 2$, quadratic ansatz: $n = 3$). The nodal values u_{cI}, w_{cI} and ψ_{cI} will be determined by inserting these interpolations into the weak form. Thus all quantities of the unknown displacement field $\boldsymbol{u} = \{u_e, w_e, \psi_e\}^T$ can be written as

$$\boldsymbol{u} = \sum_{I=1}^{n} \mathbf{N}_I(\xi)\, \mathbf{u}_I \quad \text{with} \quad \mathbf{N}_I(\xi) = \begin{bmatrix} N_I(\xi) & 0 & 0 \\ 0 & N_I(\xi) & 0 \\ 0 & 0 & N_I(\xi) \end{bmatrix} \tag{7.26}$$

where the nodal values are combined in the vector $\mathbf{u}_I = \{u_{cI}, w_{cI}, \psi_{cI}\}^T$.

Inserting (7.26) into (7.20) yields the element strains. The explicit form of the strains is obtained from (7.20)

$$\boldsymbol{\varepsilon} = \mathbf{T}(\psi_e)\, \frac{\partial \boldsymbol{u}}{\partial X_1} - \hat{\mathbf{T}}(\psi_e). \tag{7.27}$$

The angle ψ_e in the rotation matrix $\mathbf{T}$ and $\boldsymbol{n}$ is computed using $(7.25)_3$. The computation of the derivative of a variable $\boxdot$ with respect to X_1 yields

$$\boxdot_{,X_1} = [J_e(\xi)]^{-1}\,\boxdot_{,\xi}$$

where $J_e = dX_1\,/\,d\xi$ is given by $J_e = \sum_{K=1}^{n} N_K(\xi)_{,\xi}\,X_{1I}$. The *ADB* form of the strains then leads to

$$\varepsilon = \mathbf{T}(\psi_e)\left.\frac{\hat{\delta}\boldsymbol{u}}{\hat{\delta}\xi}\right|_{\frac{D\xi}{DX_1}=\left(\frac{\hat{\delta}X_1}{\hat{\delta}\xi}\right)^{-1}} - \hat{\mathbf{T}}(\psi_e). \tag{7.28}$$

In order to compute the residual and tangent stiffness matrix the strain energy function (7.24) can be written for one element as

$$\begin{aligned} \Pi &= \frac{1}{2}\int\limits_l \varepsilon\cdot\boldsymbol{D}\varepsilon\,dX - \int\limits_l \boldsymbol{n}\cdot\boldsymbol{u}\,dX \\ &\approx \sum_{g=1}^{n_g}\left[W[\varepsilon(\mathbf{p}_e,\xi_g)] - \boldsymbol{n}\cdot\boldsymbol{u}(\mathbf{p}_e,\xi_g)\right] J_e(\xi_g)\,w_g \end{aligned} \tag{7.29}$$

with

$$W[\varepsilon(\mathbf{p}_e,\xi_g)] = \frac{1}{2}\varepsilon^T(\mathbf{p}_e,\xi_g)\,\boldsymbol{D}\,\varepsilon(\mathbf{p}_e,\xi_g).$$

The vector of the nodal unknowns $\mathbf{p}_e$ is defined in (7.29) as $\mathbf{p}_e^T = \{u_1, w_1, \psi_1, u_2, w_2, \psi_2\}$. This vector contains all unknown nodal displacements and rotations related to the element.

Two-Dimensional Transformation to Global Coordinate Systems. The previous set of equations is sufficient to establish the residuals and tangent matrices needed within a solution method like the Newton scheme. These equations are related to the local coordinate system of the straight beam axis. Since beam members are used in most cases within the construction of complex structures, like multi-storey frames in which the beams are located in different positions the matrices and vectors have to be transformed to a global coordinate system. This can be performed in the same way as in the linear theory since all equations are refereed to the initial configuration. The transformation and associated matrices can be found e.g. in Hughes (1987), Bathe (1996) and Zienkiewicz and Taylor (2000b). By such transformation the local displacements and rotations $\boldsymbol{u}^L$, see (7.25), can be expressed in terms of the global deformations $\boldsymbol{u}$ via

$$\boldsymbol{u}^L = \mathbf{T}^{\alpha}\,\boldsymbol{u} = \mathbf{T}^{\alpha}\left(\sum_{I=1}^{n}\mathbf{N}_I(\xi)\,\mathbf{u}_I\right). \tag{7.30}$$

The explicit form of transformation matrix $\mathbf{T}^\alpha$ is given in the two-dimensional case by

$$\mathbf{T}^\alpha = \begin{bmatrix} \cos\alpha & \sin\alpha & 0 \\ -\sin\alpha & \cos\alpha & 0 \\ 0 & 0 & 1 \end{bmatrix}. \tag{7.31}$$

The angle α refers to the angle between the local and global coordinate axis X_1. sin and cos functions can be simply computed for a straight element by

$$\sin\alpha = \frac{(X_{2_2} - X_{2_1})}{L_e} \quad \text{and} \quad \cos\alpha = \frac{(X_{1_2} - X_{1_1})}{L_e}.$$

Here e.g. X_{2_1} is the coordinate in 2-direction at node 1, etc. and $L_e = \sqrt{(X_{2_2} - X_{2_1})^2 + (X_{1_2} - X_{1_1})^2}$ is the element length.

With this transformation the beam strains (7.28) used in (7.29) can be written in global form as

$$\boldsymbol{\varepsilon} = \mathbf{T}(\psi_e) \left. \frac{\hat{\delta}(\mathbf{T}^\alpha \boldsymbol{u})}{\hat{\delta}\xi} \right|_{\frac{D\xi}{DX_1} = \left(\frac{\hat{\delta}X_1}{\hat{\delta}\xi}\right)^{-1}} - \hat{\mathbf{T}}(\psi_e). \tag{7.32}$$

Note that $\hat{\mathbf{T}}(\psi_e)$ needs not to be transformed since the rotation is the same in local and global basis for the two-dimensional case.

Tangent and Residual. The residual of the finite beam element is now obtained using *AceGen*. Since all quantities in (7.29) depend upon the displacement field (7.25) the variation

$$\delta\Pi \approx \delta\mathbf{p}_e^T \left[\sum_{g=1}^{n_g} \frac{\partial \left[W(\boldsymbol{\varepsilon}(\mathbf{p}_e, \xi_g)) - \boldsymbol{n} \cdot \boldsymbol{u}(\mathbf{p}_e, \xi_g) \right]}{\partial \mathbf{p}_e} J_g \right] w_g \tag{7.33}$$

can be computed.

The variation of $\mathbf{p}_e$ is denoted by $\delta\mathbf{p}_e$. From (7.33) it can be recognized that the residual $\mathbf{R}_g$ at an integration point is given by

$$\mathbf{R}_g = \frac{\hat{\delta} \left[W_g(\boldsymbol{\varepsilon}_g(\mathbf{p}_e, \xi_g)) - \boldsymbol{n} \cdot \boldsymbol{u}(\mathbf{p}_e, \xi_g) \right]}{\hat{\delta}\mathbf{p}_e} J_e(\xi_g). \tag{7.34}$$

The total size of $\mathbf{R}_g$ is related to the number of entries in $\mathbf{p}_e$ which is six for the two node beam element with linear shape functions. Differentiation of this term with respect to the element displacements $\mathbf{p}_e$ yields the tangent stiffness matrix at an integration point

$$\mathbf{K}_g = \frac{\hat{\delta}\mathbf{R}_g(\mathbf{p}_e, \xi_g)}{\hat{\delta}\mathbf{p}_e}. \tag{7.35}$$

The element residual and tangent matrix are obtained by numerical integration

$$\mathbf{R}_e = \sum_{g=1}^{n_g} \mathbf{R}_g \, w_g \quad \text{and} \quad \mathbf{K}_e = \sum_{g=1}^{n_g} \mathbf{K}_g \, w_g .$$

It is well known that an under-integration of the shear term has to be used in a Timoshenko beam element. For the element with linear shape functions a one point integration is sufficient since it integrates the bending part with the right order and under-integrates the shear part in the strain energy. For the three node beam element with quadratic interpolations a two point Gauss integration is sufficient that can be applied for bending and under-integrates the shear terms.

The *AceGen* input for the beam element based on the strain (7.32) is provided in Box 7.2.

```
SMSInitialize["Beam-2D","Environment"→"AceFEM"];
SMSTemplate["SMSTopology"→"L1","SMSSymmetricTangent"→True,
  "SMSDOFGlobal"→3,"SMSDefaultIntegrationCode"→20(*one point*),
  "SMSDomainDataNames"->{"E","G","A","As","Iy"},
  "SMSDefaultData"->{1000,500,1,1,1}];
nen=SMSNoNodes;ndim=SMSNoDimensions;
SMSStandardModule["Tangent and residual"];
{Em,G,A,As,Iy}⊢SMSReal[Table[es$$["Data",i],{i,5}]];
𝕏IO⊢Table[SMSReal[nd$$[i,"X",j]],{i,nen},{j,ndim}];
𝕡eIO⊢SMSReal[Table[nd$$[i,"at",j],{i,nen},{j,3}]];
𝕡e=Flatten[𝕡eIO];
Le⊨SMSSqrt[(𝕏IO⟦1⟧-𝕏IO⟦2⟧).(𝕏IO⟦1⟧-𝕏IO⟦2⟧)];
SMSDo[Ig,1,SMSInteger[es$$["id","NoIntPoints"]]];
ξ⊢SMSReal[es$$["IntPoints",1,Ig]];
ℕh⊨1/2{(1-ξ),(1+ξ)};
X1⊢SMSFreeze[Le (ξ+1)/2];Je=SMSD[X1,ξ];
{cosα,sinα}⊨(𝕏IO⟦2⟧-𝕏IO⟦1⟧)/Le;
𝕋α⊨{{cosα,sinα,0},{-sinα,cosα,0},{0,0,1}};
𝕦⊨ℕh.𝕡eIO;𝕦L=𝕋α.𝕦;
δ𝕦⊨SMSD[𝕦L,X1,"Dependency"→{ξ,X1,1/Je}];
ψe=𝕦L⟦3⟧;𝕋a⊨{{Cos[ψe],Sin[ψe],0},{-Sin[ψe],Cos[ψe],0},{0,0,1}};
𝕋h⊨{1-Cos[ψe],Sin[ψe],0};ϵ⊨𝕋a.δ𝕦-𝕋h;
𝔻⊨DiagonalMatrix[{Em A,G As,Em Iy}];
Π⊨1/2 ϵ.𝔻.ϵ;
ℝg⊨Je SMSD[Π,𝕡e];
wgp⊢SMSReal[es$$["IntPoints",4,Ig]];
SMSExport[wgp ℝg,p$$,"AddIn"→True];
𝕂g⊨SMSD[ℝg,𝕡e];
SMSExport[wgp 𝕂g,s$$,"AddIn"→True];
SMSEndDo[];
SMSWrite[];
```

Box 7.2. *AceGen* input for the two-dimensional Reissner beam

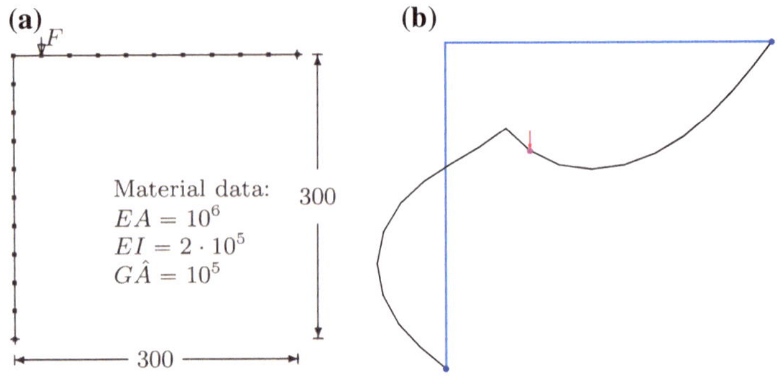

Fig. 7.6 **a** Frame structure and data, **b** deformed configuration

7.2.5 *Two-Dimensional Beam Example*

The element is now applied to a simulate the nonlinear deformation of a frame structure, depicted in Fig. 7.6a, that also documents geometrical and material data. The frame is loaded by a point force $F = \lambda \cdot 1$ acting in vertical direction and discretized by 21 finite beam elements. The computation of the load deflection curve is performed by using an arc length method. Figure 7.6b shows a deformed configuration with large deflections and rotations for a load factor $\lambda = 45$, which was computed with the geometrically exact beam model.

7.3 Three-Dimensional Geometrically Exact Beam Element

In this section we will develop a simple beam which is strictly valid only for circular cross sections. Within the present formulation warping deformations of the cross section are neglected that can occur under torsion loading. A theory that includes warping needs additional stress resultants for the higher order torsional moments, see e.g. Pimenta and Yojo (1993) and Gruttmann et al. (1998). In order to show the main features when using *AceGen* this more complicated theory is not used. However the formulation in this section is valid for large deflections and rotations. It includes shear deformation and is based on a fully three-dimensional formulation avoiding stress resultants.

7.3.1 *Beam Kinematics*

The undeformed configuration of the beam is described by the position vector of the beam axis $\boldsymbol{X}_0$ and the orthogonal basis vectors $\boldsymbol{t}_i$ that define the cross section. The latter are computed as

$$t_1 = \frac{X_{0,\xi}}{\|X_{0,\xi}\|}, \quad t_2 = \hat{Z} \times t_1, \quad t_3 = t_1 \times t_2 \tag{7.36}$$

where $\hat{\boldsymbol{Z}}$ is a given vector that defines the local $z - axis$ and ξ is the convective coordinate along the beam axis. With these definitions an arbitrary point $\boldsymbol{X}$ in the initial configuration of the beam can be computed

$$X = X_0 + \eta \, \frac{b}{2} \, t_2 + \zeta \, \frac{h}{2} \, t_3 \tag{7.37}$$

where η and ζ are the coordinates in the plane of the cross section and b is the width and h the height of the symmetric beam cross section, see Fig. 7.7.

When the beam is loaded it will undergo large displacements and rotations. The displacements are defined with respect to the beam axis. The rotations are defined around the axes of the orthogonal base vectors $\boldsymbol{t}_i$. For the update of the rotations the Rodriguez formula (7.64) is employed. Within this approach the directors $\boldsymbol{d}_2$ and $\boldsymbol{d}_3$ of the deformed beam can be computed via

$$d_2 = R\, t_2 \qquad d_3 = R\, t_3 \tag{7.38}$$

where $\boldsymbol{R} = \text{rot}(\phi)$ is the rotation tensor $\boldsymbol{R}$ representing the Rodrigues ormula and ϕ is the canonical rotation vector describing the rotations around the base vectors $\boldsymbol{t}_i$. The operator "rot" is described in Sect. 7.4.3. Now a point in the deformed configuration of the beam can be expressed by

$$x = X_0 + u + \eta \, \frac{b}{2} \, d_2 + \zeta \, \frac{h}{2} \, d_3 \tag{7.39}$$

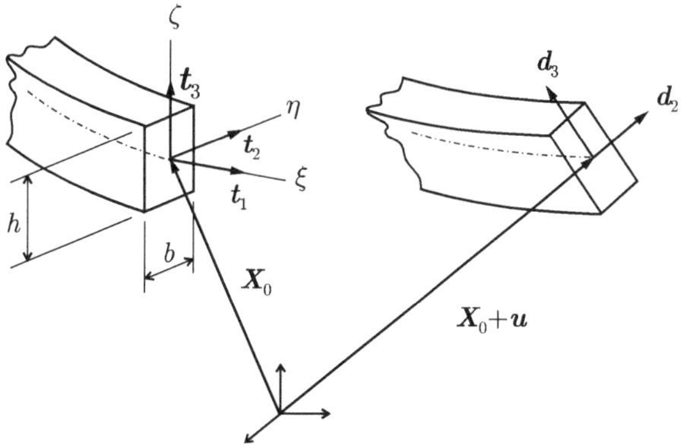

Fig. 7.7 Kinematics of the three-dimensional beam element

with the displacement vector $\boldsymbol{u}$. From this equation it is possible to compute the deformation gradient $\boldsymbol{F}$ and then a strain measure like the Green–Lagrangian strain $\boldsymbol{E}$. However one has to account for the different coordinate systems. This will be done in detail when the finite element is derived.

7.3.2 Constitutive Equation and Variational Form

In order to formulate large displacements and rotations of the beam the St. Venant constitutive equation (5.12) is used that relates the 2nd Piola–Kirchhoff stresses $\boldsymbol{S}$ to Green–Lagrangian strains $\boldsymbol{E}$ linearly

$$\boldsymbol{S} = \Lambda \text{ tr } \boldsymbol{E}\,\boldsymbol{1} + 2\,\mu\,\boldsymbol{E}. \tag{7.40}$$

This relation is valid for three-dimensional solids. In case of the beam one has to introduce two additional constraints. The stresses perpendicular to the beam axis have to be zero: $\tilde{S}_{22} = 0$ and $\tilde{S}_{33} = 0$. This condition can be incorporated directly into the formulation for the stress strain relation of the beam since both stresses can be eliminated analytically. However when using *AceGen* it is possible to do this automatically by applying the `Solve` command of *Mathematica*, see the input in Box 7.3.

The strain energy function that is needed to derive the beam element follows from the three-dimensional form (1.98) which is written here in terms of the Green–Lagrange strain tensor $\tilde{\boldsymbol{E}}$ that will be defined in a local Cartesian reference frame

$$W^{\text{beam}} = \int_B W(\tilde{\boldsymbol{E}})\,dV. \tag{7.41}$$

In case of the St. Venant constitutive equation $W(\tilde{\boldsymbol{E}})$ can be formulated as

$$W^{\text{beam}} = \frac{1}{2}\,\tilde{\boldsymbol{S}} \cdot \tilde{\boldsymbol{E}} \tag{7.42}$$

since the relation between stress and strain is linear. Together with the constraint equations for the stresses $\tilde{S}_{22}$ and $\tilde{S}_{33}$ the formulation can be employed to derive the beam element. Note that one has to integrate over the entire volume of the beam element.

7.3.3 Finite Element Formulation of the 3d-Beam

The finite element formulation will be based on an ansatz with linear shape functions. Thus the displacement vector is given by

$$\boldsymbol{u} = \begin{Bmatrix} u \\ v \\ w \end{Bmatrix} = \sum_{I=1}^{2} N_I(\xi) \begin{Bmatrix} u_{cI} \\ v_{cI} \\ w_{cI} \end{Bmatrix} \tag{7.43}$$

The rotations around the base vectors $\boldsymbol{t}_i$ are described by linear shape functions as well

$$\boldsymbol{\phi} = \begin{Bmatrix} \phi_1 \\ \phi_2 \\ \phi_3 \end{Bmatrix} = \sum_{I=1}^{2} N_I(\xi) \begin{Bmatrix} \phi_{1I} \\ \phi_{2I} \\ \phi_{3I} \end{Bmatrix} \tag{7.44}$$

with $N_1 = \frac{1}{2}(1-\xi)$ and $N_2 = \frac{1}{2}(1+\xi)$. Furthermore the discretization of the beam axis is given by

$$\boldsymbol{X}_0 = \begin{Bmatrix} X \\ Y \\ Z \end{Bmatrix} = \sum_{I=1}^{2} N_I(\xi) \begin{Bmatrix} X_{cI} \\ Y_{cI} \\ Z_{cI} \end{Bmatrix} . \tag{7.45}$$

Based on this interpolation the vector $\boldsymbol{t}_1$, tangent to the beam axis, can be computed from (7.36) using (7.45) which denotes the normal of the cross section. Now $\boldsymbol{t}_2$ and $\boldsymbol{t}_3$ follow directly from (7.36). Thus (7.37) is known and the Jacobian that maps the local beam space Ω to the reference space of a finite element $\Omega_\Box$ is given with $\Xi = \{\xi, \eta, \zeta\}$, see (4.14), by

$$\mathbf{J} = \frac{\partial \boldsymbol{X}}{\partial \boldsymbol{\Xi}} = \text{Grad}_{\Xi} \boldsymbol{X}. \tag{7.46}$$

By evaluating the rotation matrix $\boldsymbol{R}$ in terms of the rotations (7.44) the discrete form of the directors $\boldsymbol{d}_2$ and $\boldsymbol{d}_3$ is obtained from (7.38) and hence the discrete position vector of the deformed configuration follows from (7.39) as

$$\boldsymbol{x} = \boldsymbol{X}_0 + \boldsymbol{u} + \eta \frac{b}{2} \boldsymbol{d}_2 + \zeta \frac{h}{2} \boldsymbol{d}_3 . \tag{7.47}$$

The position vector $\boldsymbol{x}$ is given in terms of the unknown displacements (7.43) and rotations (7.44). From this relation the deformation gradient can be computed with respect to the coordinates of the reference configuration

$$\boldsymbol{F}_{\text{ref}} = \frac{\partial \boldsymbol{x}}{\partial \boldsymbol{\Xi}} = \frac{\hat{\delta} \boldsymbol{x}}{\hat{\delta} \boldsymbol{\Xi}} \tag{7.48}$$

The deformation gradient can now be related with the Jacobian to the global coordinate frame via

$$\boldsymbol{F} = \boldsymbol{F}_{\text{ref}} \, \mathbf{J}^{-1} = \frac{\partial \boldsymbol{x}}{\partial \boldsymbol{\Xi}} \frac{\partial \boldsymbol{\Xi}}{\partial \boldsymbol{X}} \tag{7.49}$$

By combining the local basis vectors of the cross section in the matrix

$$\mathbf{T} = [\, \boldsymbol{t}_1 \,|\, \boldsymbol{t}_2 \,|\, \boldsymbol{t}_3 \,]$$

it is possible to transform the deformation gradient to the local Cartesian reference frame

$$\tilde{\boldsymbol{F}} = \boldsymbol{F}\,\mathbf{T}. \tag{7.50}$$

Based on the last relation the discrete Green–Lagrangian strains $\tilde{\boldsymbol{E}}$ and 2nd Piola–Kirchhoff stresses $\tilde{\boldsymbol{S}}$ can be computed

$$\tilde{\boldsymbol{E}} = \frac{1}{2}\,(\tilde{\boldsymbol{F}}^T\,\tilde{\boldsymbol{F}} - \mathbf{1}), \tag{7.51}$$

$$\tilde{\boldsymbol{S}} = \Lambda\ \mathrm{tr}\,\tilde{\boldsymbol{E}}\,\mathbf{1} + 2\,\mu\,\tilde{\boldsymbol{E}}. \tag{7.52}$$

The constraint equations for the stresses $\tilde{S}_{22} = 0$ and $\tilde{S}_{33} = 0$ can then be resolved for the unknown deformations $\tilde{E}_{22}$ and $\tilde{E}_{33}$ and imposed on the strain energy for the beam

$$W^{\text{beam}} = \frac{1}{2}\tilde{\boldsymbol{S}}(\mathbf{p}_e)\cdot\tilde{\boldsymbol{E}}(\mathbf{p}_e)\bigg|_{\tilde{S}_{22}=0\wedge\tilde{S}_{33}=0}. \tag{7.53}$$

By using this form all quantities are directly related to the global coordinate system as in the two-dimensional beam, see (7.32). It may be noted that no stress resultants for normal, shear and bending modes have been defined and also no strain measures, work conjugate to the stress resultants, are needed. The additional effort for this simple formulation of the beam element is the integration over the cross sectional area and the fulfillment of the constraint equations for the stresses.

Constraint Equations for the Stresses. The constraint equations for the stresses can be enforced by the following *AceGen* input segment

```
(*1*)  Et⊨1/2 (Ftᵀ.Ft-IdentityMatrix[3]);
(*2*)  Et⟦3,3⟧=E33;Et⟦2,2⟧=E22;
(*3*)  St=λ Tr[Et] IdentityMatrix[3]+2 μ Et;
(*4*)  σcond=Solve[{St⟦2,2⟧==0,St⟦3,3⟧==0},{E22,E33}][[1]];
(*5*)  W⊨Simplify[ReplaceAll[1/2 Tr[St.Etᵀ],σcond]];
```
(7.54)

In the first line in (7.54) the Green–Lagrangian strain tensor $\tilde{\boldsymbol{E}} \equiv \mathbb{E}\mathtt{t}$ corresponding to the deformation gradient $\tilde{\boldsymbol{F}} = \mathcal{F}t$ in the local Cartesian reference frame is evaluated. The components of strain tensors $\tilde{E}_{22}$ and $\tilde{E}_{33}$ are replaced in the second line by the new independent variables $E22$ and $E33$. The stress tensor $\tilde{\boldsymbol{S}} \equiv \$\mathtt{t}$ is now evaluated for the symbolic values of $\tilde{E}_{22}$ and $\tilde{E}_{33}$ and the `Solve` command is used in line 4 to obtain an analytical solution for the constraint equations $\tilde{S}_{22} = 0$ and $\tilde{S}_{33} = 0$ with

$E22$ and $E33$ as unknowns leading to the following solution for the unknown $E22$ and $E33$

$$\left\{\mathbb{E}22 \to -\frac{\lambda\, \mathbb{E}t_{1,1}}{2\,(\lambda+\mu)},\ \mathbb{E}33 \to -\frac{\lambda\, \mathbb{E}t_{1,1}}{2\,(\lambda+\mu)}\right\}$$

Finally, in line 5, the strain energy for the beam is computed using (7.42) and the command `ReplaceAll` replaces all the occurrences of $E22$ and $E33$ in (7.42) with the obtained solution. The result is a strain energy function that implicitly fulfills the constraint equations $\tilde{S}_{22} = 0$ and $\tilde{S}_{33} = 0$.

Residual and Tangent Matrix. The residual of the three-dimensional beam element is now obtained using *AceGen*. Since all quantities in (7.53) depend upon the displacement (7.43) and rotation field (7.44) the residual at an integration point can be computed

$$\boldsymbol{R}_g = \frac{\partial W^{\text{beam}}(\mathbf{p}_e, \boldsymbol{\Xi}_g)}{\partial \mathbf{p}_e} J_e(\boldsymbol{\Xi}_g) = \frac{\hat{\delta}\, W^{\text{beam}}(\mathbf{p}_e, \boldsymbol{\Xi}_g)}{\hat{\delta} \mathbf{p}_e} J_e(\boldsymbol{\Xi}_g). \tag{7.55}$$

where $\mathbf{p}_e$ is the vector of the unknown nodal displacements and rotations with Furthermore $J_e(\boldsymbol{\Xi}_g)$ is the determinant of the Jacobian in (7.46). The total size of $\mathbf{R}_g$ is related to the number of entries in $\mathbf{p}_e$ which is 12 for the two node beam element with linear shape functions.

Differentiation of the residual $\mathbf{R}_g$ with respect to the element displacements and rotations $\mathbf{p}_e$ yields the tangent stiffness matrix at an integration point

$$\mathbf{K}_g = \frac{\partial \mathbf{R}_g(\mathbf{p}_e, \boldsymbol{\Xi}_g)}{\partial \mathbf{p}_e} = \frac{\hat{\delta} \mathbf{R}_g(\mathbf{p}_e, \boldsymbol{\Xi}_g)}{\hat{\delta} \mathbf{p}_e}. \tag{7.56}$$

The element residual and tangent matrix are obtained by numerical integration. It is well known that an under integration of the shear term has to be used in a shear deformable beam element with low order ansatz functions. Since the element is discretized by linear shape functions a one point integration is sufficient in the direction of the axis (coordinate ξ) while a two point integration across the cross section (coordinates η and ζ) is necessary to recover the normal, bending and torsional behavior of the beam.

The resulting *AceGen* input can be found in Box 7.3.

```
SMSInitialize["Beam-3D","Environment"→"AceFEM"];
SMSTemplate["SMSTopology"→"C1","SMSDOFGlobal"→6,
  "SMSDefaultIntegrationCode"→{20,21,21},
  "SMSDomainDataNames"->{"E","ν","B","H","Zx","Zt","Zz"},
  "SMSDefaultData"→{21000,.3,1,1,0,0,1}];
nen=SMSNoNodes;ndim=SMSNoDimensions;np=SMSNoDOFGlobal;
SMSStandardModule["Tangent and residual"];
{Em,ν,Bp,Hp,Zx,Zy,Zz}⊢SMSReal[Table[es$$["Data",i],{i,7}]];
XIO⊢Table[SMSReal[nd$$[i,"X",j]],{i,nen},{j,ndim}];
peIO⊢SMSReal[Table[nd$$[i,"at",j],{i,nen},{j,6}]];pe=Flatten[peIO];
Le⊨SMSSqrt[(XIO⟦1⟧-XIO⟦2⟧).(XIO⟦1⟧-XIO⟦2⟧)];
SMSDo[Ig,1,SMSInteger[es$$["id","NoIntPoints"]]];
Ξ={ξ,η,ζ}⊢Table[SMSReal[es$$["IntPoints",i,Ig]],{i,3}];
wgp⊢SMSReal[es$$["IntPoints",4,Ig]];
Nh⊨1/2 {(1-ξ),(1+ξ)};X0⊢SMSFreeze[Nh.XIO];
X0ξ=SMSD[X0,ξ];t1⊨X0ξ/SMSSqrt[X0ξ.X0ξ];t2⊨{Zx,Zy,Zz}×t1;t3⊨t1×t2;
T={t1,t2,t3}ᵀ;X=X0+η Bp/2 t2+ζ Hp/2 t3;
Je⊨SMSD[X,Ξ];Jed⊨Det[Je];uϕ⊨Nh.peIO;
R⊨Rot[uϕ⟦{4,5,6}⟧];d2=R.t2;d3=R.t3;
x=X0+uϕ⟦{1,2,3}⟧+η Bp/2 d2+ζ Hp/2 d3;
Fref⊨SMSD[x,{ξ,η,ζ}];Ft⊨Fref.SMSInverse[Je].T;
Et⊨1/2 (Ftᵀ.Ft-IdentityMatrix[3]);Et⟦3,3⟧=E33;Et⟦2,2⟧=E22;
{λ,μ}⊨SMSHookeToLame[Em,ν];St=λ Tr[Et] IdentityMatrix[3]+2 μ Et;
σcond=Solve[{St⟦2,2⟧==0,St⟦3,3⟧==0},{E22,E33}]⟦1⟧;
W⊨Simplify[ReplaceAll[1/2 Tr[St.Etᵀ],σcond]];
SMSDo[m,1,np];
  Rgm⊨Jed SMSD[W,pe,m];SMSExport[wgp Rgm,p$$[m],"AddIn"→True];
  SMSDo[n,m,np];
    Kgmn⊨SMSD[Rgm,pe,n];SMSExport[wgp Kgmn,s$$[m,n],"AddIn"→True];
  SMSEndDo[];
SMSEndDo[];
SMSEndDo[];
SMSWrite[];
```

Box 7.3. *AceGen* input for the three-dimensional beam element

7.4 General Shell Element

Basically three-dimensional solid elements, as discussed in Chap. 4, could be used to discretize shells, see the left side of Fig. 7.8. A linear interpolation through the thickness is then in close accordance with the assumption that plane cross sections remain plane during the deformation.[7] Additionally the change of thickness is taken into account in this model. It is well known that a pure displacement formulation, as

[7]Note that a plane isoparametric surface assumes in general a hyper surface after deformation and thus the cross section does not remain plane in general.

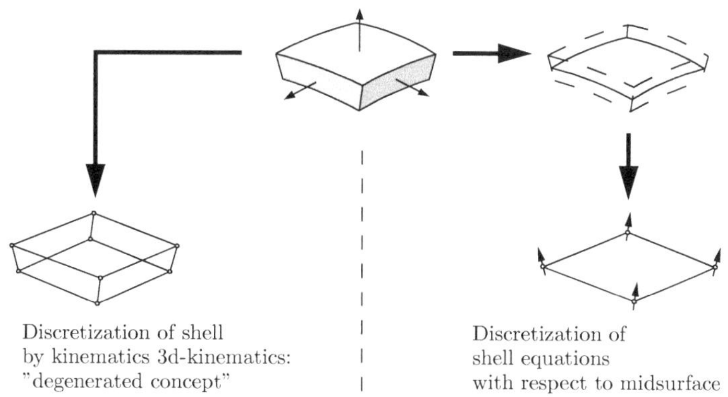

Fig. 7.8 Discretization of shells

defined in Sects. 4.1.3 and 4.2, leads to *locking* in the thin shell limit. Hence special interpolations have to be employed when three-dimensional solid formulations are used to discretize shells in order to eliminate locking, see e.g. Chap. 6.

Classically the approach described on the right side of Fig. 7.8 is followed for the development of nonlinear shell elements. However locking can also occur for specific shell formulations which are derived in that way. Here we will restrict our interest to the latter case and use the related formulation within *AceGen*

7.4.1 Introductory Remarks

The description of the shell continuum, the kinematics, the weak form of linear momentum and the constitutive equations can be stated in various ways. Some basic concepts will be discussed briefly in this introduction.

Shell Continuum and Shell Kinematics. Several possible formulations can be applied for the description of a shell continuum when finite element analysis is concerned. These are depicted in Fig. 7.8 and discussed in the following.

- Classically the description of shells is based on the definition of a middle surface, see the right path in Fig. 7.8. Based on such parametrization, kinematics, weak form and constitutive equations can be developed from the three-dimensional continuum equations. Here many different approximations for the kinematical description of the shell model in thickness direction can be employed. Different assumptions lead to equations for thin and thick shells which additionally can model deformations in thickness direction. Depending on the number of kinematical variables these approaches are denoted 5-, 6- or 7-parameter theories. Within this line of modeling different formulations were developed to construct finite elements for

shells undergoing finite deflections and rotations, see e.g. Simo et al. (1990a), Simo et al. (1990b), Onate and Cervera (1993), Sansour (1995), Eberlein and Wriggers (1999), Cirak et al. (2000), Campello et al. (2003), Pimenta et al. (2004) and Gruttmann and Wagner (2005).
- The second approach—called degenerated concept—uses the equations of a three-dimensional solid and introduces the shell kinematics at the discretization level, see left path of Fig. 7.8. As in shell theory a reference midsurface is selected, see e.g. Ramm (1976) and Bathe (1982). Within this approach no shell theory—besides the kinematical assumption—is needed for the discretization of a shell continuum. Hence this approach is conceptional simple. The introduction of stress resultants is not naturally included in this formulation. A comparison of finite elements based on classical shell theory and on the degenerated concept was investigated in Büchter and Ramm (1992), who found that both approaches lead under certain circumstances to the same finite element formulations.
- A third approach starts directly from the continuum elements discussed in Sect. 4.2. No shell midsurface is introduced explicitly but the node of the continuum elements are located at the upper and lower side of the shell continuum, see e.g. for discretizations using low order element Hughes and Liu (1981), Schoop (1986) Kühborn and Schoop (1992), Seifert (1996), Miehe (1998) and Hauptmann and Schweizerhof (1998). Higher order element formulations are discussed in Düster et al. (2001) and Benson et al. (2010).

Special measures have to be taken in all mentioned formulations to avoid locking effects.

In most approaches it is assumed that plane cross sections remain plane during the deformation of the shell continuum. This yields theories which include shear deformations that require only C^0-continuous discretizations within the finite element method.[8] In our model we will use a C^0-continuous interpolation leading to the so-called Reissner–Mindlin shell elements.

When no further assumptions besides "plane cross sections remain plane" are introduced within the derivation of the shell equations from the nonlinear continuum equations then this shell theory is called "geometrically exact". First investigations using geometrically exact shell theories can be found in Simo et al. (1989) and Wriggers and Gruttmann (1989).

[8]The classical Kirchhoff–Love hypothesis would, of course, be a natural assumption for kinematics of thin shells. However this additional constraint requires C^1-continuous interpolation functions that cannot be constructed by interpolations using only primary variables for triangular and quadrilateral finite elements. In this context, a new approach which combines interpolations of the deformations with the CAD description of the shell surfaces are of interest. These discretization employ Bezier or other C^1-continuous polynomials, see e.g. Cirak et al. (2000), Onate and Cervera (1993) and lately Hughes et al. (2005) who introduced the notion of *isogeometric* analysis.

Constitutive Equations. Not only the shell kinematics require special attention but also the formulation of constitutive equations for shells. Here different requirements have to be fulfilled, like the plane stress condition, that can complicate the shell formulation in case of nonlinear materials.

Generally the simplest model for a shell undergoing large rotations and deflections is the St. Venant constitutive equation, see (5.12). It is valid for small strains, but can capture the rigid body rotations due to finite deformations. Biomedical applications—like the analysis of skin or muscles—need the formulation of anisotropic elastic constitutive equations for large strains, see Sect. 5.1.4. Pneumatic springs made from rubber require hyper elastic constitutive equations. Here the relations of Ogden type (5.7) can be mentioned. The simulation of metal forming processes, e.g. deep drawing of sheets, needs the formulation of constitutive equations for isotropic or anisotropic finite elasto-plastic deformations. These were derived for isotropic materials in Sect. 5.3.

In general the shell equations can be derived in stress resultant form or obtained by directly using the stresses of the three-dimensional theory. Since *AceGen* can start from the strain energy function this discussion can be avoided when the kinematical equations are formulated correctly. In that case the easiest way to compute stresses is to use the direct approach, see e.g. (5.1). Thus the stresses related to the continuum are used directly, see e.g. Betsch et al. (1996) and Eberlein and Wriggers (1999). Other application of such formulations are presented in e.g. Büchter et al. (1994), Dvorkin et al. (1995), Seifert (1996) or Miehe (1998). For the special case of thin shells, the assumption of plane stress conditions have to be incorporated. Associated formulations are derived in e.g. Wriggers et al. (1996). Formulations for elasto-plastic material behaviour are documented in e.g. Roehl and Ramm (1996), Wriggers et al. (1996), Soric et al. (1997), Eberlein and Wriggers (1999) and Wagner et al. (2002).

Finite-Element Discretizations. Several new finite element discretization schemes for finite deformation analysis of shells were developed during the last years. The literature regarding this topic is quite extensive, specific approaches can be found in the papers cited so far. Different interpolation schemes were introduced that range from low order interpolations to ansatz spaces with high order.

From linear shell theory it is well known that the use of C^0 interpolation functions can lead to different stiffening effects which are known as *locking*. In shell problems different phenomena such as volume-, membrane- and shear-locking can be distinguished. In the following a special element formulation is introduced that is based on a triangular element. The interpolation of the displacements and rotations are different to avoid the aforementioned problems. Membrane locking can be eliminated by higher order interpolation. Volume and shear locking by reduced integration. Thus a quadratic interpolation is employed for the displacements to eliminate membrane locking and an incompatible linear ansatz for the rotation is used to overcome volume and shear locking. This element is simple and works very well as discussed in Campello et al. (2003) and Pimenta et al. (2004).

7.4.2 Shell Kinematics

Shells are three-dimensional curved structural members which have a small extension in thickness direction. A shell is described by introducing a shell midsurface as reference surface. Within such model it is possible to reduce the description of the shell continuum to two surface parameters ξ^α, that are defined on the shell midsurface $\mathcal{M}$, and to a coordinate ξ in thickness direction. A common choice for the parametrization of the shell midsurface are convective coordinates, which are described in detail in Appendix B.1.2. With the additional assumption that a plane cross section remains plane during the deformation the position vector of a point in the current configurations can be formulated by

$$\varphi(\xi^1, \xi^2, \xi, t) = \varphi_M(\xi^\alpha, t) + \xi\, \boldsymbol{d}(\xi^\alpha, t) \quad \text{with} \quad \xi \in \left[-\frac{h}{2}, +\frac{h}{2} \right]. \tag{7.57}$$

Here the Greek index α assumes values of 1 and 2. The director $\boldsymbol{d}$ describes the rotation of the cross section during the deformation and φ_M defines the deformation of the shell midsurface $\mathcal{M}$, see Fig. 7.9. Since as well the deformation of the shell midsurface φ_M as the director $\boldsymbol{d}$ are represented by three unknown components this formulation is called 6-Parameter theory. A point in the shell continuum is given by the position vector

$$\boldsymbol{X}(\xi^1, \xi^2, \xi, t) = \boldsymbol{X}_M(\xi^\alpha, t) + \xi\, \boldsymbol{N}(\xi^\alpha, t) \tag{7.58}$$

with respect to the initial configuration. Here $\boldsymbol{N}$ is a vector normal to the shell midsurface $\mathcal{M}$ in the initial configuration.

The co-variant base vectors related to the shell continuum are computed in the initial configuration using (1.27)

$$\boldsymbol{G}_\alpha = \frac{\partial \boldsymbol{X}}{\partial \xi^\alpha} = \boldsymbol{A}_\alpha + \xi\, \boldsymbol{N}_{,\alpha}, \qquad \boldsymbol{G}_3 = \frac{\partial \boldsymbol{X}}{\partial \xi} = \boldsymbol{N}. \tag{7.59}$$

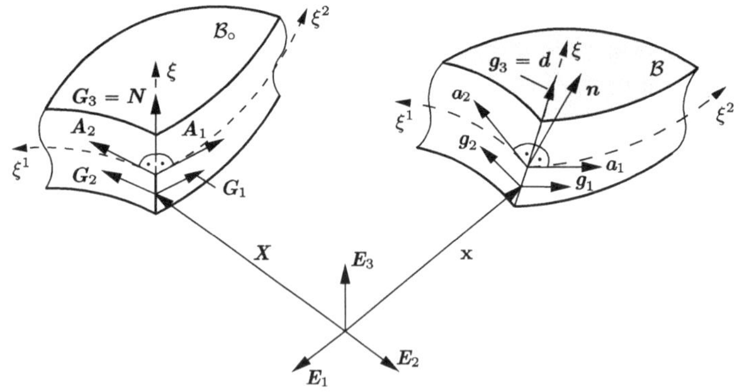

Fig. 7.9 Geometry of the shell continuum

Here the explicit dependencies of the base vectors with respect to the coordinates ξ^α and time t have been neglected to shorten notation. $\boldsymbol{A}_\alpha$ are tangent vectors at $\boldsymbol{X}_M(\xi^\alpha, t)$, defined at the shell midsurface $\mathcal{M}$, which follow from the derivative $\boldsymbol{A}_\alpha = \frac{\partial X_M}{\partial \xi^\alpha}$. Since the normal vector is perpendicular to $\boldsymbol{A}_\alpha$ the tangent vector can be used to compute the normal vector

$$\boldsymbol{N} = \frac{\boldsymbol{A}_1 \times \boldsymbol{A}_2}{\| \boldsymbol{A}_1 \times \boldsymbol{A}_2 \|}. \tag{7.60}$$

Tangent vectors can be obtained in the same manner with respect to the current configuration. From (7.57) the relations

$$\boldsymbol{g}_\alpha = \frac{\partial \boldsymbol{\varphi}}{\partial \xi^\alpha} = \boldsymbol{a}_\alpha + \xi \, \boldsymbol{d}_{,\alpha}, \qquad \boldsymbol{g}_3 = \frac{\partial \boldsymbol{\varphi}}{\partial \xi} = \boldsymbol{d}. \tag{7.61}$$

follow.

The kinematical relations are now employed to construct the deformation gradient with respect to the shell midsurface $\mathcal{M}$. From the Eq. (1.30) together with (7.61) the deformation gradient

$$\boldsymbol{F} = \boldsymbol{g}_i \otimes \boldsymbol{G}^i = (\boldsymbol{a}_\alpha + \xi \, \boldsymbol{d}_{,\alpha}) \otimes \boldsymbol{G}^\alpha + \boldsymbol{d} \otimes \boldsymbol{G}^3 \tag{7.62}$$

is obtained.

Once the deformation gradient is defined, the shell deformations are completely described. All further strain measures follow directly from (7.62) by introduction of the related three-dimensional measures such as e.g. the right Cauchy–Green tensor from $\boldsymbol{C} = \boldsymbol{F}^T \boldsymbol{F}$. Furthermore the Green–Lagrange strain tensor $\boldsymbol{E} = \frac{1}{2} (\boldsymbol{C} - \boldsymbol{1})$ can directly be computed.

The strain measures are only restricted by the kinematical assumption (7.57) and are valid for arbitrary finite strains. An additional constraint in (7.57) leads to a strain in thickness direction which is zero. This constraint is based on the assumption that the director $\boldsymbol{d}$ follows by a pure rotation $\boldsymbol{R}$ from the normal vector $\boldsymbol{N}$

$$\boldsymbol{d} = \boldsymbol{R} \, \boldsymbol{N}. \tag{7.63}$$

7.4.3 *Parametrization of the Rotations*

From many possibilities to define finite rotations of the normal vector, see e.g. Argyris (1982), we will here apply the Rodrigues formula. The director $\boldsymbol{d}$ is given by a pure rotation $\boldsymbol{d} = \boldsymbol{R} \, \boldsymbol{N}$. The rotation tensor $\boldsymbol{R}$ can be represented by the Rodrigues formula using the Euler rotation vector $\boldsymbol{\phi}$

$$\boldsymbol{R} = \boldsymbol{1} + c_1(\theta) \, \hat{\boldsymbol{\omega}} + c_2(\theta) \, \hat{\boldsymbol{\omega}} \, \hat{\boldsymbol{\omega}} \quad \text{with} \quad \theta = \| \boldsymbol{\phi} \| \tag{7.64}$$

with the vector and matrices

$$\boldsymbol{\phi} = \begin{Bmatrix} \phi_1 \\ \phi_2 \\ \phi_3 \end{Bmatrix} \quad \text{and} \quad \hat{\boldsymbol{\omega}} = \begin{bmatrix} 0 & -\phi_3 & \phi_2 \\ \phi_3 & 0 & -\phi_1 \\ -\phi_2 & \phi_1 & 0 \end{bmatrix}. \tag{7.65}$$

and the coefficients (for $\theta^2 \geqslant TOL$)

$$c_1(\theta) = \frac{\sin\theta}{\theta}, \qquad c_2(\theta) = \frac{1}{2}\left[\frac{\sin(\theta\,/\,2)}{\theta\,/\,2}\right]^2 \tag{7.66}$$

and the coefficients for very small angles (for $\theta^2 < TOL$)

$$c_1(\theta) = 1 - \frac{\theta^2}{6} + \frac{\theta^4}{120}, \qquad c_2(\theta) = \frac{1}{2} - \frac{\theta^2}{24} + \frac{\theta^4}{720} \tag{7.67}$$

where TOL has to be a small number e.g. $TOL = 10^{-8}$, however the approximation is very accurate also for larger values of θ. A detailed derivation of the Rodrigues formula can be found in Crisfield (1997). The update formula for the director vector $\boldsymbol{d}$ is singularity free within this approach. In Eq. (7.64) the three vector components of $\boldsymbol{\phi}$ are used to parameterize $\boldsymbol{R}$ and thus $\boldsymbol{d}$. This formulations is applied in many papers to describe rotations of shells undergoing finite deflections and rotations, see Simo et al. (1989), Betsch et al. (1996) and Campello et al. (2003) for the analysis of finite deformations of shells. In Box 7.4 an *AceGen* input is given that defines operator rot ≡ `Rot` in a way that $\boldsymbol{R} = \text{rot}(\boldsymbol{\phi})$.

```
Rot[ϕ_]:=Module[{ff,om,c1,c2,ϕskew,TOL},
   TOL=10^-8;ff⊧ϕ.ϕ;
   ϕskew={{0,-ϕ[[3]],ϕ[[2]]},{ϕ[[3]],0,-ϕ[[1]] },{-ϕ[[2]],ϕ[[1]] ,0}};
   SMSIf[ff<TOL
         ff   ff²        1   ff   ff²
    ,c1=1-—— +——— ;c2=— -——— +——— ;
         6   120         2   24   720
    FullSimplify[(IdentityMatrix[3]+c1 ϕskew+c2 ϕskew.ϕskew)]
    ,om = SMSSqrt[ff];
        Sin[om]       1  ⎛Sin[ om/2]⎞²
    c1=———————— ;c2=— ⎜—————————⎟  ;
          om          2  ⎝  om/2    ⎠
    FullSimplify[(IdentityMatrix[3]+c1 ϕskew+c2 ϕskew.ϕskew)]
    ,"InsertBefore"→Automatic
   ]
 ];
```

Box 7.4. Definition of operator `Rot` that calculates rotation tensor for given Euler rotation vector

7.4.4 Strain Energy Function

It is optimal to start with the derivation of a shell element from the associated strain energy function when using *AceGen* The strain energy W^{shell} has to be formulated as in the continuum form using the deformation of the shell[9] (7.57)

$$\Pi(\varphi) = \int_B W^{\text{shell}}(\varphi)\, dV - \int_B \rho_0\, \bar{b}\cdot\varphi\, dV - \int_{\partial B_\sigma} \bar{t}\cdot\varphi\, dA = 0. \tag{7.68}$$

Note that an integration over the complete shell volume has to be performed in (7.68).

The plane stress assumption in the model leads to a constraint. The constitutive equations have to include the condition $\sigma_{33} = 0$ which can be fulfilled directly for a linear relation between strains and stresses, e.g. the St. Venant constitutive equation. In case of a fully nonlinear material law a local Newton iteration has to be applied to satisfy this constraint, see e.g. Pimenta et al. (2004).

Here we will use the St. Venant material strain energy

$$W^{\text{shell}} = \frac{1}{2}\lambda\,(\text{tr}\tilde{\boldsymbol{E}})^2 - \mu\,\tilde{\boldsymbol{E}}\cdot\tilde{\boldsymbol{E}} \tag{7.69}$$

where λ and μ are the Lame material constants and $\tilde{\boldsymbol{E}}$ is the Green–Lagrange strain tensor that will be defined in a local Cartesian reference frame.

The plane stress condition can be satisfied by computing the stresses

$$\tilde{\boldsymbol{S}} = \frac{\partial W^{\text{shell}}}{\partial \tilde{\boldsymbol{E}}}$$

and solving

$$\tilde{S}_{33} = 0 \quad \text{to obtain the related } \tilde{E}_{33}.$$

This result can then be used in W^{shell} to directly enforce the plane stress condition since the dependence on $\tilde{E}_{33}$ is linear in St. Venant material equation. In *AceGen* this task is achieved by using `Solve` command in *Mathematica*that returns an analytical solution of the given set of equations as presented in Box 7.5, see also Sect. 7.3.3. Depending on the interpolation of the variables in the finite element discretization it may be necessary to include a constraint in the formulation. This is related to the drilling rotations around the normal of the shell surface which have to be zero.

[9]This formulation is contrary to classical shell theories where stress resultants are introduced with respect to the shell midsurface and related strain energies are constructed. The latter is especially complicated when finite strain constitutive models have to be considered.

7.4.5 *Finite Element Formulation for a Shear Deformable Shell*

As discussed before a triangular shell elements will be developed. The displacement field is interpolated by a quadratic conforming function while the rotations are approximated using a non conforming linear interpolation. This ansatz has the advantage that it does not depict shear locking due to the non-conforming rotations and does also not lead to membrane locking since a quadratic in plane approximation for the displacement field is selected. Furthermore the element does not depict any spurious modes and hence provides a simple but efficient discretization for shell problems that can be implemented in a straight forward manner. More details can be found in Campello et al. (2003), Pimenta et al. (2004) and Campello et al. (2007).

Isoparametric Concept. In Sect. 7.4.2 curvilinear convective coordinates ξ^α were introduced to parameterize the shell mid-surface $\mathcal{M}$. Covariant base vectors follow then from (7.59) or (7.61). These can be directly computed when using the isoparametric concept to discretize the shell geometry as well as the displacement and rotation field.

In general a unique mapping from the reference space of a finite element to the initial and current configuration is defined, see also Sect. 4.1. Within this mapping the position vector $\boldsymbol{X}_M$ and the displacement field $\boldsymbol{u}_M$ of the shell midsurface is discretized by the shape functions $\mathbf{N}_u$

$$\boldsymbol{u}_M = \sum_{I=1}^{n} N_{uI}(\xi,\eta)\,\mathbf{u}_I, \qquad \boldsymbol{X}_M = \sum_{I=1}^{n} N_{uI}(\xi,\eta)\,\mathbf{X}_I. \tag{7.70}$$

n denotes the number of nodes of the finite element and ξ and η are the convective coordinates. For the interpolation quadratic shape functions $\mathbf{N}_u$ an element with six nodes is needed ($n = 6$). For the six shape functions, see (4.32). The shape functions $\mathbf{N}_u$ are referred to the reference configuration $\Omega_\square$, see also (4.32) and Fig. 4.6. Based on (7.70) relation $\boldsymbol{X}_M = \boldsymbol{X}_M(\xi,\eta)$ is obtained. Hence the position vector $\boldsymbol{X}_M$, which describes the initial configuration B of the shell mid-surface, is given by the coordinates ξ and η of the reference element $\Omega_\square$.

In the same way non-conforming ansatz functions for the rotations ϕ are defined on a reference triangle leading to

$$\phi = \sum_{K=1}^{3} N_{\phi K}(\zeta)\,\phi_K \tag{7.71}$$

with

$$N_{\phi 1} = -1 + 2\zeta^1 + 2\zeta^2, \qquad N_{\phi 2} = 1 - 2\zeta^1, \qquad N_{\phi 3} = 1 - 2\zeta^2.$$

Here the shape functions are related to the mid nodes at the element edges (nodes with an index 4, 5 and 6 in Fig. 4.6).

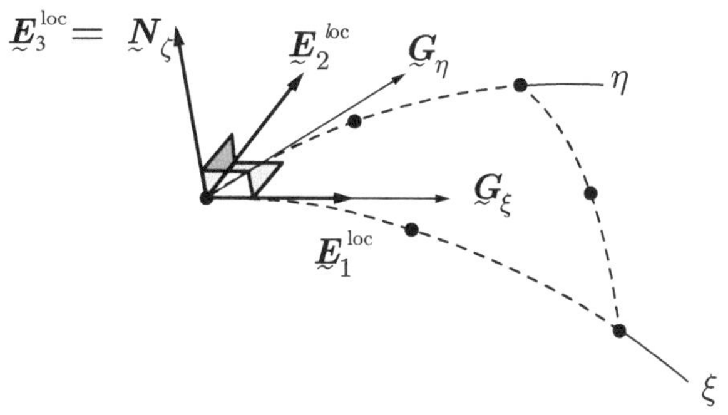

Fig. 7.10 Local cartesian basis system $\{\boldsymbol{E}_i^{\text{loc}}\}$

The isoparametric concept described in Sect. 4.1 cannot be applied in a direct way to shells since the isoparametric map of a plane reference element into three dimensional space is not singularity free. In order to avoid this singularity a modification will be used that is related to a local cartesian basis $\{\boldsymbol{E}_i^{\text{loc}} : i = 1, 2, 3\}$.[10]

The local cartesian basis is determined by the basis vectors $\boldsymbol{G}_\xi$ and $\boldsymbol{G}_\eta$ of the undeformed initial configuration of the shell midsurface. These covariant base vectors can be computed within the isoparametric formulation as, see (7.59),

$$\boldsymbol{G}_\xi = \frac{\partial \boldsymbol{X}_M}{\partial \xi} = \sum_{I=1}^{6} N_{I,\xi}\, \mathbf{X}_I = \frac{\hat{\delta} \boldsymbol{X}_M}{\hat{\delta} \xi} \tag{7.72}$$

$$\boldsymbol{G}_\eta = \frac{\partial \boldsymbol{X}_M}{\partial \eta} = \sum_{I=1}^{6} N_{I,\eta}\, \mathbf{X}_I = \frac{\hat{\delta} \boldsymbol{X}_M}{\hat{\delta} \eta} \tag{7.73}$$

The associated normal vector $\boldsymbol{N}_\zeta$ is determined by the cross product $\boldsymbol{G}_\xi \times \boldsymbol{G}_\eta$ analogous to (7.60). Now the local cartesian basis $\{\boldsymbol{E}_i^{\text{loc}}\}$—as depicted in Fig. 7.10—can be defined by

$$\boldsymbol{E}_1^{\text{loc}} = \frac{\boldsymbol{G}_1}{\|\boldsymbol{G}_1\|}, \qquad \boldsymbol{E}_3^{\text{loc}} = \frac{\boldsymbol{N}_\zeta}{\|\boldsymbol{N}_\zeta\|}, \qquad \boldsymbol{E}_2^{\text{loc}} = \boldsymbol{E}_3^{\text{loc}} \times \boldsymbol{E}_1^{\text{loc}}. \tag{7.74}$$

A point in shell space is then given by

$$\boldsymbol{X} = \boldsymbol{X}_M + \frac{h}{2} \zeta\, \boldsymbol{E}_3^{\text{loc}} \tag{7.75}$$

[10] In general this basis does not coincidence with the global cartesian coordinate system $\{\boldsymbol{E}_i\}$ of the reference configuration. Thus a transformation of vectors and tensors to the global coordinate system has to be performed in order to assemble the shell elements correctly since the FE displacements and rotations are defined in the global system.

The gradient $\mathbf{J}_e$, which belongs to the modified isoparametric map can be computed as

$$\mathbf{J}_e = \text{Grad}_{\Xi}\, \boldsymbol{X} = \frac{\partial \boldsymbol{X}}{\partial \Xi} = \frac{\hat{\delta} \boldsymbol{X}}{\hat{\delta} \Xi} \tag{7.76}$$

with its determinant $J_e = \det \mathbf{J}_e$ and reference coordinates $\Xi^T = \{\xi, \eta, \zeta\}$ describing the reference volume $[-1, 1] \times [-1, 1] \times [-1, 1]$.

Formulation of the Shell Element. The deformed configuration of the shell is described by the displacement field and the rotations. This leads with (7.57) to

$$\boldsymbol{x} = \boldsymbol{X}_M + \boldsymbol{u}_M + \frac{h}{2}\, \zeta\, \boldsymbol{d} \tag{7.77}$$

where $\boldsymbol{X}_M$ and $\boldsymbol{u}_M$ are defined in (7.70) and $\boldsymbol{d}$ is the director representing the deformed normal that is rotated using the Rodriguez formula (7.64), for the *AceGen* input, see Box 7.4,

$$\boldsymbol{d} = \boldsymbol{R}(\boldsymbol{\phi})\, \boldsymbol{E}_3^{\text{loc}} \tag{7.78}$$

Now the local deformation gradient in shell space in the reference configuration simply follows as

$$\boldsymbol{F}_{\text{ref}} = \text{Grad}_{\Xi}\, \boldsymbol{x} = \frac{\partial \mathbf{x}}{\partial \Xi} = \frac{\hat{\delta} \mathbf{x}}{\hat{\delta} \Xi}. \tag{7.79}$$

Hence $\boldsymbol{F}_{ref}$ will reduce to a unit matrix in the initial configuration with respect to $\{\mathbf{E}_i^{\text{loc}}\}$. The global deformation gradient is computed using the Jacobian which stems from the isoparametric map

$$\boldsymbol{F} = \boldsymbol{F}_{\text{ref}}\, \mathbf{J}_e^{-1}. \tag{7.80}$$

The transformation from the global to the local Cartesian reference frame is given by the transformation matrix

$$\mathbf{T} = \left[\, \boldsymbol{E}_1^{\text{loc}} \,|\, \boldsymbol{E}_2^{\text{loc}} \,|\, \boldsymbol{E}_3^{\text{loc}} \,\right] \tag{7.81}$$

that is computed from the local base vectors (7.74). With this the deformation gradient can be transformed

$$\tilde{\boldsymbol{F}} = \boldsymbol{F}\, \mathbf{T}$$

as well as the rotation matrix

$$\tilde{\boldsymbol{R}} = \boldsymbol{R}\, \mathbf{T}.$$

Now the Green–Lagrange strain $\tilde{\boldsymbol{E}}$, can be obtained from

$$\tilde{\boldsymbol{E}} = \frac{1}{2}(\tilde{\boldsymbol{C}} - \mathbf{1}) \quad \text{with} \quad \tilde{\boldsymbol{C}} = \tilde{\boldsymbol{F}}^T \tilde{\boldsymbol{F}} \tag{7.82}$$

This strain can be inserted in the strain energy function (7.69) that is needed in (7.68).

The rotation around the normal - often also called drilling degree of freedom - has to be suppressed in the formulation since in shell analysis there is no rotation around the normal. Here a penalty term is added based on the rotation tensor in the local frame and the deformation in the local frame

$$W^{\text{drill}} = \frac{1}{2}\,\varepsilon_P\,(\,\tilde{Q}_{12} - \tilde{Q}_{21}\,)^2 \tag{7.83}$$

where the components $\tilde{Q}_{12}$ and $\tilde{Q}_{21}$ describing the drilling rotations are computed from $\tilde{\boldsymbol{Q}} = \tilde{\boldsymbol{R}}^T \tilde{\boldsymbol{F}}$. A good choice for the penalty parameter is $\varepsilon_P = E\,h^3$ where E is Young's modulus and h the shell thickness. Using now (7.69) and (7.83) together with the plane stress assumption the elastic potential can be computed as

$$\Pi^*(\mathbf{p}_e,\,\Xi) = W^{\text{shell}}(\tilde{\boldsymbol{E}})\Big|_{\tilde{S}_{33}=0} + W^{\text{drill}}(\tilde{\boldsymbol{Q}}). \tag{7.84}$$

The *AceGen* procedure that can be used to fulfill the plane stress assumption automatically is equivalent to the procedure described in Sect. 7.3.3 for the fulfillment of the stress constraints of the three-dimensional beam element.

The residual $\mathbf{R}_g$ at an integration point is given for the first part in (7.68) by

$$\mathbf{R}_g = \frac{\partial \Pi^*(\mathbf{p}_e,\,\Xi_g)}{\partial \mathbf{p}_e} J_e = \frac{\hat{\delta} \Pi^*(\mathbf{p}_e,\,\Xi_g)}{\hat{\delta} \mathbf{p}_e} J_e. \tag{7.85}$$

The total size of $\mathbf{R}_g$ is related to the number of entries in $\mathbf{p}_e$. Here for the triangular shell element with quadratic and linear non-conforming shape functions the number of entries are $(3 \times 6) + (3 \times 3) = 27$.

Differentiation of this term with respect to the element displacements $\mathbf{p}_e$ yields the tangent stiffness matrix at an integration point

$$\mathbf{K}_g = \frac{\partial \mathbf{R}_g(\mathbf{p}_e,\,\Xi_g)}{\partial \mathbf{p}_e} = \frac{\hat{\delta} \mathbf{R}_g(\mathbf{p}_e,\,\Xi_g)}{\hat{\delta} \mathbf{p}_e}. \tag{7.86}$$

The differentiation in (7.85) and (7.86) is performed with respect to the set of global degrees of freedom $\mathbf{p}_e$, thus no additional transformations are needed for the proper assembly of residual vectors and tangent matrices.

Equations (7.70)–(7.86) describe the Gauss point contribution to the tangent matrix and residual of the Mindlin shell element with quadratic conforming interpolation of mid-surface displacements and linear nonconforming interpolation of rotations. All equations are stated in a form where the differentiation operator is replaced by the corresponding computational derivative operator. Thus, Eqs. (7.70)–(7.86) represent an *ADB* form of the Mindlin shell element and they can be directly translated into *AceGen* input. The corresponding *AceGen* input is provided in Box 7.5.

```
SMSInitialize["Shell-P2P1","Environment"→"AceFEM"];
SMSTemplate["SMSTopology"→"P2","SMSSymmetricTangent"→True,
  "SMSNodeID"→{"D","D","D","DFi","DFi","DFi"},
  "SMSDOFGlobal"→{3,3,3,6,6,6},"SMSDefaultIntegrationCode"→{16,21},
  "SMSDomainDataNames"->{"E","ν","h"},"SMSDefaultData"->{1,0,1}];
nen=SMSNoNodes;ndim=SMSNoDimensions;np=SMSNoDOFGlobal;
SMSStandardModule["Tangent and residual"];
{Em,ν,h}⊢SMSReal[Table[es$$["Data",i],{i,3}]];
XIO⊢Table[SMSReal[nd$$[i,"X",j]],{i,nen},{j,ndim}];
peIO⊢SMSReal[Table[nd$$[i,"at",j],{i,nen},{j,SMSDOFGlobal⟦i⟧}]];
pe=Flatten[peIO];uIO=peIO⟦All,{1,2,3}⟧;ϕIO=peIO⟦{4,5,6},{4,5,6}⟧;
SMSDo[Ig,1,SMSInteger[es$$["id","NoIntPoints"]]];
Ξ={ξ,η,ζ}⊢Table[SMSReal[es$$["IntPoints",i,Ig]],{i,3}];
wgp⊢SMSReal[es$$["IntPoints",4,Ig]];
ζc=1-ξ-η;Nu⊨{(2 ξ-1) ξ,(2 η-1) η,(2 ζc-1) ζc,4 ξ η,4 η ζc,4 ξ ζc};
Nϕ⊨{-1+2 ξ+2 η,1-2 ξ,1-2 η};
XM⊨Nu.XIO;uM⊨Nu.uIO;ϕ⊨Nϕ.ϕIO;
{Gξ,Gη}⊨SMSD[XM,{ξ,η}]ᵀ;Nζ=Gξ×Gη;
E13⊨Nζ/SMSSqrt[Nζ.Nζ];E11⊨Gξ/SMSSqrt[Gξ.Gξ];E12⊨E13×E11;
T⊨{E11,E12,E13}ᵀ;X⊨XM+h/2 ζ E13;
Je⊨SMSD[X,Ξ];Jed⊨Det[Je];R⊨Rot[ϕ];d=R.E13;
Ft⊨SMSD[XM+uM+h/2 ζ d,Ξ].SMSInverse[Je].T;
Et⊨1/2 (Ftᵀ.Ft-IdentityMatrix[3]);Et[[3,3]]=E33;
{λ,μ}⊨SMSHookeToLame[Em,ν];St=λ Tr[Et] IdentityMatrix[3]+2 μ Et;
σcond=Solve[St[[3,3]]==0,E33][[1]];
Wshell⊨Simplify[ReplaceAll[1/2 λ Tr[Et]²+μ Tr[Et.Et],σcond]];
Q⊨Tᵀ.Rᵀ.Ft;Wdrill⊨1/2 Em h³ (Q⟦1,2⟧-Q⟦2,1⟧)²;
SMSDo[m,1,np];
  Rgm⊨Jed SMSD[Wshell+Wdrill,pe,m];
  SMSExport[wgp Rgm,p$$[m],"AddIn"→True];
  SMSDo[n,m,np];
    Kgmn⊨SMSD[Rgm,pe,n];SMSExport[wgp Kgmn,s$$[m,n],"AddIn"→True];
  SMSEndDo[];
SMSEndDo[];
SMSEndDo[];
SMSWrite[];
```

Box 7.5. *AceGen* input for the Mindlin shell

7.4.6 Example

The numerical simulations in this section are performed by using the finite shell elements developed in the last section.

Buckling of a L-Shaped Plate. A L-shaped plate is considered that was already discussed in many papers, see e.g. Simo et al. (1990a), Wriggers and Gruttmann

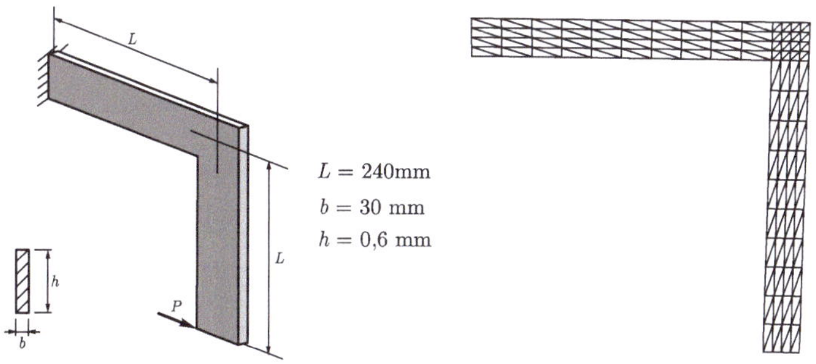

Fig. 7.11 L-shaped plate: system, data and mesh

Fig. 7.12 Load deflection curve of the l-shaped plate and deformed shape at $F = 1.35$

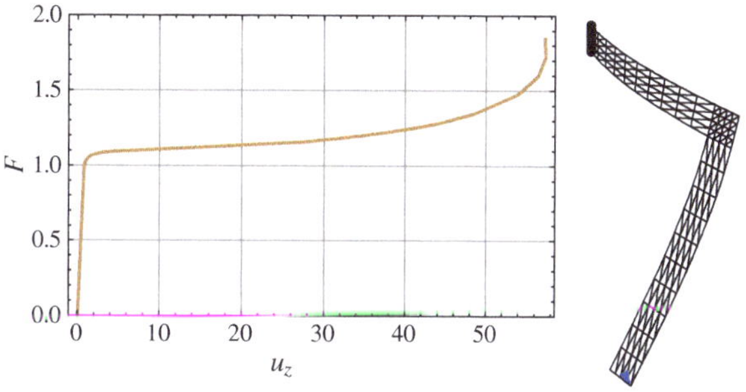

(1993) and Campello et al. (2003). The system is depicted in Fig. 7.11. Its Young's modulus is $E = 71240$ N/mm^2 the Poisson ratio is $\nu = 0.31$. The flat L-shaped plate strip of Fig. 7.11 is fully clamped in the left upper edge. It is subjected to an in-plane point load at the free end in order to investigate the lateral stability of the plate. In order to compute the load deflection curve a small perturbation load of $F_{pert} = F/1000$ is imposed at the free edge in the out-of-plane direction. This imperfection will initialize the post-critical lateral deflections. As can be seen in Fig. 7.11 on the right, a relatively coarse finite element mesh with a total of 192 triangular elements and 441 nodes has been used leading to a total of 2232 degrees of freedom.

Figure 7.12 shows the results for the chosen discretization. The out-of-plane deformation u_z is plotted versus the load F. The classical pitchfork bifurcation behaviour is observed. Furthermore the right side of Fig. 7.12 depicts the deformed configuration at load step $F = 1.348$ in true scale.

References

Argyris, J.H. 1982. An excursion into large rotations. *Computer Methods in Applied Mechanics and Engineering* 32: 85–155.

Bathe, K.J. 1982. *Finite Element Procedures in Engineering Analysis*. Englewood Cliffs: Prentice-Hall.

Bathe, K.J. 1996. *Finite Element Procedures*. Englewood Cliffs: Prentice-Hall.

Bathe, K.J., and S. Bolourchi. 1979. Large displacement analysis of three-dimensional beam structures. *International Journal for Numerical Methods in Engineering* 14: 961–986.

Benson, D., Y. Bazilevs, M. Hsu, and T. Hughes. 2010. Isogeometric shell analysis: the reissner-mindlin shell. *Computer Methods in Applied Mechanics and Engineering* 199(5): 276–289.

Betsch, P., F. Gruttmann, and E. Stein. 1996. A 4-node finite shell element for the implementation of assumed general hyperelastic 3d-elasticity at finite strains. *Computer Methods in Applied Mechanics and Engineering* 130: 57–79.

Büchter, N., E. Ramm, and D. Roehl. 1994. Three-dimensional extension of non-linear shell formulation based on the enhanced assumed strain concept. *International Journal for Numerical Methods in Engineering* 37: 2551–2568.

Büchter, N., and E. Ramm. 1992. Shell theory versus degeneration - a comparison in large rotation finite element analysis. *International Journal for Numerical Methods in Engineering* 34: 39–59.

Campello, E.M.B., P.M. Pimenta, and P. Wriggers. 2003. A triangular finte shell element based on a fully nonlinear shell formulation. *Computational Mechanics* 31: 505–518.

Campello, E.M.B., P.M. Pimenta, and P. Wriggers. 2007. Elastic-plastic analysis of metallic shells at finite strains. *Rem-Revista Escola de Minas* 60: 381–389.

Cirak, F., M. Ortiz, and P. Schröder. 2000. Subdivision surfaces: a new paradigm for thin shell finite-element analysis. *International Journal for Numerical Methods in Engineering* 47: 2039–2072.

Crisfield, M.A. 1991. *Non-linear finite element analysis of solids and structures*, vol. 1. Chichester: Wiley.

Crisfield, M.A. 1997. *Non-linear finite element analysis of solids and structures*, vol. 2. Chichester: Wiley.

Düster, A., H. Bröker, and E. Rank. 2001. The p-version of the finite element method for three-dimensional curved thin walled structures. *International Journal for Numerical Methods in Engineering* 52: 673–703.

Dvorkin, E.N., E. Onate, and J. Oliver. 1988. On a nonlinear formulation for curved Timoshenko beam elements considering large displacement/rotation increments. *International Journal for Numerical Methods in Engineering* 26: 1597–1613.

Dvorkin, E.N., D. Pantuso, and A. Repetto. 1995. A formulation of the MITC4 shell element for finite strain elasto-plastic analysis. *Computer Methods in Applied Mechanics and Engineering* 125: 17–40.

Eberlein, R., and P. Wriggers. 1999. Finite element concepts for finite elastoplastic strains and isotropic stress response in shells: theoretical and computational analysis. *Computer Methods in Applied Mechanics and Engineering* 171: 243–279.

Ehrlich, D., and F. Armero. 2005. Finite element methods for the analysis of softening plastic hinges in beams and frames. *Computational Mechanics* 35(4): 237–264.

Gruttmann, F., and W. Wagner. 2005. A linear quadrilateral shell element with fast stiffness computation. *Computer Methods in Applied Mechanics and Engineering* 194(39–41): 4279–4300.

Gruttmann, F., R. Sauer, and W. Wagner. 1998. A geometrical non-linear eccentric 3D-beam element with arbitrary cross-sections. *Computer Methods in Applied Mechanics and Engineering* 160: 383–400.

Gruttmann, F., R. Sauer, and W. Wagner. 2000. Theory and numerics of three-dimensional beams with elastoplastic material behaviour. *International Journal for Numerical Methods in Engineering* 48: 1675–1702.

Hauptmann, R., and K. Schweizerhof. 1998. A systematic development of 'solid-shell' element formulation for linear and nonlinear analyses employing only displacement degree of freedom. *International Journal for Numerical Methods in Engineering* 42: 49–69.

Hughes, T.R.J. 1987. *The Finite Element Method*. Englewood Cliffs: Prentice Hall.

Hughes, T.J.R., J.A. Cottrell, and Y. Bazilevs. 2005. Isogeometric analysis: CAD, finite elements, NURBS, exact geometry, and mesh refinement. *Computer Methods in Applied Mechanics and Engineering* 194(39–41): 4135–4195.

Hughes, T.J.R., and W.K. Liu. 1981. Nonlinear finite element analysis of shells: Part I. threedimensional shells. *Computer Methods in Applied Mechanics and Engineering* 26: 331–362.

Jelenic, G., and M. Saje. 1995. A kinematically exact space finite strain beam model - finite element formulations by generalised virtual work principle. *Computer Methods in Applied Mechanics and Engineering* 120: 131–161.

Kahn R. 1987. Finite-element-berechnungen ebener stabwerke mit flissgelenken und grossen verschiebungen. *Technical Report F 87/1*, Forschungs- und Seminarberichte aus dem Bereich der Mechanik der Universität Hannover.

Kühborn, A., and H. Schoop. 1992. A nonlinear theory for sandwich shells including the wrinkling phenomenon. *Ingenieur-Archiv* 62: 413–427.

Mäkinen, J. 2007. Total lagrangian Reissner's geometrically exact beam element without singularities. *International Journal for Numerical Methods in Engineering* 70: 1009–1048.

Lumpe G. Geometrisch nichtlineare berechnung von räumlichen stabwerken. *Technical Report 28*, Institut für Statik, Universität Hannover (1982).

Miehe, C. 1998. A theoretical and computational model for isotropic elastoplastic stress analysis in shells at large strains. *Computer Methods in Applied Mechanics and Engineering* 155: 193–233.

Onate, E., and M. Cervera. 1993. Derivation of thin plate bending elements with one degree of freedom per node: a simple three node triangle. *Engineering computations* 10(6): 543–561.

Oran, C., and A. Kassimali. 1976. Large deformations of framed structures under static and dynamic loads. *Computers and Structures* 6: 539–547.

Pimenta, P., and T. Yojo. 1993. Geometrically exact analysis of spatial frames. *Applied Mechanics Review* 46: 118–128.

Pimenta, P.M., E.M.B. Campello, and P. Wriggers. 2004. A fully nonlinear multi-parameter shell model with thickness variation and a triangular shell finite element formulation. *Computational Mechanics* 34: 181–193.

Ramm E. 1976. Geometrisch nichtlineare Elastostatik und Finite Elemente. *Technical Report Nr. 76-2*, Institut für Baustatik der Universität Stuttgart.

Rankin, C.C., and F.A. Brogan. 1984. An element independent corotational procedure for the treatment of large roations. In *Collapse analysis of structures*, ed. L.H. Sobel, and K. Thomas, 85–100. New York: ASME.

Reissner, E. 1972. On one-dimensional finite strain beam theory, the plane problem. *Journal of Applied Mathematics and Physics* 23: 795–804.

Roehl, D., and E. Ramm. 1996. Large elasto-plastic finite element analysis of solids and shells with the enhanced assumed strain concept. *International Journal of Solids & Structures* 33: 3215–3237.

Romero, I., and F. Armero. 2002. An objective finite element approximation of the kinematics of geometrically exact rods and its use in the formulation of an energy-momentum conserving scheme in dynamics. *International Journal for Numerical Methods in Engineering* 54: 1683–1716.

Sansour, C. 1995. A theory and finite element formulation of shells at finite deformations involving thickness change: circumventing the use of a rotation tensor. *Ingenieur-Archiv* 65: 194–216.

Schoop, H. 1986. Oberflächenorientierte Schalentheorien endlicher Verschiebungen. *Ingenieur-Archiv* 56: 427–437.

Seifert B. 1996. *Zur Theorie und Numerik Finiter Elastoplastischer Deformationen von Schalenstrukturen*. Dissertation, Institut für Baumechanik und Numerische Mechanik der Universität Hannover. Bericht Nr. F 96/2.

Simo, J.C. 1985. A finite strain beam formulation. The three-dimensional dynamic problem. Part I. *Computer Methods in Applied Mechanics and Engineering* 49: 55–70.
Simo, J.C., and L. Vu-Quoc. 1986. Three dimensional finite strain rod model. Part II: computational aspects. *Computer Methods in Applied Mechanics and Engineering* 58: 79–116.
Simo J.C., Fox D.D., Rifai M.S. 1989. Formulation and optimal parametrization. 1989. On a stress resultant geometrical exact shell model. Part I. *Computer Methods in Applied Mechanics and Engineering* 72: 267–304.
Simo J.C., Fox D.D., Rifai M.S. 1990a. *Computer Methods in Applied Mechanics and Engineering*. On a stress resultant geometrical exact shell model. Part III. 79: 21–70.
Simo, J.C., M.S. Rifai, and D.D. Fox. 1990b. On a stress resultant reometrically exact shell model. Part IV. Variable thickness shells with through-the-tickness stretching. *Computer Methods in Applied Mechanics and Engineering* 81: 91–126.
Soric, J., U. Montag, and W.B. Krätzig. 1997. An efficient formulation of integration algorithms for elastoplastic shell analysis based on layered finite element approach. *Computer Methods in Applied Mechanics and Engineering* 148: 315–328.
Wagner, W., S. Klinkel, and F. Gruttmann. 2002. Elastic and plastic analysis of thin-walled structures using improved hexahedral elements. *Computers and Structures* 80(9–10): 857–869.
Wempner, G. 1969. Finite elements, finite roataions and small strains of flexible shells. *International Journal of Solids & Structures* 5: 117–153.
Wriggers, P., and F. Gruttmann. 1989. Large deformations of thin shells: Theory and finite-element-discretization, analytical and computational models of shells. In *ASME, CED-Vol.3*, vol. 135–159, ed. A.K. Noor, T. Belytschko, and J.C. Simo.
Wriggers, P., R. Eberlein, and S. Reese. 1996. Comparison between shell and 3d-elements in finite plasticity. *International Journal of Solids & Structures* 33: 3309–3326.
Wriggers, P., and F. Gruttmann. 1993. Thin shells with finite rotations formulated in biot stresses theory and finite-element-formulation. *International Journal for Numerical Methods in Engineering* 36: 2049–2071.
Zienkiewicz, O.C., and R.L. Taylor. 2000b. *The Finite Element Method*, vol. 1, 5th ed. Oxford, UK: Butterworth-Heinemann.

Chapter 8
Automation of Sensitivity Analysis

Optimization of solid structures lead to better engineering designs and thus to a longer life time of e.g. technical components, engines and buildings. In an optimization process as well material parameters as shapes can be improved. However the computations needed to find optimal solutions are quite complex. Here a proper sensitivity analysis provides an efficient tool in order to compute complex derivations that are needed in the associated algorithms which finally leads to faster convergence of the underlying iterative procedures. Another applications where sensitivity analysis is needed is related to multi-scale analysis of coplex structures. Especially the so called FE^2-schemes need an accurate computation of sensitivities. Here micro and macro levels are combined in the computations to account for complex material behaviour at micro level.

This chapter will provide the necessary theoretical background of the sensitivity analysis, the numerical aspects and its integration into *AceGen* .

8.1 Introduction to Sensitivity Analysis

The aim of the sensitivity analysis is to calculate derivatives of an arbitrary response functional with respect to chosen parameters (see e.g. Kleiber et al. 1997; Keulen et al. 2005; Choi and Kim 2005a or 2005b). The response functional can depend on arbitrary analysis model inputs (material constants, load intensity and distribution, shape parameters, etc.) as well as on arbitrary intermediate or final results of the analysis (solution vectors, derived quantities such as stress tensor, integrated quantities such as damage or ductility factors, etc.). Modern finite element simulations are often coupled with optimization procedures that require additionally to the solution of the primal problem also a solution of the sensitivity problem. Sensitivity analysis has become an indispensable part of modern computational algorithms. The use of analytically exact sensitivity analysis can significantly improve optimization procedures Choi and Kim (2005b), Kristanic and Korelc (2008), multi-scale algorithms, Solinc and Korelc (2015) and implementation of nonlinear material models Korelc

J. Korelc and P. Wriggers, *Automation of Finite Element Methods*,
DOI 10.1007/978-3-319-39005-5_8

and Stupkiewicz (2014), Hudobivnik and Korelc (2016). The complete automation of the sensitivity analysis is thus only possible if the automatic differentiation is applied on the complete simulation code. However, a large variety of sensitivity problems, important for practical applications, can still be solved following the classical finite element procedure where all problem dependent quantities are evaluated at the individual element level while the global level procedures remain largely problem independent. This fact is widely exploited within the commercial FE environments where the semi-analytical sensitivity analysis is usually implemented Keulen et al. (2005). Within the semi-analytical sensitivity analysis the global sensitivity problem is solved by the direct differentiation or adjoint method, while the derivatives at the individual element level are calculated by the finite difference approximation. The automation of sensitivity analysis enables efficient evaluation of sensitivities that are exact except for the round off errors. Once the sensitivity of the basic unknowns of the problem is established, the derivatives of the response functional can also be evaluated. A comprehensive overview of the possible approaches can be found in Kleiber et al. (1997), Keulen et al. (2005), Choi and Kim (2005a), Michaleris et al. (1994) or Choi and Kim (2005b).

For the sensitivity problem, we define the residuals, vectors of unknowns and response functional as a function of a vector of design parameters

$$\boldsymbol{\phi} = \left\{\phi_I : I = 1, \ldots, n_\phi\right\} \tag{8.1}$$

where n_ϕ is the total number of sensitivity parameters. The solution algorithm of the sensitivity problem can then be obtained from the primal problem by differentiating the response functional and the residuals with respect to design parameters.

Example: stretching of an elasto-plastic bar. Let us consider a simple example, presented in Fig. 8.1, where an elastic bar of length L with the cross section area A and the elastic modulus E is subjected to a force P at the end. The goal is to minimize the response functional $F(A) = (\sigma(A) - \sigma_y)^2$ where $\sigma = E\varepsilon = E\frac{u(A)}{L}$, u is an unknown tip displacement and σ_y is the yield stress. Thus, we are looking for the cross section area at which yielding will occur for the prescribed force and yield stress. The primal problem is defined by the equilibrium equation $\frac{EA}{L}u - P = 0$. The solution (also response) of the primal problem is $u = \frac{PL}{EA}$. The solution of the primal problem depends on the shape parameter L, the material parameters E and A and the prescribed natural boundary parameter P. The sensitivity of the response of the primal problem with respect to L is called shape sensitivity and is defined by $\frac{\partial u}{\partial L} = \frac{P}{EA}$. The sensitivity with respect to the material parameters is called parameter sensitivity ($\frac{\partial u}{\partial E} = -\frac{PL}{E^2 A}$, $\frac{\partial u}{\partial A} = -\frac{PL}{EA^2}$) and with respect to the force P prescribed

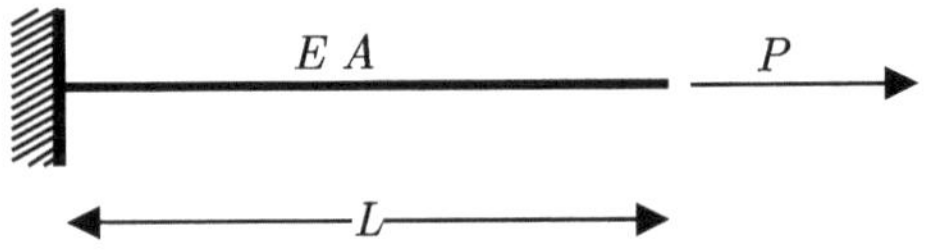

Fig. 8.1 Stretching of an elastic bar

natural boundary condition sensitivity ($\frac{\partial u}{\partial P} = \frac{L}{EA}$). The goal is to minimize the response functional F with respect to the cross sectional area A, thus only sensitivity of the response with respect to A is needed ($\boldsymbol{\phi} = \{A\}$). The gradient of the response functional is $\frac{\partial F}{\partial A} = 2(\sigma - \sigma_y)\frac{E}{L}\frac{\partial u}{\partial A}$. The optimality condition $\frac{\partial F}{\partial A} = 0$ then leads to the well known solution of the problem $A = P/\sigma_y$.

This simple example will be generalized in Sect. 8.3 to nonlinear problems as defined in Chap. 3. An example of the automatic generation of a user subroutines for sensitivity analysis is given in Sect. 8.4 followed by a general sensitivity analysis example in Sect. 8.5.

8.1.1 Parametrization of the Continuum and the Discretized Problem

The sensitivity problem is obtained from the primal problem by parametrization of any data that the primal problem implicitly or explicitly depends on, except the unknowns of the problem. Parameters used to parametrize the data become design sensitivity parameters. For example Fig. 8.2 presents a two-dimensional, solid domain B made of linear elastic material. The boundary of the domain is parametrized by its width L, thus $\Gamma : \boldsymbol{X}(L) = (X_1(L), X_2(L))$. Natural boundary conditions are prescribed at Γ_σ. Here the second component of the surface traction $\bar{\boldsymbol{t}}$ is

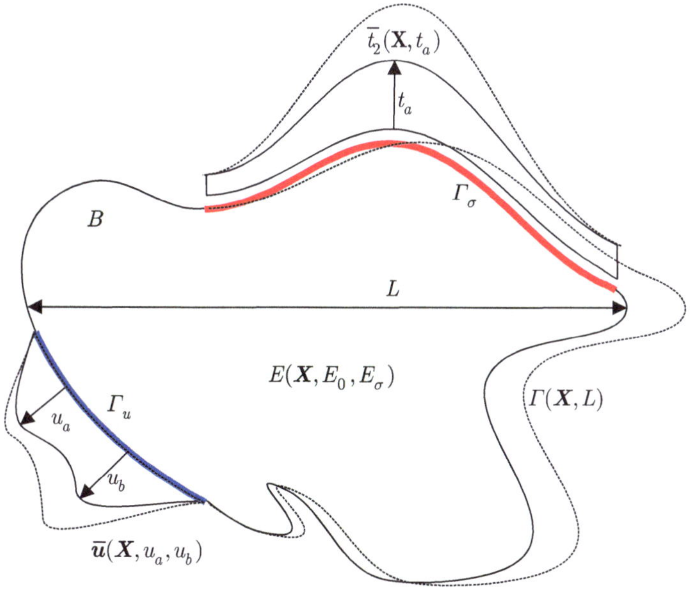

Fig. 8.2 Parametrization of a general continuum problem that is solved using the finite element model

parametrized by the parameter t_a, thus $\Gamma_\sigma : \bar{t}_2(\boldsymbol{X}, t_a)$. Essential boundary conditions are prescribed at the boundary Γ_u. Here the shape of the prescribed displacement vector $\bar{\boldsymbol{u}}$ is parametrized by parameters u_a and u_b, thus $\Gamma_u : (\bar{u}_1(\boldsymbol{X}, u_a, u_b), \bar{u}_2(\boldsymbol{X}, u_a, u_b))$. Finally, the distribution of the elastic modulus E is parametrized by the parameter E_σ, thus $E = E(\boldsymbol{X}, E_\sigma)$ and the Poisson's ratio ν takes a constant value $\nu = \nu_0$. The set of $n_\phi = 6$ sensitivity parameters of the problem is given by

$$\boldsymbol{\phi} = \{L, t_a, u_a, u_b, E_\sigma, \nu_0\} \tag{8.2}$$

as well as the set of sensitivity parameters dependent scalar fields defined on B

$$\left\{X_1, X_2, \bar{t}_2, \bar{u}_1, \bar{u}_2, E, \nu\right\}. \tag{8.3}$$

Discretized model. Figure 8.3 presents the discretized model of the continuum problem depicted in Fig. 8.2. The input parameters of the discretized model are spatial coordinates of all nodes $\mathbf{X}^J(L) = (X_1^J(L), X_2^J(L)) : J = 1, \ldots, n_{tn}$, nodal displacements $\bar{\mathbf{u}}^J(u_a, u_b) = (\bar{u}_1^J(u_a, u_b), \bar{u}_2^J(u_a, u_b)) : \forall \mathbf{X}^J \in \Gamma_u$ of all nodes at Γ_u, nodal forces $F_2^J(t_a, L) : \forall \mathbf{X}^J \in \Gamma_\sigma$ resulting from surface traction $\bar{t}_2$ in all nodes on Γ_σ and material constants $E_g = E_g(E_\sigma)$ and $\nu_g = \nu_0$ at all integration points. Note the difference between continuum and discretized model. The surface traction $\bar{t}_2 = \bar{t}_2(t_a)$ in the continuum case depends only on the parameter t_a used to parametrize the intensity of surface traction. In the discretized model the corresponding nodal forces P_2^J depend also on the shape parameter L ($P_2^J = P_2^J(t_a, L)$). The process of calculating nodal forces from surface tractions brings an additional dependency of nodal forces on the shape parameter L.

Input parameters of the finite element models can be scalars (e.g. elastic modulus E), discretized scalar fields (e.g. nodal temperatures) and discretized vector fields (e.g. nodal spatial coordinates $\mathbf{X}^J$, prescribed nodal displacements $\bar{\mathbf{u}}^J$ and nodal forces $\mathbf{P}^J$). Without loosing generality of the formulation a scalar can be considered as a constant scalar field discretized by its constant nodal values and a vector field can be considered component wise. Most of the input data of the finite element models are associated with nodes, however some quantities, such as material constants ($E_g = E_g(E_\sigma)$ and $\nu_g = \nu_0$) are associated with integration points. The following generalization will be made in order to unify nodal point based and integration point based quantities. It is assumed that integration point based quantities are obtained from the appropriate nodal based quantities using standard finite element interpolation techniques. Consequently, integration point based quantities are also represented by a discretized scalar field. Even the vector of nodal forces $\mathbf{P}^J$ that is calculated based on surface tractions and thus strictly speaking does not represent a scalar field will be considered as a discretized scalar field. Thus, all input parameters of the finite element user subroutine will be in general considered as a discretized scalar fields. A set of nodal wise ordered scalar input data fields $\boldsymbol{\psi}(\boldsymbol{\phi})$ that are input data of the discretized model in Fig. 8.3 is then given by

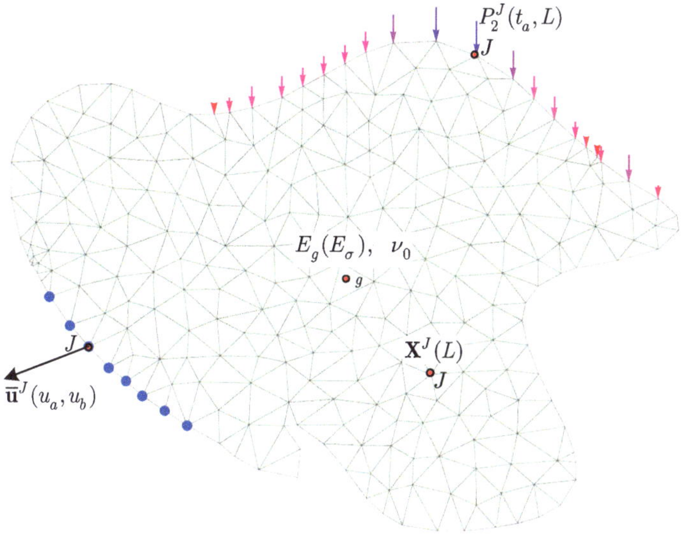

Fig. 8.3 Parametrization of input data of the discretized model

$$\boldsymbol{\psi}(\boldsymbol{\phi}) = \{\{X_1^J(L), X_2^J(L), P_2^J(t_a, L), \bar{u}_1^J(u_a, u_b), \bar{u}_2^J(u_a, u_b), E^J(E_\sigma), \nu\} \\ : J = 1, \ldots, n_{tn}\}. \tag{8.4}$$

8.1.2 *Formulation and Solution of a Simple Sensitivity Problem*

There are in general two approaches how a sensitivity problem can be formulated based on the formulation of the primal problem:

1. continuum derivatives,
2. discrete derivatives.

The continuum sensitivity equation is obtained by differentiating the continuum equations that govern response of the primal problem. Once the continuum sensitivity equation is obtained, it can be discretized in the same manner as the equations of the primal analysis.

Discrete sensitivity equations are obtained by differentiation of the governing equations of the discretized primal problem. As it will be shown later, the formulation of discrete sensitivity equations can be fully automatized for all problems, thus the discrete sensitivity approach will be adopted here.

Let us consider the most simple case of the time-independent problem that leads to the system of nonlinear equations $\mathbf{R}(\mathbf{p}) = \mathbf{0}$. The residual $\mathbf{R}$ explicitly depends on the sensitivity parameters $\boldsymbol{\phi}$, while the solution vector $\mathbf{p}$ depends on $\boldsymbol{\phi}$ implicitly, thus $\mathbf{R}(\mathbf{p}(\boldsymbol{\phi}), \boldsymbol{\phi}) = \mathbf{0}$. The direct differentiation of $\mathbf{R}(\mathbf{p}(\boldsymbol{\phi}), \boldsymbol{\phi}) = \mathbf{0}$ with respect to the Ith design parameter ϕ_I yields a set of discrete sensitivity equations

$$\frac{\partial \mathbf{R}}{\partial \mathbf{p}} \frac{D\mathbf{p}}{D\phi_I} = -\frac{\partial \mathbf{R}}{\partial \phi_I}. \tag{8.5}$$

This simple example leads to several conclusions:

- Equation (8.5) represents a system of linear equations for the unknown sensitivities $\frac{D\mathbf{p}}{D\phi_I}$,
- the number of discrete sensitivity equations is the same as the number of equations of the discretized primal problem,
- the first factor on the left hand side of (8.5) is exactly the independent tangent operator $\mathbf{K}$ of the primal problem, thus the sparsity structure of the resulting system of equations is the same as of the primal problem,
- the sensitivity problem is linear also in the case when the primal problem is nonlinear,
- if the primal problem is solved by a Newton–Raphson type iterative procedure, the already factorized tangent matrix $\mathbf{K}$ from the primal problem can be used for the solution of the sensitivity problem as well,
- the sensitivity problem can be solved within the bounds of machine precision with relatively modest additional computational cost,
- the right hand side $\frac{\partial \mathbf{R}}{\partial \phi_I}$ of the resulting system of linear equations resembles the load vector.

If $\mathbf{R}$ is obtained in an element-by-element manner using standard finite element procedures then $\frac{\partial \mathbf{R}}{\partial \phi_I}$ can also be obtained in a same way leading to

$$\frac{\partial \mathbf{R}}{\partial \phi_I} = \mathop{\mathsf{A}}_{e=1}^{n_e} \frac{\partial \mathbf{R}_e(\phi_I)}{\partial \phi_I}. \tag{8.6}$$

8.1.3 Introduction of Design Velocity Fields

The obvious problem in obtaining the right hand side $\frac{\partial \mathbf{R}}{\partial \phi_I}$ in (8.5) is that sensitivity parameters $\boldsymbol{\phi}$ in general do not appear explicitly as input parameters of the finite element solution procedure, either at the global level or at the level of user subroutines that are applied to calculate the individual finite element element contributions (e.g. residual $\mathbf{R}_e$) to the global problem.

Let us consider for example the shape parameter L. The relation between the shape parameter L and the coordinates of an arbitrary node $\mathbf{X}^J(L)$ can be an arbitrary complex function that, in general, cannot be input data of a finite element analysis. However, it is not the relation $\mathbf{X}^J(L)$ itself that is needed within the sensitivity analysis to obtain $\frac{\partial \mathbf{R}}{\partial L}$, but its derivatives. The input data of the sensitivity analysis is the rate of change of nodal coordinates with the change of sensitivity parameter L. The rate of change of X_1 coordinate of all nodes $\left\{\frac{DX_1^J}{DL} : J = 1, \ldots, n_{tn}\right\}$ are nodal values of a scalar field $\frac{DX_1}{DL}$ defined on the domain of the problem. The $\frac{DX_1}{DL}$ field is traditionally called the **"design velocity field"**. The discretized design velocity fields $\left\{\frac{DX_1^J}{DL} : J = 1, \ldots, n_{tn}\right\}$ and $\left\{\frac{DX_2^J}{DL} : J = 1, \ldots, n_{tn}\right\}$ are evaluated to values of the design sensitivity parameter L and are appropriate input data for finite element subroutines. Evaluation of the right hand side $\frac{\partial \mathbf{R}}{\partial L}$ (8.6) for a sensitivity analysis then follows as

$$\frac{\partial \mathbf{R}}{\partial L} = \mathop{\mathsf{A}}_{e=1}^{n_e} \sum_{J=1}^{n_{en}} \left(\frac{\partial \mathbf{R}_e}{\partial X_1^J} \frac{DX_1^J}{DL} + \frac{\partial \mathbf{R}_e}{\partial X_2^J} \frac{DX_2^J}{DL} \right). \tag{8.7}$$

8.1.4 Design Velocity Matrix

Let $\boldsymbol{\psi}$ be a nodal wise ordered set of all discretized scalar fields that are input parameters of the finite element solution procedure or **input data fields**. $\boldsymbol{\psi}$ must include all input data fields regardless of their dependency on sensitivity parameters of the actual problem in order to make the finite element code general and problem independent. A typical case is presented in Fig. 8.2 that leads to the following set of input data fields

$$\boldsymbol{\psi} = \left\{\left\{\bar{u}_1^J, \bar{u}_2^J, \bar{P}_1^J, \bar{P}_2^J, X_1^J, X_2^J, E^J, \nu^J\right\} : J = 1, \ldots, n_{tn}\right\}. \tag{8.8}$$

Let

$$\boldsymbol{\psi}_L = \left\{\psi_L^J : J = 1, \ldots, n_{tn}\right\} \tag{8.9}$$

be the Lth input data field and

$$\boldsymbol{\psi}^J = \left\{\psi_L^J : L = 1, \ldots, n_\psi\right\} \tag{8.10}$$

be the value of $\boldsymbol{\psi}$ at Jth node.

The rate of change $\frac{D\boldsymbol{\psi}_L}{D\phi_I}$ of the Lth input data field $\boldsymbol{\psi}_L$ with the change of the Ith sensitivity parameter ϕ_I is by definition the design velocity field. Let

$$D_{\boldsymbol{\phi}}\boldsymbol{\psi} = \left\{\frac{D\boldsymbol{\psi}^J}{D\boldsymbol{\phi}} : J = 1, \ldots, n_{tn}\right\} = \left\{\frac{D\boldsymbol{\psi}_L}{D\boldsymbol{\phi}} : L = 1, \ldots, n_\psi\right\}^T \tag{8.11}$$

be a matrix of all design velocity fields or **design velocity matrix**. $D_{\boldsymbol{\phi}}\boldsymbol{\psi}$ has dimension $n_{tn} \times n_\psi \times n_\phi$. Additionally we define $D_{\boldsymbol{\phi}}\boldsymbol{\psi}^J$ as the value of $D_{\boldsymbol{\phi}}\boldsymbol{\psi}$ at the Jth node.

The above considerations lead to the design velocity matrix $D_{\boldsymbol{\phi}}\boldsymbol{\psi}^J$ of the problem presented in Fig. 8.3.

$$
D_{\phi}\psi^{J} = \begin{array}{c} \begin{array}{cccccc} L & t_a & u_a & u_b & E_\sigma & \nu_0 \end{array} \\ \left[\begin{array}{cccccc}
0 & 0 & \frac{D\bar{u}_1^J}{Du_a} & \frac{D\bar{u}_1^J}{Du_b} & 0 & 0 \\
0 & 0 & \frac{D\bar{u}_2^J}{Du_a} & \frac{D\bar{u}_2^J}{Du_b} & 0 & 0 \\
0 & 0 & 0 & 0 & 0 & 0 \\
\frac{DP_2^J}{DL} & \frac{DP_2^J}{Dt_a} & 0 & 0 & 0 & 0 \\
\frac{DX_1^J}{DL} & 0 & 0 & 0 & 0 & 0 \\
\frac{DX_2^J}{DL} & 0 & 0 & 0 & 0 & 0 \\
0 & 0 & 0 & 0 & \frac{DE^J}{DE_\sigma} & 0 \\
0 & 0 & 0 & 0 & 0 & \frac{D\nu}{D\nu_0}
\end{array}\right] \end{array}
\begin{array}{c} \\ u_1^J \\ u_2^J \\ P_1^J \\ P_2^J \\ X_1^J \\ X_2^J \\ E^J \\ \nu^J \end{array} \tag{8.12}
$$

A complete description of the problem presented in Fig. 8.3 requires the definition of a total of 10 nonzero design velocity fields as follows

1. $\frac{DX_1^J}{DL} = \frac{DX_1^J(L)}{DL}$,
2. $\frac{DX_2^J}{DL} = \frac{DX_2^J(L)}{DL}$,
3. $\frac{D\bar{u}_1^J}{Du_a} = \begin{cases} \frac{D\bar{u}_1^J(u_a,u_b)}{Du_a} & \mathbf{X}^J \in \Gamma_u \\ 0 & \text{else} \end{cases}$,
4. $\frac{D\bar{u}_1^J}{Du_b} = \begin{cases} \frac{D\bar{u}_1^J(u_a,u_b)}{Du_b} & \mathbf{X}^J \in \Gamma_u \\ 0 & \text{else} \end{cases}$,
5. $\frac{D\bar{u}_2^J}{Du_a} = \begin{cases} \frac{D\bar{u}_2^J(u_a,u_b)}{Du_a} & \mathbf{X}^J \in \Gamma_u \\ 0 & \text{else} \end{cases}$,
6. $\frac{D\bar{u}_2^J}{Du_b} = \begin{cases} \frac{D\bar{u}_2^J(u_a,u_b)}{Du_b} & \mathbf{X}^J \in \Gamma_u \\ 0 & \text{else} \end{cases}$,
7. $\frac{D\bar{P}_2^J}{Dt_a} = \begin{cases} \frac{D\bar{P}_2^J(t_a,L)}{Dt_a} & \mathbf{X}^J \in \Gamma_\sigma \\ 0 & \text{else} \end{cases}$,
8. $\frac{D\bar{P}_2^J}{DL} = \begin{cases} \frac{D\bar{P}_2^J(t_a,L)}{DL} & \mathbf{X}^J \in \Gamma_\sigma \\ 0 & \text{else} \end{cases}$,
9. $\frac{DE^J}{DE_\sigma} = \frac{DE^I(E_\sigma)}{DE_\sigma}$,
10. $\frac{D\nu}{D\nu_0} = 1$.

Using the definition of the design velocity matrix, the right hand side of the sensitivity Eq. (8.5) for time-independent problem can simply be obtained for all sensitivity parameters as

$$
\frac{\partial \mathbf{R}}{\partial \phi} = \frac{\partial \mathbf{R}}{\partial \psi} D_{\phi}\psi . \tag{8.13}
$$

Unfortunately, as shown later, the sensitivity problem of time-dependent coupled problems cannot be formulated in that simple manner.

The design velocity matrix $D_{\phi}\psi$ is effectively used to fill in the gap in the chain rule that leads from the element residual to the global sensitivity parameters of the problem. The design velocity matrix is a sparse matrix. A proper consideration of the sparsity of the design velocity matrix is essential for the efficient numerical implementation of the sensitivity analysis. If the code is designed in a way that all input parameters of the finite element problem are by default considered to be independent variables, meaning that the associated velocity fields are zero, then only nonzero components of the velocity fields have to be stored and considered. An efficient implementation of Eq. (8.13) will be discussed in the following sections for different types of sensitivity parameters and different classes of problems.

8.2 Classification and Formulation of Problems for Sensitivity Analysis

8.2.1 Classification of Sensitivity Problems

Different nonlinear problems are classified in Chap. 3. Based on the form of the discretized equations four classes were introduced. These reflect the way how the equations are formed, assembled and solved. The classes contain uncoupled and coupled problems and, related to the dependency of the response of a system on time, they also are split into time-independent and time-dependent problems. The same classification can be adopted for sensitivity analysis and leads to the following classes:

1. sensitivity analysis of time-independent problems,
2. sensitivity analysis of time-dependent problems,
3. sensitivity analysis of time-independent coupled problems,
   (a) sensitivity analysis of time-independent locally coupled problems,
   (b) sensitivity analysis of time-independent Gauss point coupled problems,
4. sensitivity analysis of time-dependent coupled problems.
   (a) sensitivity analysis of time-dependent locally coupled problems,
   (b) sensitivity analysis of time-dependent Gauss point coupled problems,

Classification of sensitivity problems and solution procedures presented here represent extension and modification of the scheme proposed in Michaleris et al. (1994) and modified in Korelc (2009) for the purpose of automation.

In analogy to primal analysis (for definitions see Sect. 3.1), we define a general response functional F and residual $\mathbf{R}$. For the evaluation of the sensitivity of time-independent problems the form

$$F(\mathbf{p}(\boldsymbol{\phi}), \boldsymbol{\phi}), \tag{8.14}$$

$$\mathbf{R}(\mathbf{p}(\boldsymbol{\phi}), \boldsymbol{\phi}) = \mathop{\mathsf{A}}_{e=1}^{n_e} \mathbf{R}_e\left(\mathbf{p}_e(\boldsymbol{\phi}), \boldsymbol{\phi}\right) = \mathbf{0}, \tag{8.15}$$

can be introduced. For time-independent locally coupled problems the equations take the form

$$F(\mathbf{p}(\boldsymbol{\phi}), \mathbf{h}(\boldsymbol{\phi}), \boldsymbol{\phi}), \tag{8.16}$$

$$\mathbf{R}(\mathbf{p}(\boldsymbol{\phi}), \mathbf{h}(\boldsymbol{\phi}), \boldsymbol{\phi}) = \mathop{\mathsf{A}}_{e=1}^{n_e} \mathbf{R}_e\left(\mathbf{p}_e(\boldsymbol{\phi}), \mathbf{h}_e(\boldsymbol{\phi}), \boldsymbol{\phi}\right) = \mathbf{0}, \tag{8.17}$$

$$\mathbf{Q}_e(\mathbf{p}_e(\boldsymbol{\phi}), \mathbf{h}_e(\boldsymbol{\phi}), \boldsymbol{\phi}) = \mathbf{0} : e = 1, \ldots, n_e. \tag{8.18}$$

Finally, for time-independent Gauss point coupled problems one can write

$$F(\mathbf{p}(\boldsymbol{\phi}), \mathbf{h}(\boldsymbol{\phi}), \boldsymbol{\phi}), \tag{8.19}$$

$$\mathbf{R}(\mathbf{p}(\boldsymbol{\phi}), \mathbf{h}(\boldsymbol{\phi}), \boldsymbol{\phi}) = \mathop{\mathsf{A}}_{e=1}^{n_e} \sum_{g\in G_e} w_{gp} \mathbf{R}_g\left(\mathbf{p}_e(\boldsymbol{\phi}), \mathbf{h}_g(\boldsymbol{\phi}), \boldsymbol{\phi}\right) = \mathbf{0}, \tag{8.20}$$

$$\mathbf{Q}_g(\mathbf{r}_g(\mathbf{p}_e(\boldsymbol{\phi}), \boldsymbol{\phi}), \mathbf{h}_g(\boldsymbol{\phi}), \boldsymbol{\phi}) = \mathbf{0} : g = 1, \ldots, n_{tg} \tag{8.21}$$

where $G_e \subseteq \{1, \ldots, n_{tg}\}$ is a set of indices of Gauss points of eth element.

For time-dependent problems, the response functional and the residuals depend in principle explicitly on the solution vectors at all time steps. However, in the present derivation only the most common case is considered where the response functional F explicitly depends only on the last two steps $n_{\text{step}} - 1$ and n_{step} and the explicit dependency of the current residuals is limited to the last converged step n and the current step $n+1$.[1] Due to the time-dependent nature of the formulation, the response functional and the residual implicitly depend on the solution vectors at all time steps. A general response functional and the residual for time-dependent problems are then defined by

$$F(\mathbf{p}_{n_{\text{step}}}(\boldsymbol{\phi}), \mathbf{p}_{n_{\text{step}}-1}(\boldsymbol{\phi}), \boldsymbol{\phi}), \tag{8.22}$$

$$\mathbf{R}(\mathbf{p}(\boldsymbol{\phi}), \mathbf{p}_n(\boldsymbol{\phi}), \boldsymbol{\phi}) = \mathop{\mathsf{A}}_{e=1}^{n_e} \mathbf{R}_e\left(\mathbf{p}_e(\boldsymbol{\phi}), \mathbf{p}_{en}(\boldsymbol{\phi}), \boldsymbol{\phi}\right) = \mathbf{0}. \tag{8.23}$$

For time-dependent locally coupled problems the equation set

$$F(\mathbf{p}_{n_{\text{step}}}(\boldsymbol{\phi}), \mathbf{h}_{n_{\text{step}}}(\boldsymbol{\phi}), \mathbf{p}_{n_{\text{step}}-1}(\boldsymbol{\phi}), \mathbf{h}_{n_{\text{step}}-1}(\boldsymbol{\phi}), \boldsymbol{\phi}), \tag{8.24}$$

[1] The index $n+1$ is in general omitted in order to simplify the notation.

$$\mathbf{R}(\mathbf{p}(\boldsymbol{\phi}), \mathbf{h}(\boldsymbol{\phi}), \mathbf{p}_n(\boldsymbol{\phi}), \mathbf{h}_n(\boldsymbol{\phi}), \boldsymbol{\phi}) = \underset{e=1}{\overset{n_e}{\mathsf{A}}} \mathbf{R}_e \left(\mathbf{p}_e(\boldsymbol{\phi}), \mathbf{h}_e(\boldsymbol{\phi}), \mathbf{p}_{en}(\boldsymbol{\phi}), \mathbf{h}_{en}(\boldsymbol{\phi}), \boldsymbol{\phi}\right) = \mathbf{0}, \tag{8.25}$$

$$\mathbf{Q}_e(\mathbf{p}_e(\boldsymbol{\phi}), \mathbf{h}_e(\boldsymbol{\phi}), \mathbf{p}_{en}(\boldsymbol{\phi}), \mathbf{h}_{en}(\boldsymbol{\phi}), \boldsymbol{\phi}) = \mathbf{0} : g = 1, \ldots, n_e \tag{8.26}$$

has to be used and for time-dependent Gauss point coupled problems the resulting equations are

$$F(\mathbf{p}_{n_{\text{step}}}(\boldsymbol{\phi}), \mathbf{h}_{n_{\text{step}}}(\boldsymbol{\phi}), \mathbf{p}_{n_{\text{step}}-1}(\boldsymbol{\phi}), \mathbf{h}_{n_{\text{step}}-1}(\boldsymbol{\phi}), \boldsymbol{\phi}), \tag{8.27}$$

$$\mathbf{R}(\mathbf{p}(\boldsymbol{\phi}), \mathbf{h}(\boldsymbol{\phi}), \mathbf{p}_n(\boldsymbol{\phi}), \mathbf{h}_n(\boldsymbol{\phi}), \boldsymbol{\phi}) = \underset{e=1}{\overset{n_e}{\mathsf{A}}} \sum_{g \in G_e} w_{gp} \mathbf{R}_g \left(\mathbf{p}_e(\boldsymbol{\phi}), \mathbf{h}_g(\boldsymbol{\phi}), \mathbf{p}_{en}(\boldsymbol{\phi}), \mathbf{h}_{gn}(\boldsymbol{\phi}), \boldsymbol{\phi}\right) = \mathbf{0}, \tag{8.28}$$

$$\mathbf{Q}_g(\mathbf{r}_g(\mathbf{p}_e(\boldsymbol{\phi}), \mathbf{p}_{en}(\boldsymbol{\phi}), \boldsymbol{\phi}), \mathbf{h}_g(\boldsymbol{\phi}), \mathbf{p}_{en}(\boldsymbol{\phi}), \mathbf{h}_{gn}(\boldsymbol{\phi}), \boldsymbol{\phi}) = \mathbf{0} : g = 1, \ldots, n_{tg}. \tag{8.29}$$

8.2.2 Classification of Sensitivity Design Velocity Fields

The goal of automation is to preserve the standard finite element technology paradigm where all the physical problem dependent quantities are calculated at the individual finite element level and assembled at global level. Thus, the description of the relation between the parametrized continuum model and the input parameters of the finite element subroutines and the global solution procedures has to be automatized as well.

We have already mentioned the difference between the continuum model and discretized model. Surface traction $\bar{t}_2 = \bar{t}_2(t_a)$ is in continuum case independent on shape parameter L while the process of transforming the surface traction into nodal forces brings an additional dependency of nodal forces on shape parameter L. Thus, the parameter used to parametrize the shape also effects the nodal forces. From this difference it is clear that the classification of sensitivity analysis cannot be based on the classification of sensitivity parameters $\boldsymbol{\phi}$. The classification must be based on the type of the input data field $\boldsymbol{\psi}_L$ and the corresponding design velocity fields $D_{\boldsymbol{\phi}} \boldsymbol{\psi}_L$ that are input parameters of primal and sensitivity analysis.

For the purpose of automation, each input data field and the corresponding design velocity fields is classified according to its actual appearance (or lack of it) in the formulation of the finite element problem. Based on previous considerations fields can be classified into three classes:

1. general input data and general design velocity fields that include
   - parameter input data and parameter design velocity fields (e.g. E^J and $\frac{DE^J}{DE_\sigma}$),
   - spatial coordinates and shape design velocity fields (e.g. X_1^J and $\frac{DX_1^J}{DL}$),
2. essential boundary conditions and essential boundary condition velocity fields (e.g. $\bar{u}_1^J$ and $\frac{D\bar{u}_1^J}{Du_a}, \frac{D\bar{u}_1^J}{Du_b}$),
3. natural boundary conditions and natural boundary condition velocity fields (e.g. P_2^J and $\frac{DP_2^J}{Dt_a}, \frac{DP_2^J}{DL}$).

8.2.3 *General Design Velocity Fields*

General design velocity fields relate arbitrary sensitivity parameters with the input data of the finite element user subroutine that is used to evaluate the residual. Typically, this includes nodal coordinates, material constants, cross section characteristics, geometrical characteristics, etc. In fact this includes all the input parameters of the *Tangent and residual* subroutine, except the nodal unknowns. Nodal unknowns can be effected by the imposed boundary conditions, thus they have to be considered separately as presented in next section.

Let $\boldsymbol{\psi}_e$ be a set of $n_{\psi e}$ input data fields (except nodal unknowns) of the finite element user subroutine of eth element or **general input data fields** of eth element defined by

$$\boldsymbol{\psi}_e = \left\{\left\{\hat{\psi}_{eL}^J : L = 1, \ldots, n_{\psi e}\right\} : J = 1, \ldots, n_{en}\right\}. \tag{8.30}$$

$\boldsymbol{\psi}_e$ is a subset of the global set of input data fields $\boldsymbol{\psi}$ ($\boldsymbol{\psi}_e \subseteq \boldsymbol{\psi}$) evaluated at the element nodes. The dimension of $\boldsymbol{\psi}_e$ is $n_{en} \times n_{\psi e}$. $\boldsymbol{\psi}_e$ is defined at the element level.

For a typical case of 2D, linear elastic solid finite elements presented in Fig. 8.3

$$\boldsymbol{\psi}_e = \left\{\{X_1^J, X_2^J, E^J, \nu\} : J = 1, \ldots, n_{en}\right\}.$$

$\boldsymbol{\psi}_e$ defines a set input parameters for which sensitivity problem is properly defined in eth element. Since the particular finite element code is fixed, one can choose only among a finite number of input parameters that are effected by an arbitrary sensitivity parameter. The element residual $\mathbf{R}_e$ explicitly depends on $\boldsymbol{\psi}_e$ and implicitly on $\boldsymbol{\phi}$ thus, $\mathbf{R}_e = \mathbf{R}_e(\boldsymbol{\psi}_e(\boldsymbol{\phi}))$.

The derivatives of $\boldsymbol{\psi}_e$ with respect to an arbitrary sensitivity parameter ϕ_I represent **general design velocity matrix** $D_{\phi_I}\boldsymbol{\psi}_e$.

$$D_{\phi_I}\boldsymbol{\psi}_e = \frac{D\boldsymbol{\psi}_e}{D\phi_I} = \left\{\left\{\frac{D\psi^J_{eL}}{D\phi_I} : L = 1, \ldots, n_{\psi e}\right\} : J = 1, \ldots, n_{en}\right\}. \tag{8.31}$$

The term "velocity" in this case refers to the rate of change of $\boldsymbol{\psi}_e$ with the change of a sensitivity parameter ϕ_I. The $D_{\phi_I}\boldsymbol{\psi}_e$ data structure is an input data of the sensitivity problem for the *e*th element. Since $\boldsymbol{\psi}_e \subseteq \boldsymbol{\psi}$, the general design velocity matrix $D_{\phi_I}\boldsymbol{\psi}_e$ is also a subset of a global design velocity matrix $D_{\boldsymbol{\phi}}\boldsymbol{\psi}$ ($D_{\phi_I}\boldsymbol{\psi}_e \subseteq D_{\boldsymbol{\phi}}\boldsymbol{\psi}$). The components of $D_{\phi_I}\boldsymbol{\psi}_e$ can be obtained from an appropriate components of global design velocity matrix $D_{\boldsymbol{\phi}}\boldsymbol{\psi}$ as a part of a global sensitivity analysis algorithm. An example how this can be done within the *AceFEM* finite element environment is presented in Sect. 8.5. Note that, if specific design velocity field is zero then the corresponding input data field does not depend on specific sensitivity parameter and is thus independent variable for sensitivity analysis.

8.2.4 Boundary Conditions Related Sensitivity Analysis

The imposed boundary conditions can also be parametrized and thus the sensitivity of the response of the system can be analyzed with respect to the change of boundary conditions. Complex boundary conditions (e.g. symmetric boundary conditions, inclined support, etc.) can be imposed on the solution using the Lagrange multiplier method or the augmented Lagrangian method implemented as a generalized finite element. When the boundary condition is imposed by a generalized finite element and the parameters used to parametrize the intensity or the shape of the imposed boundary condition are part of input parameters of the corresponding element code, then those parameters become general input data fields and the framework introduced in the previous section for general input data fields can be applied.

However, the imposition of the boundary conditions by an additional generalized finite element is not always optimal. It can extend the set of equations, effects the conditioning of the tangent matrix and the convergence radii of iterative methods. More efficient, but less general, implementations are regularly used within the research and commercial finite element systems. All the aspects of a specific implementation of the boundary conditions have to be considered in order to obtain a correct sensitivity analysis procedure. Two specific cases are presented here.

Prescribed homogeneous essential boundary conditions. The first case considers the implementation of the prescribed homogeneous essential boundary conditions (also known as boundary conditions of the first type or Dirichlet boundary conditions) where the equation corresponding to the constraint unknown is omitted from the global system of equations or replaced by a dummy equation.

Let $\bar{\mathbf{p}}_e$ be a subset of element DOFs with prescribed essential boundary conditions and $\check{\mathbf{p}}_e$ a subset of true element DOFs. The set of global degrees of freedom of *e*th element $\mathbf{p}_e$ is thus divided into two subsets ($\mathbf{p}_e = \bar{\mathbf{p}}_e \cup \check{\mathbf{p}}_e$). Let ϕ_I be an arbitrary design parameter that effects the prescribed essential boundary conditions on the boundary of the problem, thus $\Gamma_u : \bar{\mathbf{p}} = \bar{\mathbf{p}}(\phi_I)$ and at the element level $\bar{\mathbf{p}}_e = \bar{\mathbf{p}}_e(\phi_I)$. Consequently, the element residual $\mathbf{R}_e(\mathbf{p}_e)$ explicitly depends on degrees of freedom with prescribed essential boundary conditions.

The formulation of sensitivity problem obviously requires the derivatives $\frac{D\bar{\mathbf{p}}}{D\phi_I}$ to be known and available at the element level, thus the proper parts of the global velocity matrix $D_{\phi_I}\boldsymbol{\psi}$ have to be mapped into an appropriate element input data structure. Let $D_{\phi_I}\bar{\mathbf{p}}_e$ be an **element essential boundary condition velocity field data structure** defined by

$$D_{\phi_I}\bar{\mathbf{p}}_e = \left\{\left\{D_{\phi_I}\psi^{\text{node}(J)}_{\text{EBCF}(K,\,\text{node}(J))} : K = 1, \ldots, n_{pJ}\right\} : J = 1, \ldots, n_{en}\right\} \tag{8.32}$$

where n_{pJ} is the number of DOF in Jth element node and node(J) is the global index of the Jth element node. The EBCF function defines an index of EBCF$(K, \text{node}(J))$th design velocity field taken from $D_{\phi_I}\boldsymbol{\psi}$ that belongs to Kth unknown in node(J)th global node.

Here it is assumed that there exists a subset of the global set of input data fields $\boldsymbol{\psi}_p \subseteq \boldsymbol{\psi}$ that can be uniquely mapped into nodal unknowns. For a typical case of 2D, linear elastic solid finite elements presented in Fig. 8.3 this leads to the set $\boldsymbol{\psi}_p = \left\{\{u_1^J, u_2^J\} : J = 1, \ldots, n_{tn}\right\}$.

$D_{\phi_I}\bar{p}^J_{eK}$ is then derivative of the Kth DOF in the Jth node of the eth element with respect to ϕ_I. $D_{\phi_I}\bar{\mathbf{p}}_e$ is arranged by nodes and by degrees of freedom of the nodes, thus in general it does not need to be a rectangular matrix. Following definition of the design velocity matrix $D_{\boldsymbol{\phi}}\boldsymbol{\psi}$, the sensitivity of DOF without a prescribed essential boundary condition is in $D_{\phi_I}\bar{\mathbf{p}}_e$ by default set to 0. $D_{\phi_I}\bar{p}^J_{eK}$ is an input data at the element level of the sensitivity related user subroutines.

Equation (8.32) relates design velocity matrix $D_{\boldsymbol{\phi}}\boldsymbol{\psi}$ to nodal unknowns at element level. This equation is specific for the actual problem and is fulfilled by the properly prepared input data. An example how this can be done within the *AceFEM* finite element environment is presented in Sect. 8.5.

Prescribed homogeneous natural boundary conditions. The second case considers the implementation of the prescribed homogeneous natural boundary conditions. Let assume that an arbitrary design parameter ϕ_I effects the distribution of the prescribed natural boundary conditions at the boundary of the domain, thus $\Gamma_\sigma : \Gamma_\sigma(\phi_I)$ and let $\boldsymbol{\psi}_b \subseteq \boldsymbol{\psi}$ be a subset of input data fields $\boldsymbol{\psi}$ that represent applied nodal forces. For a typical case of 2D, linear elastic solid finite elements presented in Fig. 8.3 $\boldsymbol{\psi}_b = \left\{\{P_1^J, P_2^J\} : J = 1, \ldots, n_{tn}\right\}$.

Within the standard implementation of the natural boundary conditions the element residual $\mathbf{R}_e$ does not depend explicitly on input data that defines natural boundary conditions. The resulting load vector $\lambda\,\mathbf{R}^{\text{ref}}$ is subtracted from the global residual $\mathbf{R}$ as a part of global solution algorithm as presented in Sect. 3.2. Consequently, the contribution of variation of natural boundary conditions has to be formulated within the global solution algorithm as presented later in Sect. 8.3.7.

8.3 Solution and Automation of Sensitivity Problems

If the primal problem is solved by a Newton–Raphson type iterative procedure, the corresponding sensitivity problem can also be solved within the bounds of machine precision with relatively modest additional computational cost (see Michaleris et al. 1994). In general, two approaches can be applied

- the direct differentiation sensitivity method and
- the adjoint sensitivity method.

The direct differentiation and the adjoint sensitivity methods are closely related to the forward and backward approach of automatic differentiation and they share the same advantages and drawbacks. The AD method is meant to deal with an arbitrary algorithm and arbitrary equations. The direct differentiation and adjoint sensitivity methods deal with equations of known form (8.14)–(8.26) offering the opportunity to address the drawbacks more efficiently. The sensitivity problem is solved once the primal problem is converged (see Box 8.1). Thus, in the derivation of the sensitivity problem we assume that the equations that describe the primal problem are already fulfilled.

For each of the previously defined sensitivity analysis classes, first the traditional derivation of the quantities needed within the sensitivity analysis procedure is given. The traditional notation is then transformed into automatic differentiation based notation. A numerically efficient automation can be achieved by introducing appropriate AD exceptions of type C for all element input data that can be affected by a variation of an arbitrary sensitivity parameter. The AD exceptions are used to specify the sensitivity of general input data fields $\boldsymbol{\psi}_e$ using the general design velocity matrix $D_{\phi_I}\boldsymbol{\psi}_e$ (e.g. the sensitivity of nodal coordinates or the sensitivity of material constants). These exceptions are also employed to specify the sensitivity of essential boundary conditions with respect to arbitrary parameters using the essential boundary conditions velocity field data structure $D_{\phi_I}\bar{\mathbf{p}}_e$. The variation of natural boundary conditions is accounted for by the global solution algorithm as presented later in Sect. 8.3.7, thus no AD exceptions need to be specified at the finite element level. The type C AD exception is needed because the dependency of the mentioned variables on the sensitivity parameters and the corresponding derivatives cannot be deduced by the AD procedure from the analysis of the algorithm at element level. For the presented automation procedure it is important that the derivatives of all input parameters (e.g. global unknowns $\mathbf{p}_e$, nodal coordinates $\mathbf{X}$ and material constants) are by default zero, unless the AD exception is specified.

In Chap. 3 it was shown that the automation of the implicit solution procedure for the primal analysis is essentially the same for time-independent and time-dependent problems. However, this does not hold anymore for the sensitivity analysis. In the primal analysis the response at time t_n does not depend on the response at the current time and thus time t_n quantities are considered constant within the derivation of the residual and tangent matrix. In a sensitivity analysis both, the response at time t_n

and at current time, depend on the sensitivity parameter and thus they have to be considered as variables within the derivation of sensitivity problem equations.

Imposing AD exceptions. The following formalism is used in the subsequent derivations to define proper AD exceptions for general input data ($\boldsymbol{\psi}_e$) and DOFs with prescribed essential boundary conditions ($\bar{\mathbf{p}}_e$)

$$\left.\frac{\hat{\delta}\square}{\hat{\delta}\phi_I}\right|_{\frac{D\boldsymbol{\psi}_e}{D\phi_I}=D_{\phi_I}\boldsymbol{\psi}_e;\ \frac{D\hat{\mathbf{p}}_e}{D\phi_I}=D_{\phi_I}\bar{\mathbf{p}}_e;\ \text{...terms related to implicit dependencies}} \tag{8.33}$$

At the implementational level there exists a significant difference between the general input data and prescribed essential boundary conditions. Unknowns with prescribed boundary conditions are at the element level mixed with the true unknowns. At various analysis stages (e.g. sensitivity analysis, post-processing, visualization) and for various classes of problems additional data structures related to the sensitivity of element unknowns are needed. For the purpose of clear and short definition of various AD exceptions we define the following additional data structures. The $D_{\phi_I}\check{\mathbf{p}}_e$ data structure is defined by

$$D_{\phi_I}\check{\mathbf{p}}_e = \Bigg\{ \left\{ \begin{cases} 0 & \text{if } \hat{p}^J_{eK} \in \bar{\mathbf{p}}_e \\ \frac{D\hat{p}^J_{eK}}{D\phi_I} & \text{if } \hat{p}^J_{eK} \in \mathbf{p}_e \backslash \bar{\mathbf{p}}_e \end{cases} : K = 1, \ldots, n_{pJ} \right\} : J = 1, \ldots, n_{en} \Bigg\} \tag{8.34}$$

and contains only sensitivities of true element unknowns ($\mathbf{p}_e \backslash \bar{\mathbf{p}}_e$) arranged by nodes where n_{pJ} is the number of DOFs in Jth element node. True element unknowns can be time-dependent, thus we additionally define the data structure

$$D_{\phi_I}\check{\mathbf{p}}_e = \Bigg\{ \left\{ \begin{cases} 0 & \text{if } \hat{p}^J_{eK} \in \bar{\mathbf{p}}_e \\ \frac{D\hat{p}^J_{enK}}{D\phi_I} & \text{if } \hat{p}^J_{eK} \in \mathbf{p}_e \backslash \bar{\mathbf{p}}_e \end{cases} : K = 1, \ldots, n_{pJ} \right\} : J = 1, \ldots, n_{en} \Bigg\} \tag{8.35}$$

that holds sensitivities of true element unknowns arranged by nodes at time t_n. At the end, the sensitivities of all element DOFs at time t_{n+1} and at time t_n are given by

$$D_{\phi_I}\check{\mathbf{p}}_e = \Bigg\{ \left\{ \begin{cases} D_{\phi_I}\bar{p}^J_{eK} & \text{if } \hat{p}^J_{eK} \in \bar{\mathbf{p}}_e \\ \frac{D\hat{p}^J_{eK}}{D\phi_I} & \text{if } \hat{p}^J_{eK} \in \mathbf{p}_e \backslash \bar{\mathbf{p}}_e \end{cases} : K = 1, \ldots, n_{pJ} \right\} : J = 1, \ldots, n_{en} \Bigg\} \tag{8.36}$$

$$D_{\phi_I}\hat{\mathbf{p}}_{en} = \frac{D\hat{\mathbf{p}}_{en}}{D\phi_I}. \tag{8.37}$$

Sensitivity of the response functional. The final goal of the sensitivity analysis is to calculate the sensitivity $\frac{DF}{D\phi_I}$ of the response functional F. Since the response functional F depends on the response of the system, the sensitivity of the response of the system has to be calculated before the evaluation of sensitivity of the response functional. In general, the form of the response functional F is not known in advance and no general rules can be provided for automation of the terms needed to evaluate $\frac{DF}{D\phi_I}$, such as $\frac{\partial F}{\partial \phi_I}$ and $\frac{\partial F}{\partial \mathbf{p}}$. Depending on the actual case, it might not even be practical to evaluate the sensitivity of the response functional using element-by-element procedure. The general symbolic system can be a convenient tool to form and evaluate $\frac{DF}{D\phi_I}$ for a general F. The problem independent automation procedure can be derived for a special case when F is calculated as a sum of element contributions (F_e) using the element-by-element procedure

$$F = \sum_{e=1}^{n_e} F_e. \tag{8.38}$$

The sensitivity of the response functional (8.38) is then evaluated using the element-by-element procedure as follows

$$\frac{DF}{D\boldsymbol{\phi}} = \sum_{e=1}^{n_e} \frac{DF_e}{D\boldsymbol{\phi}} \tag{8.39}$$

where $\frac{DF_e}{D\boldsymbol{\phi}}$ is the element contribution to the sensitivity of the response functional.

8.3.1 *Direct Differentiation Method for Time-Independent Problems*

The direct differentiation of the time-independent residual (8.15) with respect to Ith design parameter ϕ_I yields

$$\frac{\partial \mathbf{R}}{\partial \mathbf{p}} \frac{D\mathbf{p}}{D\phi_I} + \frac{\partial \mathbf{R}}{\partial \phi_I} = \mathbf{0} \tag{8.40}$$

where the first factor on the left hand side of (8.40) is exactly the independent tangent operator $\mathbf{K}$ of the primal problem. Equation (8.40) represents a linear system of equations

$$\mathbf{K}\frac{D\mathbf{p}}{D\phi_I} = -{}^I\tilde{\mathbf{R}} \tag{8.41}$$

for the unknown vector $\frac{D\mathbf{p}}{D\phi_I}$. The term ${}^I\tilde{\mathbf{R}}$ in (8.41) is called the **"independent sensitivity pseudo-load vector"**. It is formed using the standard FE assembly procedure

$$ {}^{I}\tilde{\mathbf{R}} = \mathop{\mathsf{A}}_{e=1}^{n_e} {}^{I}\tilde{\mathbf{R}}_e \tag{8.42} $$

where

$$ {}^{I}\tilde{\mathbf{R}}_e = \frac{\partial \mathbf{R}_e}{\partial \phi_I} \tag{8.43} $$

represents the **"element independent sensitivity pseudo-load vector"**. When the element residual is formed by numerical integration over the domain of the element, see (3.3), then the sensitivity pseudo-load vector is also formed by numerical integration over the domain of the element as follows

$$ {}^{I}\tilde{\mathbf{R}}_e = \sum_{g=1}^{n_g} w_{gp} {}^{I}\tilde{\mathbf{R}}_g \tag{8.44} $$

where

$$ {}^{I}\tilde{\mathbf{R}}_g = \frac{\partial \mathbf{R}_g}{\partial \phi_I} \tag{8.45} $$

is the **"Gauss point independent sensitivity pseudo-load vector"**.

Sensitivity analysis is numerically efficient only if the already factorized tangent matrix from the solution of the primal problem is used to solve the resulting system of linear equations. The cost of sensitivity analysis is in that case equal to the cost of evaluating n_ϕ pseudo-load vectors by the element-by element procedure. The cost of one back-substitution with n_ϕ right hand sides as presented in Box 8.1.

The element contribution to the sensitivity of the response functional is, for the known sensitivity of the response $\frac{D\mathbf{p}}{D\phi_I}$, obtained by differentiating (8.14) with respect to Ith design parameter as follows

$$ \frac{DF_e}{D\phi_I} = \frac{\partial F_e}{\partial \mathbf{p}_e}\frac{D\mathbf{p}_e}{D\phi_I} + \frac{\partial F_e}{\partial \phi_I}. \tag{8.46} $$

Automation of direct differentiation method for time-independent problems. The element independent sensitivity pseudo-load vector is evaluated at element level and thus the ADB form of Eq. (8.43) has to be derived for automation. By replacing the partial derivative in (8.43) with the computational derivative and by adding the local AD exceptions of type C for all types of sensitivity parameters as defined in (8.33) the ADB form of (8.43) leads to

$$ {}^{I}\tilde{\mathbf{R}}_e = \left. \frac{\hat{\delta} \mathbf{R}_e}{\hat{\delta} \phi_I} \right|_{\frac{D\boldsymbol{\psi}_e}{D\phi_I} = D_{\phi_I}\boldsymbol{\psi}_e;\ \frac{D\hat{\mathbf{p}}_e}{D\phi_I} = D_{\phi_I}\bar{\mathbf{p}}_e;} \tag{8.47} $$

The automation presented here is general since it accounts for all types of sensitivity parameters within a single call to the automatic differentiation procedure. Equation (8.47) can be rewritten using the global AD exceptions of type C as follows

$$\begin{aligned}
\boldsymbol{\psi}_e &\leftarrow \boldsymbol{\psi}_e \Big|_{\frac{D\boldsymbol{\psi}_e}{D\phi_I}=D_{\phi_I}\boldsymbol{\psi}_e}, \\
\hat{\mathbf{p}}_e &\leftarrow \hat{\mathbf{p}}_e \Big|_{\frac{D\hat{\mathbf{p}}_e}{D\phi_I}=D_{\phi_I}\bar{\mathbf{p}}_e}, \\
{}^{I}\tilde{\mathbf{R}}_e &= \frac{\hat{\delta}\mathbf{R}_e}{\hat{\delta}\phi_I}.
\end{aligned} \tag{8.48}$$

The global AD exceptions in Eq. (8.48) redefine the derivatives of DOFs with prescribed essential boundary conditions $\bar{\mathbf{p}}_e$ and derivatives of the general input data $\boldsymbol{\psi}_e$. Although the formulation based on the global AD exceptions appears to be less understandable than the one based on the local AD exceptions, it is much more convenient for the actual computer implementation.

The Gauss point independent sensitivity pseudo-load vector is obtained from (8.47) or (8.48) by replacing $\mathbf{R}_e$ with $\mathbf{R}_g$.

The sensitivity of the response functional (8.46) is calculated after the solution of the sensitivity of response. When the response functional is calculated as a sum of element contributions (8.38), the ADB form of the element contribution to the sensitivity of the response functional using the local AD exceptions of type C leads to

$$\frac{DF_e}{D\phi_I} = \frac{\hat{\delta} F_e}{\hat{\delta}\phi_I}\Bigg|_{\frac{D\boldsymbol{\psi}_e}{D\phi_I}=D_{\phi_I}\boldsymbol{\psi}_e;\ \frac{D\hat{\mathbf{p}}_e}{D\phi_I}=D_{\phi_I}\hat{\mathbf{p}}_e;} \tag{8.49}$$

and using the global AD exceptions of type C to

$$\begin{aligned}
\boldsymbol{\psi}_e &\leftarrow \boldsymbol{\psi}_e \Big|_{\frac{D\boldsymbol{\psi}_e}{D\phi_I}=D_{\phi_I}\boldsymbol{\psi}_e}, \\
\hat{\mathbf{p}}_e &\leftarrow \hat{\mathbf{p}}_e \Big|_{\frac{D\hat{\mathbf{p}}_e}{D\phi_I}=D_{\phi_I}\hat{\mathbf{p}}_e}, \\
\frac{DF_e}{D\phi_I} &= \frac{\hat{\delta} F_e}{\hat{\delta}\phi_I}.
\end{aligned} \tag{8.50}$$

Note that the AD exception $\frac{D\hat{\mathbf{p}}_e}{D\phi_I}=D_{\phi_I}\bar{\mathbf{p}}_e$ in (8.47) and (8.48) is replaced by the AD exception $\frac{D\hat{\mathbf{p}}_e}{D\phi_I}=D_{\phi_I}\hat{\mathbf{p}}_e$ in (8.49) and (8.50). The data structure of sensitivities of the element DOFs ($D_{\phi_I}\hat{\mathbf{p}}_e$) combines the calculated sensitivity of the true unknowns of the problems ($\frac{D\mathbf{p}}{D\phi_I}$) with the externally defined sensitivities of DOFs with prescribed essential boundary (stored in $D_{\phi_I}\bar{\mathbf{p}}_e$) into a single data structure. The algorithm for the solution and automation of the primal and sensitivity analysis of general time-independent problems is presented in Box 8.1.

8.3.2 Efficient Solution of Global Sensitivity Problem

Usually we have to solve the sensitivity problem for a set of n_ϕ sensitivity parameters. The solutions of separate sensitivity problems are, as shown in the previous section, mutually independent, however the efficiency of the actual computational algorithm can be improved by exploring the fact that some equations (e.g. shape functions) are common for all sensitivity pseudo load vectors and thus can be evaluated only once. Also the forward-elimination/back-substitution is more efficient when performed for multiple right hand sides at the same time. In the algorithm presented in Box 8.1 the element pseudo load vector is evaluated at element level for all parameters in a loop and then assembled to the matrix of element pseudo load vectors. Let $\tilde{\mathbf{R}}_e = \{{}^I\tilde{\mathbf{R}}_e : I = 1, \ldots, n_\phi\}$ be a matrix of element pseudo-load vectors and $\tilde{\mathbf{R}} = \{{}^I\tilde{\mathbf{R}} : I = 1, \ldots, n_\phi\}$ a matrix of global pseudo-load vectors obtained from the element pseudo load vector by standard FEM assembly procedure

$$\tilde{\mathbf{R}} = \mathop{\mathsf{A}}_{e=1}^{n_e} \tilde{\mathbf{R}}_e.$$

The solution of the system of linear equations with multiple right hand sides

$$\mathbf{K}\frac{D\mathbf{p}}{D\boldsymbol{\phi}} = -\tilde{\mathbf{R}}$$

then leads to the sensitivity matrix $\frac{D\mathbf{p}}{D\boldsymbol{\phi}}$.

The above consideration might not be true when the number of sensitivity parameters is comparable with an average bandwidth of the global tangent matrix. The dimension of matrix $\tilde{\mathbf{R}}$ is $n_{tp} \times n_\phi$ where n_ϕ is the number of sensitivity pseudo-load vectors. Matrix $\tilde{\mathbf{R}}$ is dense, thus the memory consumption for storing $\tilde{\mathbf{R}}$ might become a limiting factor on the size of the problem that one can solve on a given computer. The solution to the problem is to split the set of sensitivity parameters into subsets $\boldsymbol{\phi} = \bigcup \boldsymbol{\phi}_L$, $\boldsymbol{\phi}_L = \{\phi_I : I \in [\phi_{L\,\text{start}}, \phi_{L\,\text{end}}\}$ leading to the following set of equations

$$\begin{aligned}
\tilde{\mathbf{R}}_e &= \{{}^I\tilde{\mathbf{R}}_e : I = \phi_{L\,\text{start}}, \ldots, \phi_{L\,\text{end}}\}, \\
\tilde{\mathbf{R}}_L &= \mathop{\mathsf{A}}_{e=1}^{n_e} \tilde{\mathbf{R}}_e = \{{}^I\tilde{\mathbf{R}} : I = \phi_{L\,\text{start}}, \ldots, \phi_{L\,\text{end}}\}, \\
\mathbf{K}\frac{D\mathbf{p}}{D\boldsymbol{\phi}_L} &= -\tilde{\mathbf{R}}_L.
\end{aligned} \tag{8.51}$$

The second situation to be considered is parallelization of sensitivity analysis. When the analysis is parallelized on a cluster of computers or on a multi-core computer the parallel evaluation of subsets of sensitivity pseudo-load vectors ($\tilde{\mathbf{R}}_L$) on separate cluster nodes or on separate cores can be numerically highly efficient. The optimum size of a subset depends on the ratio between the number of unknowns and the number of sensitivity parameters and on the computer architecture.

8.3.3 Direct Differentiation Method for Time-Dependent Problems

The direct differentiation of the time-dependent residual (8.23) with respect to the Ith design parameter ϕ_I yields

$$\frac{\partial \mathbf{R}}{\partial \mathbf{p}} \frac{D\mathbf{p}}{D\phi_I} + \frac{\partial \mathbf{R}}{\partial \mathbf{p}_n} \frac{D\mathbf{p}_n}{D\phi_I} + \frac{\partial \mathbf{R}}{\partial \phi_I} = \mathbf{0}. \tag{8.52}$$

As in the case of time-independent problems Eq. (8.52) leads to a system of linear equations of the form

$$\mathbf{K}\frac{D\mathbf{p}}{D\phi_I} = -{}^I\tilde{\mathbf{R}} \tag{8.53}$$

with the element contribution to the independent sensitivity pseudo-load vector ${}^I\tilde{\mathbf{R}}$

$${}^I\tilde{\mathbf{R}}_e = \frac{\partial \mathbf{R}_e}{\partial \mathbf{p}_{en}} \frac{D\mathbf{p}_{en}}{D\phi_I} + \frac{\partial \mathbf{R}_e}{\partial \phi_I} \tag{8.54}$$

and its Gauss point contribution

$${}^I\tilde{\mathbf{R}}_g = \frac{\partial \mathbf{R}_g}{\partial \mathbf{p}_{en}} \frac{D\mathbf{p}_{en}}{D\phi_I} + \frac{\partial \mathbf{R}_g}{\partial \phi_I}. \tag{8.55}$$

Since the sensitivity pseudo-load vector depends on the sensitivity of the response at time t_n ($\frac{D\mathbf{p}_n}{D\phi_I}$), the $\frac{D\mathbf{p}_n}{D\phi_I}$ vector has to be stored in the memory. Thus, the sensitivity analysis of time-dependent problems requires storage of an additional $n_\phi \times n_{tp}$ real numbers.

The sensitivity of the response functional (8.22) is for the known sensitivity of the response at terminal time $t_{n_{\text{step}}}$ ($\frac{D\mathbf{p}_{n_{\text{step}}}}{D\phi_I}$) and time $t_{n_{\text{step}}-1}$ ($\frac{D\mathbf{p}_{n_{\text{step}}-1}}{D\phi_I}$) obtained by differentiating (8.22) with respect to Ith design parameter

$$\frac{DF}{D\phi_I} = \frac{\partial F}{\partial \mathbf{p}_{n_{\text{step}}}} \frac{D\mathbf{p}_{n_{\text{step}}}}{D\phi_I} + \frac{\partial F}{\partial \mathbf{p}_{n_{\text{step}}-1}} \frac{D\mathbf{p}_{n_{\text{step}}-1}}{D\phi_I} + \frac{\partial F}{\partial \phi_I}. \tag{8.56}$$

Input: $\mathbf{p}_n$ `// starting value for` $\mathbf{p}$
begin solution of primal problem
  $\mathbf{p} \leftarrow \mathbf{p}_n$
  repeat
    foreach *element e* **do**
      $\mathbf{R}_e \leftarrow \mathbf{R}_e(\mathbf{p}_e)$
      $\mathbf{K}_e \leftarrow \frac{\partial \mathbf{R}_e}{\partial \mathbf{p}_e}$ .... $\xrightarrow{\texttt{automation}}$ .... $\mathbf{K}_e \leftarrow \frac{\hat{\delta} \mathbf{R}_e}{\hat{\delta} \mathbf{p}_e}$
      add $\mathbf{R}_e$ to $\mathbf{R}$ and $\mathbf{K}_e$ to $\mathbf{K}$
    end foreach
[A]   solve $\mathbf{K}\Delta\mathbf{p} + \mathbf{R} = \mathbf{0}$ for unknown $\Delta\mathbf{p}$
    $\mathbf{p} \leftarrow \mathbf{p} + \Delta\mathbf{p}$
  until *error criterion for* $\|\mathbf{R}\|$ *and* $\|\mathbf{p}\|$ *is fulfilled*
begin solution of sensitivity problem
  foreach *element e* **do**
    foreach *sensitivity parameter* ϕ_I **do**
      ${}^I\tilde{\mathbf{R}}_e \leftarrow \frac{\partial \mathbf{R}_e}{\partial \phi_I}$ ... $\xrightarrow{\texttt{automation}}$ ... ${}^I\tilde{\mathbf{R}}_e \leftarrow \left.\frac{\hat{\delta} \mathbf{R}_e}{\hat{\delta} \phi_I}\right|_{\frac{D\boldsymbol{\psi}_e}{D\phi_I}=D_{\phi_I}\boldsymbol{\psi}_e;\ \frac{D\hat{\mathbf{p}}_e}{D\phi_I}=D_{\phi_I}\bar{\mathbf{p}}_e;}$
      add ${}^I\tilde{\mathbf{R}}_e$ to $\tilde{\mathbf{R}}_e$
    end foreach
    add $\tilde{\mathbf{R}}_e$ to $\tilde{\mathbf{R}}$
  end foreach
  solve $\mathbf{K}\frac{D\mathbf{p}}{D\boldsymbol{\phi}} + \tilde{\mathbf{R}} = \mathbf{0}$ for unknown $\frac{D\mathbf{p}}{D\boldsymbol{\phi}}$ using already factorized $\mathbf{K}$ from step [A]
begin evaluate response functional
  foreach *element e* **do**
    add $F_e(\mathbf{p}_e)$ to F
    foreach *sensitivity parameter* ϕ_I **do**
      $\frac{DF_e}{D\phi_I} \leftarrow \frac{\partial F_e}{\partial \mathbf{p}_e}\frac{D\mathbf{p}_e}{D\phi_I} + \frac{\partial F_e}{\partial \phi_I}$ .... $\xrightarrow{\texttt{automation}}$ ....
      $\frac{DF_e}{D\phi_I} \leftarrow \left.\frac{\hat{\delta} F_e}{\hat{\delta} \phi_I}\right|_{\frac{D\boldsymbol{\psi}_e}{D\phi_I}=D_{\phi_I}\boldsymbol{\psi}_e;\ \frac{D\hat{\mathbf{p}}_e}{D\phi_I}=D_{\phi_I}\hat{\mathbf{p}}_e;}$
      add $\frac{DF_e}{D\phi_I}$ to $\frac{DF}{D\boldsymbol{\phi}}$
    end foreach
  end foreach
Result: $\mathbf{p}$, $\frac{D\mathbf{p}}{D\boldsymbol{\phi}}$, F, $\frac{DF}{D\boldsymbol{\phi}}$

Box 8.1. Primal and sensitivity analysis of time-independent problem

Automation of the direct differentiation method for time-dependent problems. The ADB form of (8.54) is obtained by replacing the partial derivative with the computational derivative and by adding the AD exceptions for all types of sensitivity parameters. The dependency of the solution on the response at time t_n requires the definition of additional AD exceptions when compared to time-independent problems presented in Sect. 8.3.1. An additional vector $D_{\phi_I}\hat{\mathbf{p}}_{en}$ was introduced in Eq. (8.37) in Sect. 8.3 that stores the sensitivities of all element DOFs (true unknowns and DOFs with the prescribed essential boundary condition) at time t_n with respect to the Ith sensitivity parameter.

Using $D_{\phi_I}\hat{\mathbf{p}}_{en}$ vector the ADB form of (8.54) leads to

$$ {}^{I}\tilde{\mathbf{R}}_e = \left.\frac{\hat{\delta}\mathbf{R}_e}{\hat{\delta}\phi_I}\right|_{\frac{D\boldsymbol{\psi}_e}{D\phi_I}=D_{\phi_I}\boldsymbol{\psi}_e;\ \frac{D\hat{\mathbf{p}}_e}{D\phi_I}=D_{\phi_I}\bar{\mathbf{p}}_e;\ \frac{D\hat{\mathbf{p}}_{en}}{D\phi_I}=D_{\phi_I}\hat{\mathbf{p}}_{en};} \tag{8.57}$$

The ADB form of an element contribution to the sensitivity of the response functional F is then given by

$$\frac{DF_e}{D\phi_I} = \left.\frac{\hat{\delta}F_e}{\hat{\delta}\phi_I}\right|_{\frac{D\boldsymbol{\psi}_e}{D\phi_I}=D_{\phi_I}\boldsymbol{\psi}_e;\ \frac{D\hat{\mathbf{p}}_{en_{\text{step}}}}{D\phi_I}=D_{\phi_I}\hat{\mathbf{p}}_{en_{\text{step}}};\ \frac{D\hat{\mathbf{p}}_{en_{step-1}}}{D\phi_I}=D_{\phi_I}\hat{\mathbf{p}}_{en_{step-1}}} . \tag{8.58}$$

8.3.4 Direct Differentiation Method for Time-Independent Locally Coupled Problems

The sensitivity formulation for time-independent locally coupled problems is derived by direct differentiation of the coupled Eqs. (8.17) and (8.18) with respect to the Ith design parameter ϕ_I. The global vector of the coupled unknowns $\mathbf{h}$ is composed of n_e local (element) vectors of the unknowns $\mathbf{h}_e$. Thus the direct differentiation of Eq. (8.17) yields

$$\frac{\partial \mathbf{R}}{\partial \mathbf{p}}\frac{D\mathbf{p}}{D\phi_I} + \sum_{e=1}^{n_e}\frac{\partial \mathbf{R}}{\partial \mathbf{h}_e}\frac{D\mathbf{h}_e}{D\phi_I} + \frac{\partial \mathbf{R}}{\partial \phi_I} = \mathbf{0}. \tag{8.59}$$

where $\frac{D\mathbf{p}}{D\phi_I}$ denotes the unknown sensitivity of independent solution vector and $\frac{D\mathbf{h}_e}{D\phi_I}$ the unknown sensitivity of n_e dependent solution vectors. The sensitivity of the dependent solution vectors $\frac{D\mathbf{h}_e}{D\phi_I}$ is needed in order to solve Eq. (8.59) for the unknown $\frac{D\mathbf{p}}{D\phi_I}$. Direct differentiation of the eth dependent residual (8.18) with respect to the Ith design parameter ϕ_I yields

$$\frac{\partial \mathbf{Q}_e}{\partial \mathbf{p}_e}\frac{D\mathbf{p}_e}{D\phi_I} + \frac{\partial \mathbf{Q}_e}{\partial \mathbf{h}_e}\frac{D\mathbf{h}_e}{D\phi_I} + \frac{\partial \mathbf{Q}_e}{\partial \phi_I} = \mathbf{0}. \tag{8.60}$$

The term $\frac{\partial \mathbf{Q}_e}{\partial \mathbf{h}_e}$ in (8.60) is exactly the dependent tangent operator $\mathbf{A}_e$. The term $\frac{D\mathbf{h}_e}{D\phi_I}$ is then obtained from (8.60) and leads to

$$\frac{D\mathbf{h}_e}{D\phi_I} = -\mathbf{A}_e^{-1}\left(\frac{\partial \mathbf{Q}_e}{\partial \mathbf{p}_e}\frac{D\mathbf{p}_e}{D\phi_I} + \frac{\partial \mathbf{Q}_e}{\partial \phi_I}\right). \tag{8.61}$$

One of the assumptions of the time-independent locally coupled problems is that the local problems are mutually independent. From that assumptions relation

$$\frac{\partial \mathbf{Q}_e}{\partial \mathbf{p}_e}\frac{D\mathbf{p}_e}{D\phi_I} = \frac{\partial \mathbf{Q}_e}{\partial \mathbf{p}}\frac{D\mathbf{p}}{D\phi_I} \tag{8.62}$$

follows. The term $\frac{D\mathbf{h}_e}{D\phi_I}$ is then inserted into (8.59) to get

$$\left(\frac{\partial \mathbf{R}}{\partial \mathbf{p}} - \sum_e \frac{\partial \mathbf{R}}{\partial \mathbf{h}_e}\mathbf{A}_e^{-1}\frac{\partial \mathbf{Q}_e}{\partial \mathbf{p}}\right)\frac{D\mathbf{p}}{D\phi_I} = -\left(\frac{\partial \mathbf{R}}{\partial \phi_I} - \sum_e \frac{\partial \mathbf{R}}{\partial \mathbf{h}_e}\mathbf{A}_e^{-1}\frac{\partial \mathbf{Q}_e}{\partial \phi_I}\right). \tag{8.63}$$

The first factor on the left hand side of (8.63) is exactly the independent tangent operator $\mathbf{K}$ of the primal problem (3.48). As in previous cases, Eq. (8.63) represents a linear system of equations (8.64) for the unknown sensitivity vector $\frac{D\mathbf{p}}{D\phi_I}$

$$\mathbf{K}\frac{D\mathbf{p}}{D\phi_I} = -{}^I\tilde{\mathbf{R}} \tag{8.64}$$

where the element contribution ${}^I\tilde{\mathbf{R}}_e$ to independent sensitivity pseudo-load vector ${}^I\tilde{\mathbf{R}}$ is defined by

$${}^I\tilde{\mathbf{R}}_e = \frac{\partial \mathbf{R}_e}{\partial \phi_I} - \frac{\partial \mathbf{R}_e}{\partial \mathbf{h}_e}\mathbf{A}_e^{-1}\frac{\partial \mathbf{Q}_e}{\partial \phi_I}. \tag{8.65}$$

With the known sensitivity of the independent solution vector ($\frac{D\mathbf{p}}{D\phi_I}$), the system of linear equations

$$\mathbf{A}_e\frac{D\mathbf{h}_e}{D\phi_I} = -{}^I\tilde{\mathbf{Q}}_e \ : e = 1, \ldots, n_e \tag{8.66}$$

can be formed for the unknown sensitivity of dependent solution vectors ($\frac{D\mathbf{h}_e}{D\phi_I}$) from (8.61). The term ${}^I\tilde{\mathbf{Q}}_e$ in (8.66) is called the **"dependent sensitivity pseudo-load vector"** and defined by

$${}^I\tilde{\mathbf{Q}}_e = \frac{\partial \mathbf{Q}_e}{\partial \mathbf{p}_e}\frac{D\mathbf{p}_e}{D\phi_I} + \frac{\partial \mathbf{Q}_e}{\partial \phi_I} \ : e = 1, \ldots, n_e. \tag{8.67}$$

Note that for a primal analysis the independent problem is solved after the dependent problem. For the sensitivity analysis the independent sensitivity problem is solved first and then the dependent sensitivity problem.

The sensitivity of the response functional in the case of time-independent locally coupled problems is obtained by differentiating (8.16) with respect to ϕ_I as follows

$$\frac{DF}{D\phi_I} = \frac{\partial F}{\partial \mathbf{p}}\frac{D\mathbf{p}}{D\phi_I} + \frac{\partial F}{\partial \mathbf{h}}\frac{D\mathbf{h}}{D\phi_I} + \frac{\partial F}{\partial \phi_I}. \tag{8.68}$$

Automation of the direct differentiation method for time-independent locally coupled problems. The element level quantities for which the ADB form have to be derived include the independent sensitivity pseudo-load vector (8.65), the dependent sensitivity pseudo-load vector (8.67) and the local tangent matrix $\mathbf{A}_e$. For the solution of the linear system of equations (8.64) the already derived and triangulated tangent matrix from the primal analysis can be used. The ADB form of local tangent matrix is

$$\mathbf{A}_e = \frac{\hat{\delta}\mathbf{Q}_e}{\hat{\delta}\mathbf{h}_e}. \tag{8.69}$$

The ADB form of (8.65) and (8.67) is obtained by replacing the partial derivative with the computational derivative and by adding the AD exceptions for general and essential boundary condition input data (8.33). The dependency between the dependent $\mathbf{h}_e$ and independent $\mathbf{p}$ solution vectors requires definition of additional AD exceptions when compared to uncoupled problems as presented in (Sect. 8.3.1). Let $D_{\phi_I}\mathbf{h}_e$ be an additional vector that stores the sensitivity of the eth dependent solution vector with respect to the Ith sensitivity parameter. Additionally, the vector $D_{\phi_I}\hat{\mathbf{p}}_e$ was introduced by Eq. (8.36) in Sect. 8.3 that stores the sensitivities of all element DOFs (true unknowns and DOFs with prescribed essential boundary conditions) with respect to the Ith sensitivity parameter.

Definition (8.65) of the independent sensitivity pseudo-load vector includes the term $\frac{\partial \mathbf{R}_e}{\partial \mathbf{h}_e}\mathbf{A}_e^{-1}\frac{\partial \mathbf{Q}_e}{\partial \phi_I}$ that cannot be reduced to a single call of the automatic differentiation procedure. Thus, the evaluation of (8.65) and (8.67) is divided into two steps. First the auxiliary quantity ${}^I\mathbf{Z}_e$ is defined

$${}^I\mathbf{Z}_e = -\mathbf{A}_e^{-1} \left.\frac{\hat{\delta}\mathbf{Q}_e}{\hat{\delta}\phi_I}\right|_{\frac{D\boldsymbol{\psi}_e}{D\phi_I}=D_{\phi_I}\boldsymbol{\psi}_e;\ \frac{D\hat{\mathbf{p}}_e}{D\phi_I}=D_{\phi_I}\bar{\mathbf{p}}_e;}. \tag{8.70}$$

The ADB form of the element independent sensitivity pseudo-load vector then follows from (8.65) as

$${}^I\tilde{\mathbf{R}}_e = \left.\frac{\hat{\delta}\mathbf{R}_e}{\hat{\delta}\phi_I}\right|_{\frac{D\boldsymbol{\psi}_e}{D\phi_I}=D_{\phi_I}\boldsymbol{\psi}_e;\ \frac{D\hat{\mathbf{p}}_e}{D\phi_I}=D_{\phi_I}\bar{\mathbf{p}}_e;\ \frac{D\mathbf{h}_e}{D\phi_I}={}^I\mathbf{Z}_e}. \tag{8.71}$$

In (8.70) and (8.71) the local AD exceptions of type C are used. With the use of the global AD exceptions of type C Eqs. (8.70) and (8.71) are rewritten

$$\begin{aligned}
\boldsymbol{\psi}_e &\leftarrow \left.\boldsymbol{\psi}_e\right|_{\frac{D\boldsymbol{\psi}_e}{D\phi_I}=D_{\phi_I}\boldsymbol{\psi}_e}, \\
\hat{\mathbf{p}}_e &\leftarrow \left.\hat{\mathbf{p}}_e\right|_{\frac{D\hat{\mathbf{p}}_e}{D\phi_I}=D_{\phi_I}\bar{\mathbf{p}}_e}, \\
{}^I\mathbf{Z}_e &\leftarrow -\mathbf{A}_e^{-1}\frac{\hat{\delta}\mathbf{Q}_e}{\hat{\delta}\phi_I}, \\
{}^I\tilde{\mathbf{R}}_e &= \left.\frac{\hat{\delta}\mathbf{R}_e}{\hat{\delta}\phi_I}\right|_{\frac{D\mathbf{h}_e}{D\phi_I}={}^I\mathbf{Z}_e}.
\end{aligned} \tag{8.72}$$

By comparing Eqs. (8.70) and (8.71) with (8.72), the true advantage of the global AD exceptions becomes apparent. While the local AD exceptions have to be defined twice, for each call of AD procedure separately, the global AD exceptions require only one definition.

By using the local AD exceptions the automation of the dependent sensitivity pseudo-load vector (8.67) leads to

$$
{}^{I}\tilde{\mathbf{Q}}_e = \left.\frac{\hat{\delta}\mathbf{Q}_e}{\hat{\delta}\phi_I}\right|_{\frac{D\boldsymbol{\psi}_e}{D\phi_I}=D_{\phi_I}\boldsymbol{\psi}_e;\ \frac{D\hat{\mathbf{p}}_e}{D\phi_I}=D_{\phi_I}\hat{\mathbf{p}}_e;} \tag{8.73}
$$

and by using the global AD exceptions it follows

$$
\begin{aligned}
\boldsymbol{\psi}_e &\leftarrow \boldsymbol{\psi}_e\Big|_{\frac{D\boldsymbol{\psi}_e}{D\phi_I}=D_{\phi_I}\boldsymbol{\psi}_e}, \\
\hat{\mathbf{p}}_e &\leftarrow \hat{\mathbf{p}}_e\Big|_{\frac{D\hat{\mathbf{p}}_e}{D\phi_I}=D_{\phi_I}\hat{\mathbf{p}}_e}, \\
{}^{I}\tilde{\mathbf{Q}}_e &= \frac{\hat{\delta}\mathbf{Q}_e}{\hat{\delta}\phi_I}.
\end{aligned} \tag{8.74}
$$

The formulation presented so far does not require any additional memory storage, which is consistent with the time-independent nature of the problem. However, if the auxiliary quantity ${}^{I}\mathbf{Z}_g$ is stored in the memory during the evaluation of ${}^{I}\tilde{\mathbf{R}}_e$ and used later during the evaluation of the sensitivity of the dependent solution vectors, even more numerically efficient formulation can be derived. This possibility will be explored in the next section for the automation of time-dependent locally coupled problems.

From (8.68) follows the ADB form of an element contribution F_e to the sensitivity of the response functional F

$$
\frac{DF_e}{D\phi_I} = \left.\frac{\hat{\delta}F_e}{\hat{\delta}\phi_I}\right|_{\frac{D\boldsymbol{\psi}_e}{D\phi_I}=D_{\phi_I}\boldsymbol{\psi}_e;\ \frac{D\hat{\mathbf{p}}_e}{D\phi_I}=D_{\phi_I}\hat{\mathbf{p}}_e;\ \frac{D\mathbf{h}_e}{D\phi_I}=D_{\phi_I}\mathbf{h}_e}. \tag{8.75}
$$

Furthermore, the ADB form based on global AD exceptions of (8.75) is given by

$$
\begin{aligned}
\boldsymbol{\psi}_e &\leftarrow \boldsymbol{\psi}_e\Big|_{\frac{D\boldsymbol{\psi}_e}{D\phi_I}=D_{\phi_I}\boldsymbol{\psi}_e}, \\
\hat{\mathbf{p}}_e &\leftarrow \hat{\mathbf{p}}_e\Big|_{\frac{D\hat{\mathbf{p}}_e}{D\phi_I}=D_{\phi_I}\hat{\mathbf{p}}_e}, \\
\mathbf{h}_e &\leftarrow \mathbf{h}_e\Big|_{\frac{D\mathbf{h}_e}{D\phi_I}=D_{\phi_I}\mathbf{h}_e}, \\
\frac{DF_e}{D\phi_I} &= \frac{\hat{\delta}F_e}{\hat{\delta}\phi_I}.
\end{aligned} \tag{8.76}
$$

8.3.5 Direct Differentiation Method for Time-Independent Gauss Point Coupled Problems

The Gauss point coupled problems differ from locally coupled problems in two key points. The dependent residual $\mathbf{Q}_g$ does not directly depend on $\mathbf{p}_e$ but on a set of intermediate variables $\mathbf{r}_g(\mathbf{p}_e)$. Direct differentiation of the gth dependent residual (8.21) with respect to the Ith design parameter ϕ_I yields

$$\frac{\partial \mathbf{Q}_g}{\partial \mathbf{r}_g}\left(\frac{\partial \mathbf{r}_g}{\partial \mathbf{p}_e}\frac{D\mathbf{p}_e}{D\phi_I}+\frac{\partial \mathbf{r}_g}{\partial \phi_I}\right)+\frac{\partial \mathbf{Q}_e}{\partial \mathbf{h}_e}\frac{D\mathbf{h}_e}{D\phi_I}+\frac{\partial \mathbf{Q}_e}{\partial \phi_I}=\mathbf{0}. \tag{8.77}$$

After rearrangement we obtain

$$\underline{\frac{\partial \mathbf{Q}_g}{\partial \mathbf{r}_g}\frac{\partial \mathbf{r}_g}{\partial \mathbf{p}_e}}\frac{D\mathbf{p}_e}{D\phi_I}+\frac{\partial \mathbf{Q}_e}{\partial \mathbf{h}_e}\frac{D\mathbf{h}_e}{D\phi_I}+\underline{\frac{\partial \mathbf{Q}_g}{\partial \mathbf{r}_g}\frac{\partial \mathbf{r}_g}{\partial \phi_I}+\frac{\partial \mathbf{Q}_g}{\partial \phi_i}}=\mathbf{0}. \tag{8.78}$$

If the procedure presented for locally coupled problems would be repeated here, then the first underlined term would become a part of the global tangent matrix. The global tangent is, for the sensitivity analysis, taken from the primal analysis, thus the first underlined term does not need to be derived again for sensitivity analysis. The second underlined term is, from the point of automation, equal to $\frac{\hat{\delta}\mathbf{Q}_g}{\hat{\delta}\phi_I}$. Consequently, within the sensitivity analysis it holds: $\mathbf{r}_g = \mathbf{p}_e$, and the expressions derived for locally coupled problems are directly applicable within Gauss point coupled problems.

Automation of time-independent Gauss point coupled problems follows from the additive nature of the element assembly procedure, from Gauss integration and from the assumption that Gauss point problems are mutually independent. The formulation of locally coupled problems (Eqs. (8.70)–(8.74)) yields the ADB form of a Gauss point contribution to the independent sensitivity pseudo-load vector ${}^I\tilde{\mathbf{R}}_g$

$${}^I\mathbf{Z}_g = -\mathbf{A}_g^{-1}\left.\frac{\hat{\delta}\mathbf{Q}_g}{\hat{\delta}\phi_I}\right|_{\frac{D\boldsymbol{\psi}_e}{D\phi_I}=D_{\phi_I}\boldsymbol{\psi}_e;\ \frac{D\hat{\mathbf{p}}_e}{D\phi_I}=D_{\phi_I}\bar{\mathbf{p}}_e;}, \tag{8.79}$$

$${}^I\tilde{\mathbf{R}}_g = \left.\frac{\hat{\delta}\mathbf{R}_g}{\hat{\delta}\phi_I}\right|_{\frac{D\boldsymbol{\psi}_e}{D\phi_I}=D_{\phi_I}\boldsymbol{\psi}_e;\ \frac{D\hat{\mathbf{p}}_e}{D\phi_I}=D_{\phi_I}\bar{\mathbf{p}}_e;\ \frac{D\mathbf{h}_g}{D\phi_I}={}^I\mathbf{Z}_g}, \tag{8.80}$$

and the dependent sensitivity pseudo-load vector ${}^I\tilde{\mathbf{Q}}_g$

$${}^I\tilde{\mathbf{Q}}_g = \left.\frac{\hat{\delta}\mathbf{Q}_g}{\hat{\delta}\phi_I}\right|_{\frac{D\boldsymbol{\psi}_e}{D\phi_I}=D_{\phi_I}\boldsymbol{\psi}_e;\ \frac{D\hat{\mathbf{p}}_e}{D\phi_I}=D_{\phi_I}\hat{\mathbf{p}}_e;}. \tag{8.81}$$

By using global AD exceptions Eqs. (8.79)–(8.81) can be rewritten to

$$
\begin{aligned}
\boldsymbol{\psi}_e &\leftarrow \boldsymbol{\psi}_e\Big|_{\frac{D\boldsymbol{\psi}_e}{D\phi_I}=D_{\phi_I}\boldsymbol{\psi}_e}, \\
\hat{\mathbf{p}}_e &\leftarrow \hat{\mathbf{p}}_e\Big|_{\frac{D\hat{\mathbf{p}}_e}{D\phi_I}=D_{\phi_I}\bar{\mathbf{p}}_e}, \\
{}^I\mathbf{Z}_g &\leftarrow -\mathbf{A}_g^{-1}\frac{\hat{\delta}\mathbf{Q}_g}{\hat{\delta}\phi_I}, \\
{}^I\tilde{\mathbf{R}}_g &= \frac{\hat{\delta}\mathbf{R}_g}{\hat{\delta}\phi_I}\Bigg|_{\frac{D\mathbf{h}_g}{D\phi_I}={}^I\mathbf{Z}_g}
\end{aligned}
\tag{8.82}
$$

and

$$
\begin{aligned}
\boldsymbol{\psi}_e &\leftarrow \boldsymbol{\psi}_e\Big|_{\frac{D\boldsymbol{\psi}_e}{D\phi_I}=D_{\phi_I}\boldsymbol{\psi}_e}, \\
\hat{\mathbf{p}}_e &\leftarrow \hat{\mathbf{p}}_e\Big|_{\frac{D\hat{\mathbf{p}}_e}{D\phi_I}=D_{\phi_I}\hat{\mathbf{p}}_e}, \\
{}^I\tilde{\mathbf{Q}}_g &= \frac{\hat{\delta}\mathbf{Q}_g}{\hat{\delta}\phi_I}.
\end{aligned}
\tag{8.83}
$$

8.3.6 *Direct Differentiation Method for Time-Dependent Locally Coupled Problems*

In case of time-dependent locally coupled problems the direct differentiation of Eq. (8.25) with respect to the Ith design parameter ϕ_I yields

$$
\begin{aligned}
&\frac{\partial \mathbf{R}}{\partial \mathbf{p}}\frac{D\mathbf{p}}{D\phi_I} + \sum_{e=1}^{n_e}\frac{\partial \mathbf{R}}{\partial \mathbf{h}_e}\frac{D\mathbf{h}_e}{D\phi_I} + \frac{\partial \mathbf{R}}{\partial \mathbf{p}_n}\frac{D\mathbf{p}_n}{D\phi_I} \\
&\quad + \sum_{e=1}^{n_e}\frac{\partial \mathbf{R}}{\partial \mathbf{h}_{en}}\frac{D\mathbf{h}_{en}}{D\phi_I} + \frac{\partial \mathbf{R}}{\partial \phi_I} = \mathbf{0}
\end{aligned}
\tag{8.84}
$$

where $\frac{D\mathbf{p}}{D\phi_I}$ denotes the unknown sensitivity of the independent solution vector at time t_{n+1}. The unknown sensitivity of the dependent solution vectors $\frac{D\mathbf{h}_e}{D\phi_I}$ at time t_{n+1} is needed in order to solve Eq. (8.84) for the unknown $\frac{D\mathbf{p}}{D\phi_I}$. The differentiation of the eth dependent residual (8.26) with respect to the Ith design parameter ϕ_I yields

$$
\begin{aligned}
\frac{D\mathbf{h}_e}{D\phi_I} = -\mathbf{A}_e^{-1}\Bigg(&\frac{\partial \mathbf{Q}_e}{\partial \mathbf{p}_e}\frac{D\mathbf{p}_e}{D\phi_I} + \frac{\partial \mathbf{Q}_e}{\partial \mathbf{p}_{en}}\frac{D\mathbf{p}_{en}}{D\phi_I} \\
&+ \frac{\partial \mathbf{Q}_e}{\partial \mathbf{h}_{en}}\frac{D\mathbf{h}_{en}}{D\phi_I} + \frac{\partial \mathbf{Q}_e}{\partial \phi_I}\Bigg).
\end{aligned}
\tag{8.85}
$$

By assuming the relation (8.62) and inserting the term $\frac{D\mathbf{h}_e}{D\phi_I}$ into (8.84) we obtain

$$\left(\frac{\partial \mathbf{R}}{\partial \mathbf{p}} - \sum_e \frac{\partial \mathbf{R}}{\partial \mathbf{h}_e}\mathbf{A}_e^{-1}\frac{\partial \mathbf{Q}_e}{\partial \mathbf{p}}\right)\frac{D\mathbf{p}}{D\phi_I}$$
$$= -\left(\frac{\partial \mathbf{R}}{\partial \mathbf{p}_n}\frac{D\mathbf{p}_n}{D\phi_I} + \sum_e \frac{\partial \mathbf{R}}{\partial \mathbf{h}_{en}}\frac{D\mathbf{h}_{en}}{D\phi_I} + \frac{\partial \mathbf{R}}{\partial \phi_I}\right.$$
$$\left. - \sum_e \frac{\partial \mathbf{R}}{\partial \mathbf{h}_e}\mathbf{A}_e^{-1}\left(\frac{\partial \mathbf{Q}_e}{\partial \mathbf{p}_{en}}\frac{D\mathbf{p}_{en}}{D\phi_I} + \frac{\partial \mathbf{Q}_e}{\partial \mathbf{h}_{en}}\frac{D\mathbf{h}_{en}}{D\phi_I} + \frac{\partial \mathbf{Q}_e}{\partial \phi_I}\right)\right). \tag{8.86}$$

The first factor on the left hand side of (8.86) is the independent tangent operator $\mathbf{K}$ of the primal problem (3.48). As in previous cases, Eq. (8.86) represents a linear system of equation (8.87) for the unknown sensitivity vector $\frac{D\mathbf{p}}{D\phi_I}$

$$\mathbf{K}\frac{D\mathbf{p}}{D\phi_I} = -{}^I\tilde{\mathbf{R}}. \tag{8.87}$$

Here the element contribution (${}^I\tilde{\mathbf{R}}_e$) to the independent sensitivity pseudo-load vector (${}^I\tilde{\mathbf{R}}$) is defined by

$${}^I\tilde{\mathbf{R}}_e = \frac{\partial \mathbf{R}_e}{\partial \mathbf{p}_{en}}\frac{D\mathbf{p}_{en}}{D\phi_I} + \frac{\partial \mathbf{R}_e}{\partial \mathbf{h}_{en}}\frac{D\mathbf{h}_{en}}{D\phi_I} + \frac{\partial \mathbf{R}_e}{\partial \phi_I} + \frac{\partial \mathbf{R}_e}{\partial \mathbf{h}_e}{}^I\mathbf{Z}_e \tag{8.88}$$

where ${}^I\mathbf{Z}_e$ is an additional auxiliary quantity. It is defined for each locally coupled problem as

$${}^I\mathbf{Z}_e = -\mathbf{A}_e^{-1}\left(\frac{\partial \mathbf{Q}_e}{\partial \mathbf{p}_{en}}\frac{D\mathbf{p}_{en}}{D\phi_I} + \frac{\partial \mathbf{Q}_e}{\partial \mathbf{h}_{en}}\frac{D\mathbf{h}_{en}}{D\phi_I} + \frac{\partial \mathbf{Q}_e}{\partial \phi_I}\right). \tag{8.89}$$

The dimension of ${}^I\mathbf{Z}_e$ is the same as the dimension of $\frac{D\mathbf{h}_e}{D\phi_I}$ and the independent sensitivity problem is solved before the dependent sensitivity problem, thus the ${}^I\mathbf{Z}_g$ term can be stored temporarily during the assembly of the pseudo-load vector at the place of $\frac{D\mathbf{h}_e}{D\phi_I}$ and be overwritten by the true $\frac{D\mathbf{h}_e}{D\phi_I}$ during the solution of dependent sensitivity problem. With the term $\frac{D\mathbf{p}_e}{D\phi_I}$ and the auxiliary vector ${}^I\mathbf{Z}_e$ being known, the sensitivity of the dependent solution vectors leads to

$$\frac{D\mathbf{h}_e}{D\phi_I} = {}^I\mathbf{Z}_e - \mathbf{A}_e^{-1}\frac{\partial \mathbf{Q}_e}{\partial \mathbf{p}_e}\frac{D\mathbf{p}_e}{D\phi_I}. \tag{8.90}$$

The introduction of the auxiliary quantity ${}^I\mathbf{Z}_e$ splits the solution of the dependent sensitivity problem into two parts. Consequently, the dependent sensitivity pseudo-load vector ${}^I\tilde{\mathbf{Q}}_e$ is not explicitly formed as it was the case for the sensitivity analysis of time-independent locally coupled problems presented in the previous section.

The sensitivity of the response functional (8.24) is obtained by differentiating (8.24) with respect to the Ith design parameter

$$\frac{DF}{D\phi_I} = \frac{\partial F}{\partial \mathbf{p}_{n_{\text{step}}}} \frac{D\mathbf{p}_{n_{\text{step}}}}{D\phi_I} + \frac{\partial F}{\partial \mathbf{h}_{n_{\text{step}}}} \frac{D\mathbf{h}_{n_{\text{step}}}}{D\phi_I} + \frac{\partial F}{\partial \mathbf{p}_{n_{\text{step}}-1}} \frac{D\mathbf{p}_{n_{\text{step}}-1}}{D\phi_I} + \frac{\partial F}{\partial \mathbf{h}_{n_{\text{step}}-1}} \frac{D\mathbf{h}_{n_{\text{step}}-1}}{D\phi_I} + \frac{\partial F}{\partial \phi_I}. \tag{8.91}$$

Automation of the direct differentiation method for time-dependent locally coupled problems. The ADB form for time-dependent locally coupled problems combines the definitions for the automation of time-dependent problems presented in Sect. 8.3.3 with definitions used for automation of time-independent locally coupled problems in Sect. 8.3.4. As in Sect. 8.3.4 the quantities at element level for which the ADB form has to be derived include the independent sensitivity pseudo-load vector (8.65), the dependent sensitivity pseudo-load vector (8.67) and the local tangent matrix $\mathbf{A}_e$. The dependency between the dependent $\mathbf{h}_e$ and independent $\mathbf{p}$ solution vectors and the dependency on the solution at time t_n require the definition of additional AD exceptions when compared to (Sect. 8.3.1). Let $D_{\phi_I}\mathbf{h}_e$ and $D_{\phi_I}\mathbf{h}_{en}$ be additional vectors that store the sensitivity of the eth dependent solution vector with respect to the Ith sensitivity parameter at times $t_n + 1$ and t_n, respectively. Additionally, the vectors $D_{\phi_I}\hat{\mathbf{p}}_e$ and $D_{\phi_I}\hat{\mathbf{p}}_{en}$ were introduced in Sect. 8.2.2 that store the sensitivities of all element DOFs (true unknowns and DOFs with a prescribed essential boundary condition) with respect to the Ith sensitivity parameter at times $t_n + 1$ and t_n, respectively.

Definitions (8.89) and (8.88) can now be rewritten as

$$^I\mathbf{Z}_e = -\mathbf{A}_e^{-1} \left. \frac{\hat{\delta}\mathbf{Q}_e}{\hat{\delta}\phi_I} \right|_{\frac{D\boldsymbol{\psi}_e}{D\phi_I}=D_{\phi_I}\boldsymbol{\psi}_e;\ \frac{D\hat{\mathbf{p}}_e}{D\phi_I}=D_{\phi_I}\bar{\mathbf{p}}_e;\ \frac{D\hat{\mathbf{p}}_{en}}{D\phi_I}=D_{\phi_I}\hat{\mathbf{p}}_{en};\ \frac{D\mathbf{h}_{en}}{D\phi_I}=D_{\phi_I}\mathbf{h}_{en}}, \tag{8.92}$$

$$^I\tilde{\mathbf{R}}_e = \left. \frac{\hat{\delta}\mathbf{R}_e}{\hat{\delta}\phi_I} \right|_{\substack{\frac{D\boldsymbol{\psi}_e}{D\phi_I}=D_{\phi_I}\boldsymbol{\psi}_e;\ \frac{D\hat{\mathbf{p}}_e}{D\phi_I}=D_{\phi_I}\bar{\mathbf{p}}_e; \\ \frac{D\hat{\mathbf{p}}_{en}}{D\phi_I}=D_{\phi_I}\hat{\mathbf{p}}_{en};\ \frac{D\mathbf{h}_{en}}{D\phi_I}=D_{\phi_I}\mathbf{h}_{en};\ \frac{D\mathbf{h}_e}{D\phi_I}={}^I\mathbf{Z}_e}} . \tag{8.93}$$

The derivation of the ADB form of the sensitivity of the dependent solution vector (8.90) requires an additional consideration. The DOFs with prescribed essential boundary conditions are in the traditional notation not unknowns, thus their contribution to $\frac{D\mathbf{h}_e}{D\phi_I}$ in (8.90) does not come from the $\frac{\partial \mathbf{Q}_e}{\partial \mathbf{p}_e}$ term but from the $\frac{\partial \mathbf{Q}_e}{\partial \phi_I}$ term and is accounted for in the derivation of $^I\mathbf{Z}_e$. Thus, the $\bar{\mathbf{p}}_e$ DOFs have to be kept constant when the term $\frac{\partial \mathbf{Q}_e}{\partial \mathbf{p}_e}\frac{D\mathbf{p}_e}{D\phi_I}$ in (8.90) is evaluated. This can be achieved by the introduction of an additional field of sensitivities of true element unknowns $D_{\phi_I}\check{\mathbf{p}}_e$ defined by Eq. (8.34). With Eq. (8.34), the automation of the evaluation of the sensitivity of the dependent solution vectors $\frac{D\mathbf{h}_e}{D\phi_I}$, given by (8.90), leads to

$$\frac{D\mathbf{h}_e}{D\phi_I} = {}^I\mathbf{Z}_e - \mathbf{A}_e^{-1}\left(\left.\frac{\hat{\delta}\mathbf{Q}_e}{\hat{\delta}\phi_I}\right|_{\frac{D\hat{\mathbf{p}}_e}{D\phi_I}=D_{\phi_I}\check{\mathbf{p}}_e}\right). \tag{8.94}$$

The ADB form of the element contribution to the sensitivity of the response functional F is given by

$$\frac{DF_e}{D\phi_I} = \left.\frac{\hat{\delta}F_e}{\hat{\delta}\phi_I}\right|_{\substack{\frac{D\boldsymbol{\psi}_e}{D\phi_I}=D_{\phi_I}\boldsymbol{\psi}_e; \\ \frac{D\hat{\mathbf{p}}_{en_{\text{step}}}}{D\phi_I}=D_{\phi_I}\hat{\mathbf{p}}_{en_{\text{step}}};\ \frac{D\hat{\mathbf{p}}_{en_{\text{step}}-1}}{D\phi_I}=D_{\phi_I}\hat{\mathbf{p}}_{en_{\text{step}}-1}; \\ \frac{D\mathbf{h}_{en_{\text{step}}}}{D\phi_I}=D_{\phi_I}\mathbf{h}_{en_{\text{step}}};\ \frac{D\mathbf{h}_{en_{\text{step}}-1}}{D\phi_I}=D_{\phi_I}\mathbf{h}_{en_{\text{step}}-1}}} \tag{8.95}$$

Equations (8.92) and (8.93) can be rewritten by using the global AD exceptions of type C

$$\begin{aligned}
\boldsymbol{\psi}_e &\leftarrow \boldsymbol{\psi}_e\Big|_{\frac{D\boldsymbol{\psi}_e}{D\phi_I}=D_{\phi_I}\boldsymbol{\psi}_e}, \\
\hat{\mathbf{p}}_e &\leftarrow \hat{\mathbf{p}}_e\Big|_{\frac{D\hat{\mathbf{p}}_e}{D\phi_I}=D_{\phi_I}\bar{\mathbf{p}}_e}, \\
\hat{\mathbf{p}}_{en} &\leftarrow \hat{\mathbf{p}}_{en}\Big|_{\frac{D\hat{\mathbf{p}}_{en}}{D\phi_I}=D_{\phi_I}\hat{\mathbf{p}}_{en}}, \\
\mathbf{h}_{en} &\leftarrow \mathbf{h}_{en}\Big|_{\frac{D\mathbf{h}_{en}}{D\phi_I}=D_{\phi_I}\mathbf{h}_{en}}, \\
{}^I\mathbf{Z}_e &\leftarrow -\mathbf{A}_e^{-1}\frac{\hat{\delta}\mathbf{Q}_e}{\hat{\delta}\phi_I}, \\
{}^I\tilde{\mathbf{R}}_e &= \left.\frac{\hat{\delta}\mathbf{R}_e}{\hat{\delta}\phi_I}\right|_{\frac{D\mathbf{h}_e}{D\phi_I}={}^I\mathbf{Z}_e}.
\end{aligned} \tag{8.96}$$

Additionally Eq. (8.94) is rewritten as

$$\begin{aligned}
\hat{\mathbf{p}}_e &\leftarrow \hat{\mathbf{p}}_e\Big|_{\frac{D\hat{\mathbf{p}}_e}{D\phi_I}=D_{\phi_I}\check{\mathbf{p}}_e}, \\
\frac{D\mathbf{h}_e}{D\phi_I} &= {}^I\mathbf{Z}_e - \mathbf{A}_e^{-1}\frac{\hat{\delta}\mathbf{Q}_e}{\hat{\delta}\phi_I}.
\end{aligned} \tag{8.97}$$

Automation of the direct differentiation method for time-dependent Gauss point coupled problems. Following the considerations presented in Sect. 8.3.5 the direct differentiation method for time-dependent Gauss point coupled problems can be obtained from the direct differentiation method for time-dependent locally coupled problems simply by replacing the index e of the local problem by the index g of the Gauss point problem in all equations. The resulting automation of the direct differentiation method for time-dependent Gauss point coupled problems is presented in Box 8.2. Only constant time stepping is considered. The independent sensitivity pseudo load vector is evaluated in Algorithm 8.2 for all sensitivity parameters, as discussed in Sect. 8.3.2.

8.3.7 Natural Boundary Condition Sensitivity Analysis

Problems in solid mechanics and nonlinear structural mechanics, subjected to quasi-static proportional load, are frequently formulated as

$$\mathbf{R} = \mathbf{R}^{\text{int}} - \lambda\,\mathbf{R}^{\text{ref}} = \mathbf{0} \tag{8.98}$$

where $\mathbf{R}^{\text{int}}$ denotes the contribution of the internal forces to the global residual vector. Vector $\mathbf{R}^{\text{ref}}$ is the reference load vector associated with the pattern of the applied nodal forces (natural boundary condition input data) and λ is the loading parameter. The load vector $\lambda\ \mathbf{R}^{\text{ref}}$ is subtracted from the internal force vector and thus does not effect directly the residual vectors of the finite elements at local element level. This is opposed to the way how the variation of prescribed essential boundary conditions is treated. As shown in the previous sections the variation of prescribed essential boundary conditions can be handled at the element level. If the contribution of the natural boundary conditions to the global residual $\mathbf{R}$ is accounted for by a special generalized finite elements then the natural boundary condition input data can be considered as a part of general input data $\boldsymbol{\psi}_e$ and treated accordingly.

The general equation (8.98) leads for an arbitrary time-dependent problem and for an arbitrary sensitivity parameter ϕ_I to

$$\mathbf{R}^{\text{int}}(\mathbf{p}(\phi_I), \mathbf{p}_n(\phi_I), \phi_I) - \lambda\,\mathbf{R}^{\text{ref}}(\phi_I) = \mathbf{0}. \tag{8.99}$$

Direct differentiation of (8.99) with respect to ϕ_I yields

$$\frac{\partial \mathbf{R}^{\text{int}}}{\partial \mathbf{p}} \frac{D\mathbf{p}}{D\phi_I} + \frac{\partial \mathbf{R}^{\text{int}}}{\partial \mathbf{p}_n} \frac{D\mathbf{p}_n}{D\phi_I} + \frac{\partial \mathbf{R}^{\text{int}}}{\partial \phi_I} - \lambda \frac{D\mathbf{R}^{\text{ref}}}{D\phi_I} = \mathbf{0}. \tag{8.100}$$

Using the independent sensitivity pseudo-load vector ${}^I\tilde{\mathbf{R}}$

$${}^I\tilde{\mathbf{R}} = \frac{\partial \mathbf{R}^{\text{int}}}{\partial \mathbf{p}_n} \frac{D\mathbf{p}_n}{D\phi_I} + \frac{\partial \mathbf{R}^{\text{int}}}{\partial \phi_I} - \lambda \frac{D\mathbf{R}^{\text{ref}}}{D\phi_I}, \tag{8.101}$$

leads to the sensitivity of the response $\frac{D\mathbf{p}}{D\phi_I}$ by the solution of the linear equation system

$$\mathbf{K} \frac{D\mathbf{p}}{D\phi_I} = -{}^I\tilde{\mathbf{R}}. \tag{8.102}$$

The first two terms in (8.101) represent contribution of an internal forces to the sensitivity pseudo load vector as described in detail in Sect. 8.3.3. Thus, we only need to evaluate the contribution of variation of imposed homogeneous natural boundary conditions $\frac{D\mathbf{R}^{\text{ref}}}{D\phi_I}$ to the sensitivity pseudo-load vector ${}^I\tilde{\mathbf{R}}$.

The actual dimension and the contents of the sensitivity pseudo-load vector $\frac{D\mathbf{R}^{\text{ref}}}{D\phi_I}$ is not known before the analysis. It can change during the analysis (e.g. in the case of contact problems). Thus, the term $\frac{D\mathbf{R}^{\text{ref}}}{D\phi_I}$ can not be an actual input data for sensitivity analysis.

Let assume that there exists a subset of the global set of input data fields $\boldsymbol{\psi}_b \subseteq \boldsymbol{\psi}$ that can be uniquely mapped into nodal forces and that the reference load vector $\mathbf{R}^{\text{ref}}$ is composed of the components of $\boldsymbol{\psi}$. The formulation of sensitivity problem obviously requires the derivatives $\frac{D\mathbf{R}^{\text{ref}}}{D\phi_I}$ to be known and available at the global level, thus the proper parts of the global velocity matrix $D_{\phi_I}\boldsymbol{\psi}$ have to be mapped into an appropriate parts of $\frac{D\mathbf{R}^{\text{ref}}}{D\phi_I}$. The mapping form $D_{\phi_I}\boldsymbol{\psi}$ to $\frac{D\mathbf{R}^{\text{ref}}}{D\phi_I}$ is given by

$$\frac{D\mathbf{R}^{\text{ref}}}{D\phi_I} = \left\{ D_{\phi_I}\psi^{\text{node}(K)}_{\text{NBCF}(\text{dof}(K),\text{node}(K))} : K = 1, \ldots, n_{tp} \right\} \tag{8.103}$$

where n_{tp} is the total number of global unknowns, node(K) is the global index of the node with Kth global unknown, dof(K) is relative position of the Kth global unknown within the unknowns of the node(K)th node. The NBCF function defines an index of the design velocity field taken from $D_{\phi_I}\boldsymbol{\psi}$ that belongs to dof(K)th unknown in node(K)th global node.

Equation (8.103) relates design velocity matrix $D_{\phi_I}\boldsymbol{\psi}$ to components of the pseudo load vector $\frac{D\mathbf{R}^{\text{ref}}}{D\phi_I}$. Note that the mapping function NBCF in (8.103) has the same purpose as the mapping function EBCF in Eq. (8.32) that relates design velocity matrix $D_{\phi_I}\boldsymbol{\psi}$ to the nodal unknowns at element level. This equation is specific for the actual problem solved and is fulfilled by the properly prepared input data for sensitivity analysis. An example how this can be done within the *AceFEM* finite element environment is presented in Sect. 8.5.

8.3.8 Implicit Dependency and the Cases with Multiple Domains

The finite element model can be composed of several domains that are discretized by different types of elements. In general, a specific design parameter would not effect all domains in the same way. For example, in the model of reinforced concrete the change of yield stress of steel does not effect directly the residual vectors of the elements used to discretize the concrete. However, in the final result of the sensitivity analysis,

Input: $\mathbf{p}_0$ // starting value for $\mathbf{p}$ (e.g. $\mathbf{0}$)
Input: $\mathbf{h}_0$ // starting value for $\mathbf{h}$ (e.g. $\mathbf{0}$)
$t_n \leftarrow 0$; $\mathbf{p}_n \leftarrow \mathbf{p}_0$; $\mathbf{h}_n \leftarrow \mathbf{h}_0$
$D_{\boldsymbol{\phi}}\hat{\mathbf{p}} \leftarrow \mathbf{0}$; $D_{\boldsymbol{\phi}}\mathbf{h}_n \leftarrow \mathbf{0}$ // starting value for all sensitivity data is 0
repeat
  $t \leftarrow t_n + \Delta t$ // update time
  begin solution of primal problem
    solution procedure of primal problem is presented in Box 3.5
  begin solution of independent sensitivity problem
    foreach *element e* **do**
      foreach *sensitivity parameter* ϕ_I **do**
        $\boldsymbol{\psi}_e \leftarrow \boldsymbol{\psi}_e\Big|_{\frac{D\boldsymbol{\psi}_e}{D\phi_I}=D_{\phi_I}\boldsymbol{\psi}_e}$; $\hat{\mathbf{p}}_e \leftarrow \hat{\mathbf{p}}_e\Big|_{\frac{D\hat{\mathbf{p}}_e}{D\phi_I}=D_{\phi_I}\bar{\mathbf{p}}_e}$; $\hat{\mathbf{p}}_{en} \leftarrow \hat{\mathbf{p}}_{en}\Big|_{\frac{D\hat{\mathbf{p}}_{en}}{D\phi_I}=D_{\phi_I}\hat{\mathbf{p}}_{en}}$
        foreach *Gauss point g* **do**
          $\mathbf{h}_{gn} \leftarrow \mathbf{h}_{gn}\Big|_{\frac{D\mathbf{h}_{gn}}{D\phi_I}=D_{\phi_I}\mathbf{h}_{gn}}$; ${}^I\mathbf{Z}_g \leftarrow -\mathbf{A}_g^{-1}\frac{\hat{\delta}\mathbf{Q}_g}{\hat{\delta}\phi_I}$; ${}^I\tilde{\mathbf{R}}_g \leftarrow \frac{\hat{\delta}\mathbf{R}_g}{\hat{\delta}\phi_I}\Bigg|_{\frac{D\mathbf{h}_g}{D\phi_I}={}^I\mathbf{Z}_g}$
          add $w_{gp}\,{}^I\tilde{\mathbf{R}}_g$ to ${}^I\tilde{\mathbf{R}}_e$
          export ${}^I\mathbf{Z}_g \rightarrow D_{\phi_I}\mathbf{h}_g$ // ${}^I\mathbf{Z}_g$ is stored into $D_{\phi_I}\mathbf{h}_g$ for use in line [A]
        end foreach
        add ${}^I\tilde{\mathbf{R}}_e$ to $\tilde{\mathbf{R}}_e$
      end foreach
      add $\tilde{\mathbf{R}}_e$ to $\tilde{\mathbf{R}}$
    end foreach
    solve $\mathbf{K}\frac{D\mathbf{p}}{D\boldsymbol{\phi}} + \tilde{\mathbf{R}} = \mathbf{0}$ for unknown $\frac{D\mathbf{p}}{D\boldsymbol{\phi}}$ and store $\frac{D\mathbf{p}}{D\boldsymbol{\phi}}$ into $D_{\boldsymbol{\phi}}\hat{\mathbf{p}}$
  begin solution of dependent sensitivity problem
    foreach *element e* **do**
      foreach *sensitivity parameter* ϕ_I **do**
        $\hat{\mathbf{p}}_e \leftarrow \hat{\mathbf{p}}_e\Big|_{\frac{D\hat{\mathbf{p}}_e}{D\phi_I}=D_{\phi_I}\bar{\mathbf{p}}_e}$
        foreach *Gauss point g* **do**
[A] ${}^I\mathbf{Z}_g \leftarrow D_{\phi_I}\mathbf{h}_g$; $\frac{D\mathbf{h}_g}{D\phi_I} \leftarrow {}^I\mathbf{Z}_g - \mathbf{A}_g^{-1}\frac{\hat{\delta}\mathbf{Q}_g}{\hat{\delta}\phi_I}$
          export $\frac{D\mathbf{h}_g}{D\phi_I}$ to $D_{\phi_I}\mathbf{h}_g$ data structure
        end foreach
      end foreach
    end foreach
  $\mathbf{p}_n \leftarrow \mathbf{p}$; $\mathbf{h}_n \leftarrow \mathbf{h}$; // update solution data for next time step
  $D_{\boldsymbol{\phi}}\hat{\mathbf{p}}_n \leftarrow D_{\boldsymbol{\phi}}\hat{\mathbf{p}}$; $D_{\boldsymbol{\phi}}\mathbf{h}_n \leftarrow D_{\boldsymbol{\phi}}\mathbf{h}$ // update sensitivity data
until *terminal time* $\bar{t}$ *is reached*
Result: $\mathbf{p}_n$, $\frac{D\mathbf{p}_n}{D\boldsymbol{\phi}}$, $\mathbf{h}_n$, $\frac{D\mathbf{h}_n}{D\boldsymbol{\phi}}$, $\mathbf{p}$, $\frac{D\mathbf{p}}{D\boldsymbol{\phi}}$, $\mathbf{h}$, $\frac{D\mathbf{h}}{D\boldsymbol{\phi}}$

Box 8.2. Automation of the direct differentiation method for time-dependent Gauss point coupled problems using global AD exceptions

unknowns used to discretize the concrete domain will exhibit nonzero sensitivity due to the implicit dependency of the global vector of unknowns $\mathbf{p}$ on all design parameters. Additionally, in the case of time-dependent problems (elasto-plastic model of steel and concrete), the sensitivity pseudo-load vectors would always be nonzero due to the implicit dependency on the history dependent variables. Consequently, although some domains might not be effected by the chosen sensitivity parameter directly, one still has to consider their contribution to the sensitivity pseudo-load vectors.

8.4 Generation of Sensitivity Analysis Related Subroutines

Implementation of sensitivity analysis in a finite element environments is not straightforward. In fact, most of the commercial finite element environments do not support analytical sensitivity analysis. Instead they resort to more easily implementable semi-analytical sensitivity analysis where the sensitivity pseudo-load vector is evaluated by the finite difference method. Automatic generation of the sensitivity analysis related subroutines by the *AceGen* automatic code generator is described here. Subroutines are generated for the *AceFEM* finite element environment. However, the symbolic description is universal and with a proper symbolic-numeric interface the generated subroutines can be used also within other finite element environments.

Design velocity fields can be classified into three types, see Sect. 8.2.2. Only general velocity fields and essential boundary condition velocity fields have to be considered at the individual finite element level. The natural boundary condition velocity fields are considered at the global level of the finite element environment.

A general case is considered where the finite element model is composed of n_B domains $B = B_1 \cup B_2, \ldots, \cup B_{n_B}$ that are discretized by different types of finite elements. The set of sensitivity parameters $\boldsymbol{\phi}$ is split into subsets $\boldsymbol{\phi} = \bigcup \boldsymbol{\phi}_L$, $\boldsymbol{\phi}_L = \{\phi_I : I \in [\phi_{L\,\mathrm{start}}, \ldots, \phi_{L\,\mathrm{end}}\}$ for efficiency reasons, as discussed in Sect. 8.3.2.

The algorithm presented in Box 8.2 reveals that a combined primal and sensitivity analysis of a time-dependent coupled problem contains three steps:

- solution of the primal analysis,
- solution of the independent sensitivity problem and
- solution of the dependent sensitivity problem.

At least a part of each of these steps has to be performed at finite element level and consequently three different user subroutines have to be generated for those tasks:

`"Tangent and residual"` user subroutine is a part of the solution algorithm for the primal analysis and returns the element tangent matrix $\mathbf{K}_e$ and the residual vector $\mathbf{R}_e$.

`"Sensitivity pseudo-load"` user subroutine is a part of the solution algorithm of the independent sensitivity problem and returns the contribution of an individual finite element ($\tilde{\mathbf{R}}_e$) to a matrix of sensitivity pseudo-load vectors $\tilde{\mathbf{R}}_L$ that belongs to the Lth subset of sensitivity parameters $\boldsymbol{\phi}_L$ as defined in Sect. 8.3.2.

`"Dependent sensitivity"` user subroutine is a part of the solution algorithm for the dependent sensitivity problem and updates the sensitivity of locally or Gauss point coupled variables. It is only needed for the sensitivity analysis of locally and Gauss point coupled problems.

Some derivations (e.g. interpolation of unknown fields and definition of the hyper-elastic strain energy) is needed in all of the above subroutines. In order to avoid repeating the same or similar symbolic inputs at several points, three auxiliary functions will be defined:

`IO[user-subroutine]` auxiliary function that creates definitions of the input/output parameters needed within the specific user subroutine.
`GP[]` auxiliary function that evaluates the Gauss point related quantities such as the interpolation of coordinates and unknown fields and strain energy function.
`fQW[task,hg]` auxiliary function in the case of coupled problems evaluates a coupled set of equations as described in Sect. 5.2.4 for elasto-plastic problems.

8.4.1 Axisymmetric, Hyper-Elastic Element for Primal and Sensitivity Analysis

In this section we derive user subroutines that can be applied to perform primal and sensitivity analysis of axisymmetric, hyper-elastic problems. The general description of the axisymmetric, hyper-elastic element was already given in Sect. 6.2.2 and will not be repeated here. The *AceGen* input closely follows the general algorithm for primal and sensitivity analysis of time-independent problems presented in Box 8.1.

Step 1: *AceGen* and Template initialization

```
SMSInitialize["Q1AX-sens","Environment"→"AceFEM"];
SMSTemplate["SMSTopology"→"Q1","SMSSymmetricTangent"→True,
  "SMSNodeID"→"D","SMSDomainDataNames"->{"E","ν"},
  "SMSDefaultData"->{21 000,0.3},
  "SMSSensitivityNames"→{"X","Y","E","ν"}];
nen=SMSNoNodes;
ndim=SMSNoDimensions;
np=SMSNoDOFGlobal;
ndB=2;
nψe=ndim+ndB;
```

At the beginning of the session the `SMSInitialize` function starts *AceGen* and the `SMSTemplate` function initializes constants that are needed to create the interface to the chosen FE environment. The constants `"Environment"`, `"SMSTopology"`, `"SMSDomainDataNames"`, `"SMSDefaultData"` and `"SMSSymmetricTangent"` were already described in Sect. 2.7. Additionally we need to specify the `"SMSSensitivityNames"` template constant that specifies the ordering of the general input data fields $\boldsymbol{\psi}_e$ for which sensitivity is defined within the element and unique node identification "SMSNodeID" → "D" for the 2D nodes with 2 unknowns. As a rule, the first parameters are spatial coordinates (for the axisymmetric case "X", "Y") followed by the material parameters (for the present material law E and ν). The total number of general input data fields for the present element is then $n_{\psi e} = 4$. This ordering has to be used when the input data that describes the actual problem is prepared.

Step 2: Definition of IO auxiliary function. In addition to the already defined input/output parameters used for primal analysis (see Sects. 2.7 and 5.2.6) we need to define:

nd$$[J, "SDVF", Iϕ, L] is sensitivity of the Lth general input data field (ordering is defined by "SMSSensitivityNames" template constant) with respect to the Ith sensitivity parameter in the Jth element node. Data is used to construct the general design velocity matrix $D_{\phi_I}\boldsymbol{\psi}_e \equiv$ DψeDϕIO as defined by Eq. (8.31).

nd$$[J, "st", Iϕ, K] is sensitivity of the Kth nodal unknown with respect to the Ith sensitivity parameter in the Jth element node at time t_{n+1}. The data is used to set up the $D_{\phi_I}\hat{\mathbf{p}}_e \equiv$ DpeDϕIO data structure as defined in Eq. (8.36).

```
IO[subroutine_]:=Block[{},
  Ξ={ξ,η}⊢Table[SMSReal[es$$["IntPoints",i,Ig]],{i,2}];
  wgp⊢SMSReal[es$$["IntPoints",4,Ig]];
  Nh⊨1/4 {(1-ξ) (1-η),(1+ξ) (1-η),(1+ξ) (1+η),(1-ξ) (1+η)};
  Switch[subroutine
   ,"Tangent and residual",
   XIO⊢Table[SMSReal[nd$$[J,"X",K]],{J,nen},{K,ndim}];
   uIO⊢Table[SMSReal[nd$$[J,"at",K]],{J,nen},{K,ndim}];
   dB⊢Table[SMSReal[es$$["Data",i]],{i,ndB}];
   ,"Sensitivity pseudo-load",
   DψeDϕIO⊢Table[SMSReal[nd$$[J,"SDVF",Iϕ,L]],{J,nen},{L,nψe}];
   XIO⊢Table[SMSReal[nd$$[J,"X",K],
       "Dependency"→{ϕ,DψeDϕIO⟦J,K⟧}],{J,nen},{K,ndim}];
   dB⊢Table[SMSReal[es$$["Data",i],
       "Dependency"→{ϕ,Nh.DψeDϕIO⟦All,ndim+i⟧}],{i,ndB}];
   DpeDϕIO⊨SMSReal[Table[nd$$[J,"st",Iϕ,K],{J,nen},{K,ndim}]];
   uIO⊢Table[SMSReal[nd$$[J,"at",K],
       "Dependency"→{ϕ,DpeDϕIO[[J,K]]}],{J,nen},{K,ndim}];
  ];
  pe=Flatten[uIO];
 ]
```

The general design velocity matrix $D_{\phi_I}\boldsymbol{\psi}_e$ is used within the IO auxiliary function to establish the dependency of nodal coordinates and material parameters on sensitivity parameters. The global AD exceptions are imposed on a set of nodal displacements $\mathbf{u}$ and material parameters $\mathbf{d}_B$ according to Eq. (8.48).

Step 3: Definition of Gauss point quantities.

```
GP[]:=Block[{},
   X⊢SMSFreeze[Nh.XIO];
   u⊨Nh.uIO;
   Je⊨SMSD[X,Ξ];
   Jed⊨2 π X[[1]] Det[Je];
   H⊨SMSD[u,X,"Dependency"→{Ξ,X,SMSInverse[Je]}];
   F⊨{{1+H[[1,1]],H[[1,2]],0},{H[[2,1]],1+H[[2,2]],0},{0,0,1+u[[1]]/X[[1]]}};
   Ct⊨F^T.F;JF⊨Det[F];{λ,μ}⊨SMSHookeToLame[dB[[1]],dB[[2]]];
   W⊨λ/4 (JF^2-1)-λ/2 Log[JF]-μ Log[JF]+1/2 μ (Tr[Ct]-3);
  ];
```

The general formulation of the axisymmetric element is presented in Sect. 6.2.2.

Step 4: Derivation of `"Tangent and residual"` user subroutine.

```
SMSStandardModule["Tangent and residual"];
SMSDo[Ig,1,SMSInteger[es$$["id","NoIntPoints"]]];
  IO["Tangent and residual"];GP[];
  SMSDo[m,1,np];
    Rgm⊨Jed SMSD[W,pe,m];
SMSExport[wgp Rgm,p$$[m],"AddIn"→True];
     SMSDo[n,m,np];
       Kgmn⊨SMSD[Rgm,pe,n];
SMSExport[wgp Kgmn,s$$[m,n],"AddIn"→True];
     SMSEndDo[];
  SMSEndDo[];
SMSEndDo[];
```

The formulation of the element tangent matrix and residual vector for the axisymmetric element is presented in Sect. 6.2.2.

Step 5: Derivation of `"Sensitivity pseudo-load"` user subroutine.

```
SMSStandardModule["Sensitivity pseudo-load"];
{ϕLstart,ϕLend}⊢
  SMSInteger[{idata$$["SensIndexStart"],idata$$["SensIndexEnd"]}];
SMSDo[Ig,1,SMSInteger[es$$["id","NoIntPoints"]]];
  SMSDo[Iϕ,ϕLstart,ϕLend];
    ϕ⊢SMSFictive[];IO["Sensitivity pseudo-load"];GP[];
    SMSDo[m,1,np];
      Rgm⊨Jed SMSD[W,pe,m];Rtgm⊨SMSD[Rgm,ϕ];
      SMSExport[wgp Rtgm,s$$[Iϕ-ϕLstart+1,m],"AddIn"→True];
    SMSEndDo[];
  SMSEndDo[];
SMSEndDo[];
```

Table 8.1 Statistics of the axisymmetric, hyper-elastic element for primal and sensitivity analysis

Subroutine	No. of formulae	No. of leafs
`Tangent and residual`	110	1709
`Sensitivity pseudo-load`	168	2900

File size: 16150 bytes

Additional data are needed for the derivation of the `"Sensitivity pseudo-load"` user subroutine:

`idata$$["SensIndexStart"]` $\phi_{L\,\text{start}} \equiv \phi$`Lstart` is an index of the first sensitivity parameter within the currently calculated subset of sensitivity parameters $\boldsymbol{\phi}_L$ as defined in Sect. 8.3.2.

`idata$$["SensIndexEnd"]` $\phi_{L\,\text{end}} \equiv \phi$`Lend` is an index of the last sensitivity parameter within $\boldsymbol{\phi}_L$.

`s$$` is a matrix of sensitivity pseudo-load vectors ($\tilde{\mathbf{R}}_e^T \equiv$ `s$$`) with dimension $(\phi_{L\,\text{end}} - \phi_{L\,\text{start}} + 1) \times n_p$.

Step 6: Automatic generation of the element source code file.

```
SMSWrite[];
```

The essential statistics related to the derivations of the formulas and the generated source code file are provided in Table 8.1. The number of leafs is a measure of the size of the derived formulas and represents roughly the number of items in a graph representing the formulas.

8.4.2 Three-Dimensional, Elasto-Plastic Element for Primal and Sensitivity Analysis

A three-dimensional, von Mises small strain elasto-plastic, eight nodes brick element for primal and sensitivity analysis is derived here. The general description of the *AceGen* input for the generation of the three-dimensional elasto-plastic element for primal analysis was provided in Sect. 5.2.6. The general description of the *AceGen* input for sensitivity analysis of time-independent problems is given in Sect. 8.4.1. Only the differences related to the sensitivity analysis of time-dependent coupled problems are explained here. The elasto-plastic problem is a coupled problem, thus as presented in the introduction, three auxiliary functions `IO`, `GP` and `fQW` have to be defined.

Step 1: *AceGen* and Template initialization

```
SMSInitialize["ElastoPlastic","Environment"→"AceFEM"];
nhg=7;
lhg=nhg+1+nhg idata$$["NoSensParameters"];
leh=lhg es$$["id","NoIntPoints"];
SMSTemplate["SMSTopology"→"H1","SMSSymmetricTangent"→True,
  "SMSDomainDataNames"→{"E","ν","Y0","H"},
  "SMSDefaultData"->{21 000,0.3,24,0},"SMSNoTimeStorage"→leh,
  "SMSPostIterationCall"→True,
  "SMSSensitivityNames"→{"X","Y","E","ν","Y0","H"}];
nen=SMSNoNodes;
ndim=SMSNoDimensions;
ndB=4;
nψe=ndim+ndB;
np=SMSNoDOFGlobal;
```

The initialization phase requires the additional definition of the length of the history data vector per element. The internal organization of the element history data vector $\mathbf{d}_{he}$ at time t_{n+1} can be arbitrary. In the present element is the element history data vector organized Gauss point wise as follows

$$\mathbf{d}_{he} = \bigcup_{g=1}^{n_g} \mathbf{d}_{hg}$$

with the Gauss point history vector $\mathbf{h}_g$ that is composed of

$$\mathbf{d}_{hg} = \mathbf{h}_g \bigcup \{state\} \bigcup_{I=1}^{n_\phi} \frac{D\mathbf{h}_g}{D\phi_I}.$$

Here $state$ is set to 0 when a Gauss point is in the elastic regime and to 1 when a Gauss point is in the plastic regime. The total length of the element history data vector (l_{eh}) is then

$$l_{eh} = l_{hg}\, n_g \quad \text{and} \quad l_{hg} = n_{hg} + 1 + n_{hg}\, n_\phi$$

where n_{hg} is the number of Gauss point unknowns. The organization of the element history data vector $\mathbf{d}_{hen}$ at time t_n is analogous to the organization of $\mathbf{d}_{he}$. The Gauss point history vector $\mathbf{h}_g$ is stored in the element history data vector $\mathbf{d}_{he}$ starting with the position $I_{hg} = (I_g - 1)\, l_{hg}$.

Step 2: Definition of `IO` auxiliary function. In addition to the already defined input/output parameters (see Sects. 2.7, 5.2.6 and 8.4.1) we need to define sensitivity of the Gauss point unknowns $D_{\phi_I}\mathbf{h}_g \equiv$ `DhgDϕIO` at time t_{n+1} and $D_{\phi_I}\mathbf{h}_{gn} \equiv$ `DhgnDϕIO` at time t_n. As shown in the previous step, the sensitivity of the Gauss point unknowns in the gth point with respect to the I_ϕth sensitivity parameter is stored in the element history data vector $\mathbf{d}_{he}$ starting with the position $I_{sg} = I_{hg} + n_{hg} + 1 + (I_\phi - 1)\, n_{hg}$.

Appropriate AD exceptions are imposed on a set of nodal displacements $\mathbf{u}$, material parameters $\mathbf{d}_B$ and Gauss point unknowns at times t_n and t_{n+1} in subroutines `"Sensitivity pseudo-load"` and `"Dependent sensitivity"`. Global AD exceptions are imposed in the `"Sensitivity pseudo-load"` user subroutine according to Eq. (8.96).

In the `"Dependent sensitivity"` user subroutine it is more convenient to use local AD exceptions, thus AD exceptions in the `"Dependent sensitivity"` user subroutine are imposed later in the definition of the subroutine according to Eq. (8.94). This can be achieved by the introduction of an additional field of sensitivities of true element unknowns $D_{\phi_I}\check{\mathbf{p}}_e \equiv$ `DpceDϕIO` defined by (8.34).

```
IO[subroutine_]:=Block[{},
  Ξ={ξ,η,ζ}⊢Table[SMSReal[es$$["IntPoints",i,Ig]],{i,3}];
  wgp⊢SMSReal[es$$["IntPoints",4,Ig]];
  Ξn={{-1,-1,-1},{1,-1,-1},{1,1,-1},{-1,1,-1},
    {-1,-1,1},{1,-1,1},{1,1,1},{-1,1,1}};
  Nh⊨Table[1/8 (1+ξ Ξn⟦J,1⟧) (1+η Ξn⟦J,2⟧) (1+ζ Ξn⟦J,3⟧),{J,1,nen}];
  Ihg⊢SMSInteger[(Ig-1) lhg];
  Switch[subroutine
   ,"Tangent and residual"|"Dependent sensitivity",
   XIO⊢Table[SMSReal[nd$$[J,"X",K]],{J,nen},{K,ndim}];
   uIO⊢Table[SMSReal[nd$$[J,"at",K]],{J,nen},{K,ndim}];
   hgnIO⊢Table[SMSReal[ed$$["hp",Ihg+i]],{i,nhg}];
   dB⊢Table[SMSReal[es$$["Data",i]],{i,ndB}];
   If[subroutine=="Dependent sensitivity",
    hgIO⊢Table[SMSReal[ed$$["ht",Ihg+i]],{i,nhg}];
    Isg⊢SMSInteger[Ihg+nhg+1+(Iϕ-1) nhg];
    Zg⊢Table[SMSReal[ed$$["ht",Isg+i]],{i,nhg}];
    DpceDϕIO⊢
     Flatten[Table[SMSIsDOFConstrained[SMSInteger[nd$$[J,"DOF",K]],
        0,SMSReal[nd$$[J,"st",Iϕ,K]]],{J,nen},{K,ndim}]];
   ];
   ,"Sensitivity pseudo-load",
   DψeDϕIO⊢Table[SMSReal[nd$$[J,"SDVF",Iϕ,L]],{J,nen},{L,nψe}];
   XIO⊢Table[SMSReal[nd$$[J,"X",K],
      "Dependency"→{ϕ,DψeDϕIO⟦J,K⟧}],{J,nen},{K,ndim}];
   dB⊢Table[SMSReal[es$$["Data",i],
      "Dependency"→{ϕ,Nh.DψeDϕIO⟦All,ndim+i⟧}],{i,ndB}];
   DpeDϕIO⊢Table[SMSReal[nd$$[J,"st",Iϕ,K]],{J,nen},{K,ndim}];
   uIO⊢Table[SMSReal[nd$$[J,"at",K],
```

```
    "Dependency"→{ϕ,DpeDϕIO[[J,K]]}],{J,nen},{K,ndim}];
  Isg ⊢SMSInteger[Ihg+nhg+1+ (Iϕ-1) nhg];
  DhgnDϕIO⊢Table[SMSReal[ed$$["hp",Isg+i]],{i,nhg}];
  hgnIO⊢
   Table[SMSReal[ed$$["hp",Ihg+i],"Dependency"→{ϕ,DhgnDϕIO[[i]]}],
    {i,nhg}];
  hgIO⊢Table[SMSReal[ed$$["ht",Ihg+i]],{i,nhg}];
 ];
 pe=Flatten[uIO];
]
```

Note that the elasto-plastic evolution equations do not depend on the nodal displacements $\mathbf{u}_n$ at time t_n, thus $\mathbf{u}_n$ is not defined within the `IO` auxiliary function.

Step 3: Definition of Gauss point quantities.

```
GP[]:=Block[{},
   X⊢SMSFreeze[Nh.XIO];
   u⊨Nh.uIO;
   Je⊨SMSD[X,Ξ];
   Jed⊨Det[Je];
   H⊨SMSD[u,X,"Dependency"→{Ξ,X,SMSInverse[Je]}];
   SMSFreeze[ϵ,1/2 (H+Transpose[H]),"Symmetric"→True];
   rg=SMSVariables[ϵ];
  ];
```

The extension of the formulation to sensitivity analysis has no effect on interpolations of coordinates, displacement gradient and definition of a small strain tensor. The equations given in Sect. 5.2.6 are summarized in presented definition of the `GP` auxiliary function.

Step 4: Definition of `fQW` auxiliary function.

The extension of the formulation to sensitivity analysis has no effect on the `fQW` auxiliary function. Thus, the `fQW` function defined in Box 5.3 can be used here as well. The `fQW` function is essential for the transparent representation of the operator split procedure for the integration of the elasto-plastic evolution equations.

Step 5: Derivation of `"Tangent and residual"` user subroutine.

```
SMSStandardModule["Tangent and residual"];
SMSDo[Ig,1,SMSInteger[es$$["id","NoIntPoints"]]];
IO["Tangent and residual"];GP[];
TOL⊢SMSReal[rdata$$["SubIterationTolerance"]];
iNR⊢SMSInteger[idata$$["Iteration"]];
staten⊢SMSReal[ed$$["hp",Ihg+1hg]];
ftr⊨fQW["f",hgnIO];
SMSIf[(iNR==1 && staten==0)||(iNR>1 && ftr<TOL)];(*elastic*)
  hg⊣hgnIO;
  SMSExport[Join[hg,{0}],Table[ed$$["ht",Ihg+i],{i,nhg+1}]];
SMSElse[];(*plastic*)
  hgb⊣hgnIO;
  SMSDo[jNR,1,30,1,hgb];
    Qg⊨fQW["Q",hgb];
    Ag⊨SMSD[Qg,hgb];
    LU⊨SMSLUFactor[Ag];
    Δh⊨SMSLUSolve[LU,-Qg];
    SMSIf[Sqrt[Δh.Δh]<TOL,
      DhgDr⊣SMSLUSolve[LU,-SMSD[Qg,rg,"Constant"→hgb]];
      SMSExport[Join[hgb+Δh,{1}],Table[ed$$["ht",Ihg+i],{i,nhg+1}]];
      SMSBreak[];];
    hgb⊣hgb+Δh;
    SMSIf[jNR=="29",SMSExport[{1,2},{idata$$["SubDivergence"],
        idata$$["ErrorStatus"]}];SMSBreak[];];
  SMSEndDo[hgb,DhgDr];
  hg⊣SMSFreeze[hgb,"Dependency"→{rg,DhgDr}];
SMSEndIf[hg];
W⊨fQW["W",hg];
SMSDo[m,1,np];
  Rgm⊨Jed SMSD[W,pe,m,"Constant"->hg];
  SMSExport[wgp Rgm,p$$[m],"AddIn"→True];
  SMSDo[n,m,np];
    Kgmn⊨SMSD[Rgm,pe,n];
    SMSExport[wgp Kgmn,s$$[m,n],"AddIn"→True];
  SMSEndDo[];
SMSEndDo[];
SMSEndDo[];
```

A detailed description of the *ADB* formulation of the element tangent matrix and residual vector for elasto-plastic elements is presented in Sect. 5.2.6.

Step 6: Derivation of "`Sensitivity pseudo-load`" user subroutine.

```
SMSStandardModule["Sensitivity pseudo-load"];
{ϕLstart,ϕLend}⊢
  SMSInteger[{idata$$["SensIndexStart"],idata$$["SensIndexEnd"]}];
SMSDo[Ig,1,SMSInteger[es$$["id","NoIntPoints"]]];
  SMSDo[Iϕ,ϕLstart,ϕLend];
    ϕ⊢SMSFictive[];IO["Sensitivity pseudo-load"];GP[];
    state⊢SMSReal[ed$$["ht",Ihg+lhg]];
    Qg⊨fQW["Q",hgIO];
    Zg⊨SMSIf[SMSReal[state]==0
        ,DhgnDϕIO
        ,Ag⊨SMSD[Qg,hgIO];
        SMSLinearSolve[Ag,-SMSD[Qg,ϕ]]
        ];
    SMSExport[Zg,Table[ed$$["ht",Isg+i],{i,nhg}]];
    SMSDo[m,1,np];
      Rgm⊨Jed  SMSD[W,pe,m,"Constant"→hgIO];
      Rtgm⊨SMSD[Rgm,ϕ,"Dependency"→{{hgIO,ϕ,Zg}}];
      SMSExport[wgp Rtgm,s$$[Iϕ-ϕLstart+1,m],"AddIn"→True];
    SMSEndDo[];
  SMSEndDo[];
SMSEndDo[];
```

The *AceGen* input that generates the `"Sensitivity pseudo-load"` user subroutine closely follows the second step "solution of independent sensitivity problem" of the general algorithm for the automation of the direct differentiation method for time-dependent Gauss point coupled problems presented in Box 8.2.

Step 7: Derivation of `"Dependent sensitivity"` user subroutine.

```
SMSStandardModule["Dependent sensitivity"];
{ϕLstart,ϕLend}⊢
  SMSInteger[{idata$$["SensIndexStart"],idata$$["SensIndexEnd"]}];
SMSDo[Ig,1,SMSInteger[es$$["id","NoIntPoints"]]];
  SMSDo[Iϕ,ϕLstart,ϕLend];
    ϕ⊢SMSFictive[];IO["Dependent sensitivity"];
    state⊢SMSReal[ed$$["ht",Ihg+lhg]];
    SMSIf[SMSReal[state]!=0];
      GP[];Qg⊨fQW["Q",hgIO];Ag⊨SMSD[Qg,hgIO];
      DhgDϕ⊨SMSLinearSolve[Ag,
            -SMSD[Qg,ϕ,"Dependency"→{{pe,ϕ,DpceDϕIO}}]]+Zg;
      SMSExport[DhgDϕ,Table[ed$$["ht",Isg+i],{i,nhg}]];
    SMSEndIf[];
  SMSEndDo[];
SMSEndDo[];
```

Table 8.2 Statistics related to the generation of the three-dimensional, elasto-plastic element for primal and sensitivity analysis

Subroutine	No. of formulae	No. of leafs
`Tangent and residual`	538	8663
`Sensitivity pseudo-load`	537	11030
`Dependent sensitivity`	325	5889

File size: 76953 bytes

The "solution of dependent sensitivity problem" of the general algorithm presented in Box 8.2 is implemented in this step. The only difference is that instead of the global AD exceptions the local AD exceptions defined in Eq. (8.34) are used to define the missing variations of the true element unknowns ($D_{\phi_I}\check{\mathbf{p}}_e \equiv$ `DpceDϕIO`).

Step 8: Automatic generation of the element source code file.

```
SMSWrite[];
```

The essential statistics of the derived formulas and the generated source code file is provided in Table 8.2.

8.5 Sensitivity Analysis of a Double-Layered Axisymmetric Cone by *AceFEM*

The element derived in Sect. 8.4.1 is applied here to perform the primal and sensitivity analysis of a double layered axisymmetric cone depicted in Fig. 8.4.

The cone is composed of two layers B_1 and B_2 with different material characteristics as presented in Table 8.3. The position of the points $\boldsymbol{T}_1 - \boldsymbol{T}_6$ defining the geometry is described by the axisymmetric coordinates (R, Z) in Table 8.3. The specimen is clamped at the bottom. A uniform surface traction $q_p = 10^5$ is imposed in Z direction at the outer surface of the cone, thus $\bar{t}_2 = q_p$ at $\overline{T_3 T_6}$. Additionally, the prescribed displacement $u_p = 0.1$ is enforced towards the center in point $\boldsymbol{T}_4$, thus $u_R^4 = -u_p$. The layers are meshed using a structured mesh of $m_h \times m_h$ elements where $m_h = 10$. Within the primal analysis the problem converges in only one load step. The contours of the norm of the nodal displacements $\|\mathbf{u}_I\|$ are depicted in Fig. 8.4.

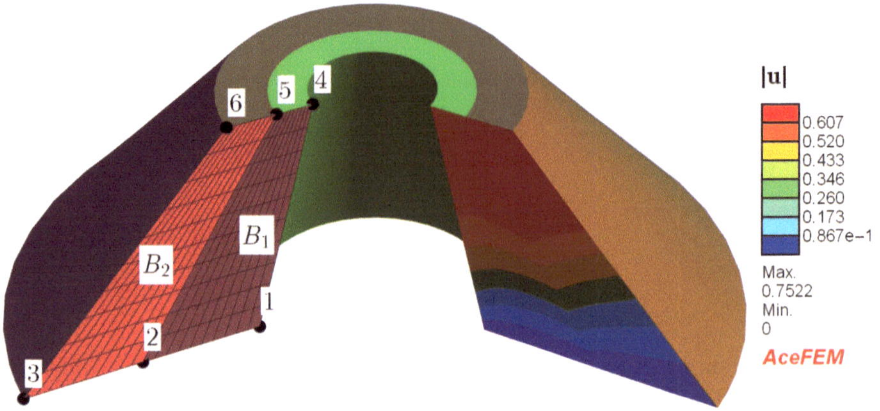

Fig. 8.4 Layout of double-layered axisymmetric cone

Table 8.3 Data of the double-layered axisymmetric cone example

Mesh points (R, Z)	Boundary and material
$\boldsymbol{T}_1(1, 0)$	$z = 0 : u_1 = u_2 = 0$
$\boldsymbol{T}_2(2, 0)$	$\boldsymbol{T}_4 : u_1 = -0.1$
$\boldsymbol{T}_3(3, 0)$	$B_1 : E = 1000, \nu = 0.3$
$\boldsymbol{T}_4(0.5, 2)$	$B_2 : E = 5000, \nu = 0.2$
$\boldsymbol{T}_5(0.8, 2)$	$\overline{T_3 T_6}$: $\bar{t}_2 = 10^5$
$\boldsymbol{T}_6(1.2, 2)$	mesh $2m \times m$ with $m = 10$

8.5.1 Definition of Sensitivity Parameters

Sensitivity problem is formed in a way that includes all major cases defined in Sect. 8.2.2. The following sensitivity parameters are considered:

Elastic modulus: $\phi_1 = E_1$

In the elements derived in Sects. 8.4.1 and 8.4.2 it is assumed that the material parameters are constant over the element domain Ω_e. However, even if the material parameter is constant, the design velocity field does not need to be constant. For example, let us assume that the elastic modulus is interpolated over the global domain of the problem by

$$E = E_0 + \sum_I N_I(\boldsymbol{X}) E_I$$

where E_0 is an average value of the elastic modulus and $\sum N_I E_I$ is its variable part interpolated as linear combination of the shape function N_I and the weights

E_I. A linear expansion of the response of the system with respect to weights E_I around $E_I = 0$ is sought. The coefficients of the expansion can be obtained by sensitivity analysis. The velocity field for the change of E is then defined by

$$\frac{DE}{DE_I} = N_I(\boldsymbol{X}).$$

The above situation frequently appears. For example, in imperfection sensitivity analysis the $\mathbf{N}_I$ can be the base vectors used to model material imperfection (Kristanic and Korelc 2008). In stochastic analysis the $\mathbf{N}_I$ can be the deterministic part of a Karhunen-Loève decomposition (Melink and Korelc 2014) of the stochastic field that represents the elastic modulus.
In the present example the elastic modulus of the material in layer B_1 is changing in radial direction in a form of a bubble as follows

$$E(R, Z) = E_{B_1} + E_1 N_1 \tag{8.104}$$

where

$$N_1(R, Z) = (R + 0.25(Z - 2) - 0.5)(R + 0.6(Z - 2) - 0.8). \tag{8.105}$$

The distribution of the elastic modulus is parametrized by (8.104) with the general sensitivity parameter $\phi_1 = E_1$. The general velocity field that belongs to sensitivity parameter ϕ_1 at the Jth global node is then defined by

$$\left.\frac{DE}{DE_1}\right|_{\mathbf{X}^J} = \begin{cases} N_1(\mathbf{X}^J) & \text{if } \mathbf{X}^J \in B_1 \\ 0 & \text{else} \end{cases}. \tag{8.106}$$

Surface traction: $\phi_2 = q_p$
The second design parameter is the intensity of the surface traction in Z direction $\bar{t}_2 = q_p$ imposed on the outer surface of the cone. The sensitivity analysis with respect to the natural boundary conditions is described in Sect. 8.3.7. The reference nodal force in nodes along line $\overline{T_3 T_6}$ is for quadrilateral mesh with $m_h + 1$ equidistant nodes $\frac{1}{m_h} q_p L_{36}$ where $L_{36} = \|\boldsymbol{T}_6 - \boldsymbol{T}_3\|$. The reference nodal force in nodes at the edge is $\frac{1}{2}\frac{1}{m_h} q_p L_{36}$. The natural boundary condition velocity field at the Jth node then leads to

$$\frac{DP_2^J}{Dq_p} = \begin{cases} \frac{L_{36}}{m_h} & \text{if } \mathbf{X}^J \in]\boldsymbol{T}_3, \boldsymbol{T}_6[\\ \frac{L_{36}}{2m_h} & \text{if } \mathbf{X}^J = \boldsymbol{T}_3 \vee \mathbf{X}^J = \boldsymbol{T}_6 \\ 0 & \text{else} \end{cases}. \tag{8.107}$$

Prescribed displacement: $\phi_3 = u_p$

The third sensitivity parameter is the intensity of the prescribed displacement in R direction at point $\boldsymbol{T}_4$. The corresponding essential boundary conditions design velocity field is defined by

$$\left.\frac{D\bar{u}_1}{Du_p}\right|_{\mathbf{X}^J} = \begin{cases} -1 & \text{if } \mathbf{X}^J = \boldsymbol{T}_4 \\ 0 & \text{else} \end{cases}. \tag{8.108}$$

where $\left.\frac{D\bar{u}_1}{Du_p}\right|_{\mathbf{X}^J}$ is the derivative of the first DOF at the Jth node with respect to ϕ_3. Observe that the proposed methodology for the sensitivity analysis imposes no restriction on the form of the input data fields $\boldsymbol{\psi}$. The input data field can be discontinuous or even be a Dirac delta function as in the present case where the displacement is prescribed for a single node.

Shape: $\phi_4 = T_R^3$

The R coordinate of the point $\boldsymbol{T}_3$ is used to change the shape of the mesh, thus $\phi_4 = T_R^3$. With this the shape of the cone is transformed by changing the outer radius of the cone. The change of the shape of the mesh effects nodal coordinates as well as the nodal forces resulting from the surface traction $\bar{t}_2$. Consequently, we need to define three nonzero design velocity fields. The shape design velocity fields $\left.\frac{DR}{DT_R^3}\right|_{\mathbf{X}^J}$ and $\left.\frac{DZ}{DT_R^3}\right|_{\mathbf{X}^J}$ that are associated with the change of the spatial coordinates R and Z and natural boundary condition velocity field $\frac{DP_2^J}{DT_R^3}$ associated with the change of nodal force P_2^J resulting from the nodal traction $\bar{t}_2 = q_p$ in Jth node. First two velocity fields can be obtained by direct differentiation of the meshing algorithm implemented in *AceFEM* as presented and explained in Sect. 8.5.3. In analogy with (8.107) the natural boundary condition velocity field at the Jth global node leads to

$$\frac{DP_2^J}{DT_R^3} = \begin{cases} \frac{q_p}{m_h}\frac{\partial L_{36}}{\partial T_R^3} & \text{if } \mathbf{X}^J \in]\boldsymbol{T}_3, \boldsymbol{T}_6[\\ \frac{q_p}{2m_h}\frac{\partial L_{36}}{\partial T_R^3} & \text{if } \mathbf{X}^J = \boldsymbol{T}_3 \vee \mathbf{X}^J = \boldsymbol{T}_6 \\ 0 & \text{else} \end{cases}. \tag{8.109}$$

8.5.2 Design Velocity Matrix

The sensitivity problem described in Sect. 8.5.1 results in a total set of 4 sensitivity parameters

$$\boldsymbol{\phi} = \{\phi_1 = E_1,\ \phi_2 = q_p,\ \phi_3 = u_p,\ \phi_4 = T_R^3\}. \tag{8.110}$$

Hyperelastic, axisymmetric problem discretized by elements derived in Sect. 8.4.1 is defined by a total of 8 input data fields as follows

$$\boldsymbol{\psi} = \left\{\left\{\bar{u}_1^J, \bar{u}_2^J, \bar{P}_1^J, \bar{P}_2^J, R^J, Z^J, E^J, \nu^J\right\} : J = 1, \ldots, n_{tn}\right\}. \tag{8.111}$$

The corresponding general input data fields of axisymmetric element at the individual element level are

$$\boldsymbol{\psi}_e = \left\{\left\{R^J, Z^J, E^J, \nu^J\right\} : J = 1, \ldots, n_{en}\right\}. \tag{8.112}$$

where index J in this case refers to the numbering of the element nodes. The global design velocity matrix at the Jth global node is then given by

$$D_{\boldsymbol{\phi}}\boldsymbol{\psi}^J = \begin{array}{|c|c|c|c|l}
\multicolumn{1}{c}{E_1} & \multicolumn{1}{c}{q_p} & \multicolumn{1}{c}{u_p} & \multicolumn{1}{c}{T_R^3} & \\
\hline
0 & 0 & \left.\frac{D\bar{u}_1}{Du_p}\right|_{\mathbf{X}^J} & 0 & \bar{u}_1^J \\
\hline
0 & 0 & 0 & 0 & \bar{u}_2^J \\
\hline
0 & 0 & 0 & 0 & P_1^J \\
\hline
0 & \frac{DP_2^J}{Dq_p} & 0 & \frac{DP_2^J}{DT_R^3} & P_2^J \\
\hline
0 & 0 & 0 & \left.\frac{DR}{DT_R^3}\right|_{\mathbf{X}^J} & R^J \\
\hline
0 & 0 & 0 & \left.\frac{DZ}{DT_R^3}\right|_{\mathbf{X}^J} & Z^J \\
\hline
\left.\frac{DE}{DE_1}\right|_{\mathbf{X}^J} & 0 & 0 & 0 & E^J \\
\hline
0 & 0 & 0 & 0 & \nu^J \\
\hline
\end{array}. \tag{8.113}$$

Only the nonzero components of design velocity matrix $D_{\boldsymbol{\phi}}\boldsymbol{\psi}$ represent an actual input data for the sensitivity analysis procedures implemented in *AceFEM* . The vector of sensitivity parameters (8.110) is in *AceFEM* defined by the `SMTSensitvity Problem` command and the nonzero components of design velocity matrix (8.113) by the `SMTSetVelocityFields` command as shown in the next section.

8.5.3 AceFEM Input for Primal and Sensitivity Analysis

The *AceFEM* input used to analyze the above problem is presented here in order to demonstrate how an advanced symbolic-numeric environment such as *Mathematica* simplifies formulation of complex computational problems. The analysis is divided into several phases.

```
(*Mesh and element input data, φ4 = T3R is left in a symbolic form.*)
mh=10;T1={1,0};T2={2,0};T3={T3R,0};
T4={0.5,2};T5={0.8,2};T6={1.2,2};
SMTInputData[];
SMTAddDomain[{"B1","Q1AX-sens",{}},{"B2","Q1AX-sens",{}}];
SMTMesh["B1","Q1",{mh,mh},{{T1,T2},{T4,T5}}];
SMTMesh["B2","Q1",{mh,mh},{{T2,T3},{T5,T6}}];
SMTAnalysis["SearchFunction"→(#/.T3R→3&)];
(*Evaluation of shape velocity fields DR^J/DT3R and DZ^J/DT3R by differentiation of
symbolically generated mesh.*)
{DRDT3R,DZDT3R}=D[SMTNodes⟦All,{2,3}⟧,T3R]ᵀ/.T3R→3;
(*Evaluation of an auxiliary quantities L36 and ∂L36/∂T3R required in (10.109) *)
L36=Sqrt[(T6-T3).(T6-T3)]/.T3R→3;
δL36=D[Sqrt[(T6-T3).(T6-T3)],T3R]/.T3R→3
```

Box 8.3. *AceFEM* input for the evaluation of shape velocity fields of hyper-elastic cone problem.

Dual symbolic-numeric spproach. First the shape velocity fields are evaluated. The corresponding *AceFEM* input is presented in Box 8.3. The code in Box 8.3 starts with the creation of the two-dimensional mesh for the double-layered cone presented in Fig. 8.4. However, while most of the finite element environments require purely numerical data as an input, the *AceFEM* finite element environment explores the advantages of the general symbolic system *Mathematica* and allows some input data to be left in a symbolic form (Korelc 2011b). The actual mesh is structured, thus the nodal coordinates are obtained by the isoparametric mapping from the uniformly meshed unit square to an actual quadrilateral shape. This transformation can be based on purely numerical data or on some data left in a symbolic form. In the present case, the coordinate $T_R^3 \equiv$ `T3R` is left in a symbolic form. The results are nodal coordinates expressed as a symbolic function of the design parameter T_R^3. The shape velocity fields $\frac{DR^J}{D\phi_4} \equiv$ `DRDT3R` and $\frac{DZ^J}{D\phi_4} \equiv$ `DZDT3R` that belong to the shape parameter T_R^3 can then be obtained by a symbolic differentiation of the symbolically expressed nodal coordinates. A *Mathematica* command `D[x,y]` is used for that purpose. The result is evaluated for the true value of the shape parameter $T_R^3 = 3$ by the *Mathematica* command *expr* `/. T3R→3`. An auxiliary quantities $L_{36} \equiv$ `L36` and $\frac{\partial L_{36}}{\partial T_R^3} \equiv$ `δL36` required in definition of velocity fields (8.107) and (8.109) are also evaluated together with the shape velocity fields.

```
(*Step 1: Definition of mesh and material data.*)
mh=10;qp=10 000;up=0.1;T1={1,0};T2={2,0};T3={3,0};
T4={0.5,2};T5={0.8,2};T6={1.2,2};
SMTInputData[];
SMTAddDomain[{"B1","Q1AX-sens",{"E"->1000,"ν"->0.3}},
  {"B2","Q1AX-sens",{"E"->5000,"ν"->0.2}}];
SMTAddEssentialBoundary[Line[{T1,T3}],1->0,2→0];
SMTAddNaturalBoundary[ Line[{T3,T6}],2→Line[{qp}]];
SMTAddEssentialBoundary[ Point[T4],1→-up];
SMTMesh["B1","Q1",{mh,mh},{{T1,T2},{T4,T5}}];
SMTMesh["B2","Q1",{mh,mh},{{T2,T3},{T5,T6}}];
(*Step 2: Definition of the sensitivity parameters.*)
SMTSensitivityProblem[{{"E1",{"P","DE/DE1",(*E on B1*)"B1"→3}},
   {"qp",{"NBC","DP2/Dqp",(*2nd dof*)"D"→2}},
   {"up",{"EBC","Du/Dup",(*1st dof*)"D"→1}},
   {"L",{"S","DR/DT3R",All->(*R*)1},{"S","DZ/DT3R",All->(*Z*)2},
    {"NBC","DP2/DT3R",(*2nd dof*)"D"→2}}}];
(*Step 3: This creates actual mesh and initializes data structures.*)
SMTAnalysis[];
(*Step 4: This sets the nonzero components of design velocity matrix DΦψ *)
SMTSetVelocityFields[{"DE/DE1"→{{"DomainID","B1"},
     Function[{n,R,Z},(R+0.25 (Z-2)-0.5) (R+0.6 (Z-2)-0.8)]},
   "DP2/Dqp"→{Line[{T3,T6}],Function[{n,R,Z},L36/mh]},
   "DP2/Dqp"→{Point[T3]||Point[T6],{L36/mh/2,L36/mh/2}},
   "Du/Dup"→{Point[T4],{-1}},"DR/DT3R"→{All,DRDT3R},
   "DZ/DT3R"→{All,DZDT3R},
   "DP2/DT3R"→{Line[{T3,T6}],Function[{n,R,Z},qp/mh δL36]},
   "DP2/DT3R"→{Point[T3]||Point[T6],{qp/mh/2 δL36,qp/mh/2 δL36}}}];
(*Step 5: Here is the primal analysis executed.*)
SMTNextStep[1,1];
While[SMTConvergence[10^-9,10],SMTNewtonIteration[]];
(*Step 6: This performs the sensitivity analysis with respect to all parameters.*)
SMTSensitivity[];
```

Box 8.4. *AceFEM* input for primal and sensitivity analysis of hyper-elastic cone problem.

Primal and sensitivity analysis. The complete *AceFEM* input for primal and sensitivity analysis is provided in Box 8.4. Here are all calculations done using purely numerical data. The *AceFEM* module for numerical calculations is separated from *Mathematica* and written in the numerically efficient compiled language C. The above procedure efficiently combines two worlds, the symbolic capabilities of *Mathematica* and the numerical efficiency of compiled languages such as C.

The complete analysis is divided into 6 steps:

1. definition of mesh and material data,
2. definition of the sensitivity parameters,
3. data base initialization,
4. definition of design sensitivity matrix,

5. primal analysis,
6. sensitivity analysis.

The mesh generation phase in Box 8.4 is followed by the description of the sensitivity problem (`SMTSensitivityProblem` command), the initialization of the general data structures (`SMTAnalysis` command) and the definition of nonzero components of design velocity matrix (`SMTSetVelocityFields` command). A Newton iterative procedure then solves the problem in one load step. After convergence of the primal problem has been achieved the `SMTSensitivity` command executes the sensitivity analysis with respect to all four parameters. The results of the primal and sensitvity analysis are presented in Table 8.5 and in Fig. 8.5. Figure 8.5 depicts the sensitivity of the displacement in Z direction with respect to all sensitivity parameters.

Definition of sensitivity parameters by `SMTSensitivityProblem` command. For each of four sensitivity parameters an unique identification is defined followed by the corresponding definitions of nonzero terms of design velocity matrix that belong to particular sensitivity parameter. The definition of the design velocity field is composed of the type specification, unique identification and the range of validity as presented in Box 8.4 and explained in Table 8.4.

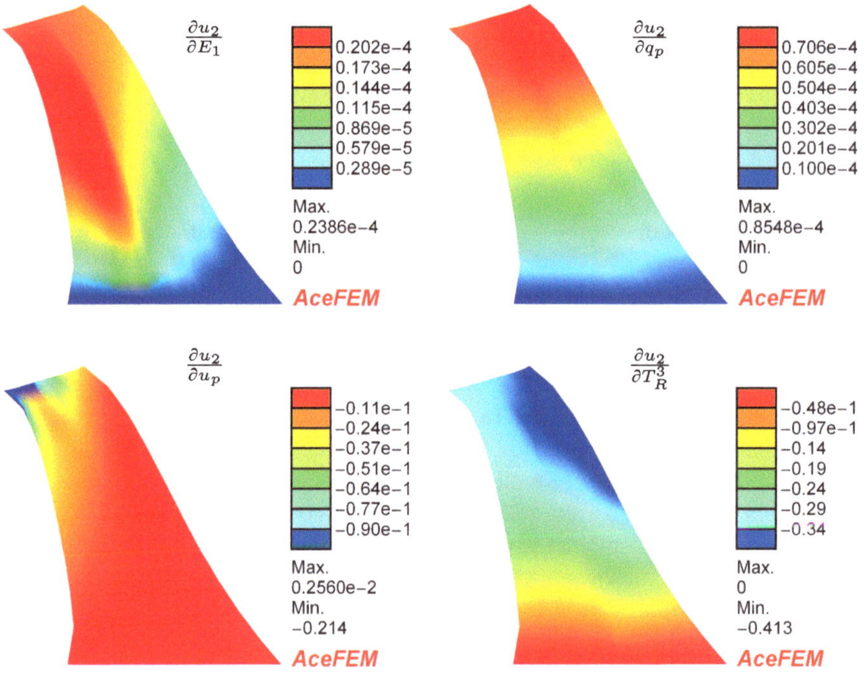

Fig. 8.5 The sensitivity of displacement in Z direction with respect to all sensitivity parameters

Table 8.4 Classification of sensitivity parameters

Type	Description	Field/DOF index
`"P"`	Parameter design velocity field	`"DomainID"`$\rightarrow L$
`"S"`	Shape design velocity field	`"DomainID"`$\rightarrow L$
`"EBC"`	Essential boundary conditions velocity field	`"NodeID"`$\rightarrow K$
`"NBC"`	Natural boundary conditions velocity field	`"NodeID"`$\rightarrow K$

In the case of general input data fields (type `"P"` and `"S"`), the term `"Domain ID"`$\rightarrow L$ in Table 8.4 specifies that velocity field belongs to the Lth general input data field $\boldsymbol{\psi}_{eL}$ on domain with given domain identification `"DomainID"`. For example, `{"E1",{"P", "DE/DE1","B1"→3}}` specifies that sensitivity parameter `"E1"`$\equiv E_1 \equiv \phi_1$ has one nonzero parameter velocity field assigned. Corresponding general velocity field is identified by `"DE/DE1"`$\equiv \frac{DE}{DE_1}$, defined only on subdomain of the problem B_1 and represents the derivative of the third general input data field (elastic modulus E) with respect to E_1. The ordering of the general input data fields is uniquely defined by the `"SMSSensitivityNames"` template constant as a part of *AceGen* input for the generation of axisymmetric element as presented in Sect. 8.4.1.The data is used to form the element general design velocity matrix $D_{\phi_I}\boldsymbol{\psi}_e$.

In the case of natural and essential boundary condition velocity fields (type `"EBC"` and `"NBC"`), the term `"NodeID"`$\rightarrow K$ in Table 8.4 specifies the mapping between the given velocity field and the nodal unknowns. For example, `{"up", {"EBC", "Du/Dup","D"→1}}` means that sensitivity parameter `"up"` has one nonzero essential boundary condition velocity field assigned. Corresponding essential boundary condition velocity field is identified by `"Du/Dup"` and represents the derivative of the first unknown in nodes with node identification `"D"` with respect to `"up"`. The node identification `"D"` is defined by the `"SMSNodeID"` template constant as a part of *AceGen* input for the generation of axisymmetric element as presented in Sect. 8.4.1. The data is used to construct the EBCVF function needed in formulation of essential boundary condition velocity field data structure $D_{\phi_I}\bar{\mathbf{p}}_e$ (8.32). In the case of natural boundary conditions velocity field the `"NodeID"`$\rightarrow K$ gives the necessary data to construct the NBCVF function needed in formulation of the derivatives of reference load vector $\frac{D\mathbf{R}^{\text{ref}}}{D\phi_I}$ (8.103).

Definition of design velocity matrix by `SMTSetVelocityFields` command. Only the nonzero components of the design velocity matrix have to be specified. Velocity field is defined by its nodal values in selected nodes or by a function that is applied on selected nodes as shown in Box 8.4. For example, `"DP2/Dqp"→{Line[{T3, T6}], Function[{n, R, Z},L36/mh]}` specifies constant value $\frac{L_{36}}{m_h}$ of velocity field `"DP2/Dqp"` in all nodes on line $\overline{T_3T_6}$. Other ways how to select nodes are: with `{"DomainID", "B1"}` all nodes in domain B_1 are selected, with `Point[T3]` the node at point T_3 is selected and with `"All"` all nodes of the problem are selected.

Table 8.5 Analytical sensitivity of displacement u_2 in point T_6 and its finite difference approximation with $\delta\phi = 10^{-6}$

Parameter	Analytical sensitivity	Finite difference approximation
E_1	1.88653×10^{-5}	Not applicable
q_p	8.44073×10^{-5}	8.44073×10^{-5}
u_p	-2.35153×10^{-2}	-2.35153×10^{-2}
T_R^3	-3.89401×10^{-1}	-3.89401×10^{-1}

Visualization and post-processing. It is a general assumption among many scientists that the finite difference approximation can always be performed and applied within arbitrary problems while the analytical methods have limitations. However, the Table 8.5 shows a case where the finite difference method cannot be applied while analytical methods give result. The finite difference problem for differentiation with respect to parameter E_1 cannot be formulated. The derived finite element assumes constant distribution of elastic modulus over the sub-domain B_1, thus the variable part of the elastic modulus cannot be modeled. Consequently, the finite difference scheme cannot be applied.

Table 8.5 presents the results of sensitivity analysis for displacement u_2 at point $\boldsymbol{T}_6$. The corresponding displacement at point $\boldsymbol{T}_6$ is $u_2 = 0.734739$. The analytical sensitivity obtained by the proposed automation procedure is compared with the results of the finite difference approximation $\frac{Du_2}{D\phi_I} \approx \frac{u_2(\phi_I+\delta\phi_I)-u_2(\phi_I)}{\delta\phi_I}$ using a perturbation $\delta\phi_I = 10^{-6}$. The results agree up to 5 significant digits.

References

Choi, K.K., and N.H. Kim. 2005a. *Structural sensitivity analysis and optimization 1, Linear systems*. New York: Springer Science+Business Media.

Choi, K.K., and N.H. Kim. 2005b. *Structural sensitivity analysis and optimization 2, Nonlinear systems and applications*. New York: Springer Science+Business Media.

Hudobivnik, B., and J. Korelc. 2016. Closed-form representation of matrix functions in the formulation of nonlinear material models. *Finite Elements in Analysis and Design* 111: 19–32.

Keulen, F., R. Haftka, and N. Kim. 2005. Review of options for structural design sensitivity analysis. part 1: Linear systems. *Computer Methods in Applied Mechanics and Engineering* 194: 3213–3243.

Kleiber, M., H. Antunez, T. Hien, and P. Kowalczyk. 1997. *Parameter sensitivity in nonlinear mechanics*. New York: Wiley.

Korelc, J. 2009. Automation of primal and sensitivity analysis of transient coupled problems. *Computational Mechanics* 44: 631–649.

Korelc, J. 2011b. Semi-analytical solution of path-independed nonlinear finite element models. *Finite Elements in Analysis and Design* 47: 281–287.

Korelc, J., and S. Stupkiewicz. 2014. Closed-form matrix exponential and its application in finite-strain plasticity. *International Journal for Numerical Methods in Engineering* 98: 960–987.

Kristanic, N., and J. Korelc. 2008. Optimization method for the determination of the most unfavorable imperfection of structures. *Computational Mechanics* 42: 859–872.

Melink, T., and J. Korelc. 2014. Stability of karhunen-love expansion for the simulation of gaussian stochastic fields using galerkin scheme. *Probabilistic Engineering Mechanics* 37: 7–15.

Michaleris, P., D. Tortorelli, and C. Vidal. 1994. Tangent operators and design sensitivity formulations for transient non-linear coupled problems with applications to elastoplasticity. *International Journal for Numerical Methods in Engineering* 37: 2471–2499.

Solinc, U., and J. Korelc. 2015. A simple way to improved formulation of fe2 analysis. *Computational Mechanics* 56: 905–915.

Erratum to: Automation of Finite Element Methods

Jože Korelc and Peter Wriggers

Erratum to:
J. Korelc and P. Wriggers, *Automation of Finite Element Methods*, DOI 10.1007/978-3-319-39005-5

An error in the production process unfortunately led to publication of this book prematurely, before incorporation of the final corrections. The version supplied here has been corrected and approved by the authors.

The updated original online version for this book can be found at
DOI 10.1007/978-3-319-39005-5

J. Korelc (✉)
Faculty of Civil and Geodetic Engineering, University of Ljubljana,
Ljubljana, Slovenia
e-mail: jkorelc@fgg.uni-lj.si

P. Wriggers
Leibniz University Hannover, Hannover, Germany
e-mail: wriggers@ikm.uni-hannover.de

J. Korelc and P. Wriggers, *Automation of Finite Element Methods*,
DOI 10.1007/978-3-319-39005-5_9

Appendix A
Mathematica and *AceGen* Syntax

Mathematica is one of the largest of and most complex software system. Thus it is impossible to describe all the aspects of *Mathematica* here. Only a short explanation and the syntax of the commands that are used in examples presented in the book and are essential for the understanding of the *AceGen* is given here. The same is valid for the *AceGen* automatic code generator and the *AceFEM* finite element environment. All three software systems used in this book come with an extensive manual where an additional information can be found.

A.1 Highlights of *Mathematica* Syntax

Constants

`1`	integer number
`1.0`	real number
`"string"`	string constant
π, `E`	mathematical constants (symbols representing mathematical constant can not be used as a name of a variable)

Structure of expressions

`1+2, 3-4, 5*6, 7/8`	arithmetic operations (empty space is also interpreted as multiplication, e.g. `2 * 3 ≡ 2 3`)
`a=5`	assigns the value 5 to a symbol `a`
`a=b+1; c=2a; d=a+c;`	compound expression
`a == b`	is true if `a` and `b` are identical expressions
`a === b`	is true if expressions `a` and `b` are identically the same

Brackets

`2*(3+4)`	the `(...)` brackets define the precedence of operators
`y[x]`	the `[...]` brackets envelope the arguments of a function or *Mathematica* command, e.g. $\sin x \equiv$ `Sin[x]`
`a={a1,a2,a3}`	the `{...}` brackets envelope one-dimensional set

J. Korelc and P. Wriggers, *Automation of Finite Element Methods*,
DOI 10.1007/978-3-319-39005-5

a〚i〛	the 〚...〛 ≡ [[...]] brackets are used to select the i th element of a (note that index i has to have an integer value at the time of execution of *Mathematica* session and that the 〚 and 〛 characters are inserted by the <ESC>[[<ESC> and <ESC>]]<ESC> keyboard sequences)

Manipulations of symbolic expressions

a → b	rule (the → character is inserted by the <ESC>-><ESC> keyboard sequence)
exp/.a → b	transform expression accordingly to the given rule (also ReplaceAll[expr, a → b]
Simplify[expr]	command tries to simplify the expression by using various algebraic transformations
N[exp,n]	gives the numerical value of expression with n-digit precision
Solve[f==0, x]	attempts to find a symbolic solution of equation $f(x) = 0$ for the variable x
D[f, x]	gives partial derivative $\frac{\partial f(x)}{\partial x}$ (note that f has to be explicitly expressed by x)

Arrays and tensors

S={a,{b},{{c,6}}}	the {...} brackets can also be used to create an arbitrarily nested sets, e.g. a two-dimensional matrix T = {{T11, T12}, {T21, T22}}.
S〚i, j, ...〛	i, j, ...-th element of a set S
a〚i;; j〛	take a subset of a composed of elements from position i to position j
Table[exp, {i, i_{max}}]	generates a vector of the values of exp when i runs from 1 to i_{max}
Table[exp, {i, i_{max}}, {j, j_{max}}]	generates a $i_{max} \times j_{max}$ matrix of the values of of exp when i runs from 1 to i_{max} and j runs from 1 to j_{max}
Table[exp, {i, i_{max}}, {j, j_{max}[i]}]	generates a two-dimensional nested set of values of exp when i runs from 1 to i_{max} and j runs from 1 to j_{max}[i]
Sum[exp, {i, i_{max}}]	calculates a sum of values of exp when i runs from 1 to i_{max}
a^{T}	transposes the first two levels in a (the ^{T} character is inserted by the <ESC>tr<ESC> keyboard sequence)
Flatten[S]	flattens out nested lists
a . b	scalar product of vectors a and b
A . B	matrix product of matrices A and B
Det[A]	determinant of matrix A
Inverse[A]	inverse of matrix A
Tr[A]	trace of matrix A

Conditionals and loops

`If[cond, a, b]` gives `a` if condition evaluates to True, and `b` if it evaluates to False (note that `cond` has to have a constant True or False value at the time when the statement is executed).

`Do[exp, {i, imax}]` evaluates expression `exp` with the variable `i` successively taking on the values 1 through i_{max} (note that i_{max} has to have a constant integer value at the time when the statement is executed).

A.2 Highlights of *AceGen* Syntax

In principle we can get *AceGen* input simply by replacing the assignment operator = in standard *Mathematica* input by an appropriate AceGen assignment operator (⊢, ⊨, ⊣ or ⫤), the standard *Mathematica* conditional statement `If` by the *AceGen* command `SMSIf` and the standard *Mathematica* loop statement `Do` by the *AceGen* `SMSDo` statement. Various aspects of *AceGen* syntax have already been explained throughout the book on examples. Here they are summarized in a concise form.

A.2.1 AceGen Session

The general composition of a typical *AceGen* automatic code generation procedure is given in Sect. 2.5.2 and a procedure for automatic generation of finite element user subroutines in Sect. 2.7. The most important commands are summarized here.

`SMSInitialize[`*filename*`, "Environment"→` *env*`]` function initializes the *AceGen* system and specifies the name of the generated source code file. Additional option `"Environment"` specifies the finite element environment for which the element source code will be generated. The supported finite element environment are `"AceFEM"`, `"AceFEM-MDriver"`, `"FEAP"`, `"ELFEN"`, `"ABAQUS"` or a generic `"User"` finite element environment.

`SMSTemplate["const1"→` val_1`, "const2"→` val_2`, ...]` function initializes constants `"const`$_i$`"` that are needed to create proper interface between the generated user subroutines and the finite element environment. Template constants are described in Sect. A.3.

`SMSStandardModule[task]` command starts the definition of the finite element user subroutine meant to perform a specific `task` (e.g. calculate element tangent matrix and residual vector) with predefined set of input/output parameters that is adjusted for the needs of a specified finite element environment. The input parameters of the user subroutines are described in Sect. A.4. The typical user subroutines and their output parameter are as follows.

`"Tangent and residual"` is subroutine that returns the tangent matrix (stored in variable `s$$` with dimensions $(n_p + n_{he}) \times (n_p + n_{he})$) and residual or internal load vector (stored in variable `p$$` with dimension $n_p + n_{he}$) for the current state of element related data.

`"Sensitivity pseudo-load"` is subroutine that returns a matrix of pseudo-load vectors (stored in variable `s$$` with dimensions $(\phi_{L\text{end}} - \phi_{L\text{start}} + 1) \times (n_p + n_{he})$) that is used in sensitivity analysis to calculate the sensitivity of the global unknowns with respect to arbitrary parameter.

`"Dependent sensitivity"` is subroutine that is used to calculate sensitivity of locally coupled unknowns at the element level. Details about sensitivity analysis are presented in Chap. 8.

`"Postprocessing"` is subroutine that returns two arrays with arbitrary number of post-processing quantities as follows: `gpost$$` is array of integration point post-processing quantities with dimensions *number of integration points* × *number of integration point quantities*; `npost$$` is array of the nodal point quantities with dimensions *number of nodes* × *number of nodal point quantities*.

The "Tangent and residual" user subroutine is supported by all finite element environments and used within the iterative solution of the primal problem. The use of other user subroutines depends on the chosen finite element environment.

`p⊢SMSReal[rp$$]` introduces the input/output parameter `rp$$` from the list of available input/output parameters of the particular user subroutine as an independent real type variable and assigns the value of the parameter to the symbol `p`. Double $ sign in the name of the parameter indicates that it is an input/output parameter of the subroutine that have to be properly interpreted accordingly to the chosen finite element environment and chosen programing language (C, Fortran, etc.). Within the *AceGen* session, the `p` have to be used exclusively. A list of standard input/output parameters is given in Sect. A.4.

`p⊢SMSReal[rp$$,"Dependency"→ {y,` $\frac{Dp}{Dy}$`}]` introduces the input parameter `rp$$` as a real type variable that depends on variable `y` and defines the total derivative of `p` with respect to `y` to be $\frac{Dp}{Dy}$.

`i⊢SMSInteger[ip$$]` introduces the input/output parameter `ip$$` from the list of available input/output parameters as an independent integer type variable and assigns the value of the parameter to the symbol `i`.

`SMSExport[`*exp*`, out$$]` command exports the value of the expression *exp* to the output parameter `out$$` of the user subroutine. A list of standard output parameters is given in Sect. A.4.

`SMSWrite[]` function writes the generated formulas together with the code that provides interface to the chosen finite element environment to the file in a program language of the chosen finite element environment (e.g. C, Fortran, Fortran90, C#, Matlab language or Mathematica language).

A.2.2 Assignment Operators

During the description of the problem the special operators (⊢, ⊨, ⊣ and ⫤) are used to perform the simultaneous optimization of expressions and the creation of new auxiliary variables as described in Sect. 2.5.3. For example, the expression `x⊢exp` first evaluates right-hand side expression `exp`, creates new auxiliary variable, and assigns the new auxiliary variable to be the value of the left-hand side symbol `x`. From then on, `exp` is replaced by a new auxiliary variable whenever it appears. Operators ⊢, ⊨, ⊣ and ⫤ have the following characteristics:

`x` ⊢ *exp* A new auxiliary variable is created, regardless on the contents of *exp*. The primal functionality of this form is to force creation of the new auxiliary variable.

`x` ⊨ *exp* A new auxiliary variable is created if *AceGen* finds out that the introduction of the new variable is necessary, otherwise `x=`*exp*. This is the basic form for introducing new expressions in an optimized way. Ordinary *Mathematica* input can be converted to the *AceGen* input by replacing the set operator = with *AceGen* operator ⊢.

`x` ⫤ *exp* A new auxiliary variable is created, regardless on the contents of exp. The primal functionality of this form is to create variable which will appear more than once on a left-hand side of equation (multi-valued variables). For example, the following snippet of *AceGen* input

```
y⫤0;
SMSDo[
   y⊣y+1/i;
  ,{i,1,n,1,y}];
```

would create a code segment that evaluates the sum $y = \sum_{i=1}^{n} \frac{1}{i}$ for an arbitrary n.

`x` ⊣ *exp* A new value (*exp*) is assigned to the previously created auxiliary variable `v`. At the input `x` has to be auxiliary variable created as the result of `x` ⫤ *exp* command.

`x` ⊢ `SMSFreeze[`*exp*`]` Differentiation with respect to intermediate auxiliary variable can lead to incorrect results due to the interaction of automatic differentiation algorithm and expression optimization algorithm. With the `SMSFreeze` operator a new auxiliary variable is created that can be safely used as independent variable for differentiation.

`SMSFreeze[M, B, "Symmetric" → True]` A symmetric matrix `A` of auxiliary variables is created from a symmetric matrix of expressions `B`.

`x` ⊢ `SMSFreeze[`*exp*`, "Dependency" → {y,` $\frac{Dx}{Dy}$`}]` An independent auxiliary variable `x` is created with the value of *exp*. The true dependencies of `x` are neglected. Instead, it is assumed that `x` depends on variable `y` and that the total derivative of `x` with respect to `y` is $\frac{Dx}{Dy}$.

As a rule, the ⊢ and ⊨ operators are used for variables that will appear only once on the left-hand side of equation and for variables that will appear more that once on the left-hand side the operators ⊣ and ⫤ have to be used.

A.2.3 Program Flow Control

AceGen can automatically generate conditionals (`SMSIf` construct) and loops (`SMSDo` construct). The program structure specified by the conditionals and loops is created simultaneously during the *AceGen* session and it will appear as a part of automatically generated code in a specified language. All other conditional and loop structures have to be manually replaced by the equivalent forms consisting only of `SMSIf` and `SMSDo` statements. It is important to notice that only the replaced conditionals and loops produce corresponding conditionals and loops in the generated code and are evaluated when the generated program is executed. The conditional and loops that are left unchanged are evaluated directly in *Mathematica* during the *AceGen* session.

`lhs` ⊢ `SMSIf[condition, a, b]` creates source code that evaluates `a` if condition evaluates to True, and `b` if it evaluates to False. The value assigned to `lhs`.

`SMSDo[exp, {i, i`$_{min}$`, i`$_{max}$`, i}]` creates source code that evaluates `exp` with the variable `i` successively taking on the values `i`$_{min}$ through `i`$_{max}$ in steps of Δi (note that `i`$_{min}$, `i`$_{max}$ and `i` does not need to be an integer constants)

`SMSIf[condition];`
True branch
`SMSElse[];`
False branch
`SMSEndIf[];`
is an alternative form of `SMSIf` construct.

`SMSDo[i, i`$_{min}$`, i`$_{max}$`, i];`
loop body
`SMSEndDo[];`
is an alternative form of `SMSDo` construct.

A.2.4 Generalized Automatic Differentiation with AceGen

Essential for the automation of computational procedures via generalized automatic differentiation exceptions is straightforward transformation of the notations introduced in Sect. 2.4 into *AceGen* input code. Table A.1 introduces snippets of *AceGen* input that corresponds to the local definition of AD exception of all types and Table A.2 the corresponding snippets for the global definition of AD exceptions.

Table A.1 Snippets of *AceGen* input for the typical local definitions of AD exceptions

Type	Local AD exception	*AceGen* snippet

A	$\nabla f_A := \frac{\hat{\delta} f(\mathbf{a},\mathbf{b}(\mathbf{a}))}{\hat{\delta}\mathbf{a}}\Big\vert_{\frac{D\mathbf{b}}{D\mathbf{a}}=\mathbf{M}}$	`a⊢SMSReal[a$$]`
		`b⊢SMSFreeze[fb[a]]`
		`∇fA⊨SMSD[f[a,b],a,`
		`"Dependency"→{b,a,M}]`

B	$\nabla f_B := \frac{\hat{\delta} f(\mathbf{a},\mathbf{b}(\mathbf{a}))}{\hat{\delta}\mathbf{a}}\Big\vert_{\frac{D\mathbf{b}}{D\mathbf{a}}=\mathbf{0}}$	`a⊢SMSReal[a$$]`
		`b⊢SMSFreeze[fb[a]]`
		`∇fB⊨SMSD[f[a,b],a,"Constant"→b]`

C	$\nabla f_C := \frac{\hat{\delta} f(\mathbf{a},\mathbf{b})}{\hat{\delta}\mathbf{a}}\Big\vert_{\frac{D\mathbf{b}}{D\mathbf{a}}=\mathbf{M}}$	`a⊢SMSReal[a$$]`
		`b⊢SMSReal[b$$]`
		`∇fC⊨SMSD[f[a,b],a,`
		`"Dependency"→{b,a,M}]`

D	$\nabla f_D := \frac{\hat{\delta} f(\mathbf{a},\mathbf{b}(\mathbf{c}(\mathbf{a})))}{\hat{\delta}\mathbf{a}}\Big\vert_{\frac{D\mathbf{b}}{D\mathbf{c}}=\mathbf{M}}$	`a⊢SMSReal[a$$]`
		`c⊢SMSFreeze[fc[a]]`
		`b⊢SMSFreeze[fb[c]]`
		`∇fD⊨SMSD[f[a,b],a,`
		`"Dependency"→{b,c,M}]`

Table A.2 Snippets of *AceGen* input for the typical global definitions of AD exceptions

Type	Global AD exception	*AceGen* snippet

A	$\mathbf{b} := \mathbf{f}_b(\mathbf{a})\vert_{\frac{D\mathbf{b}}{D\mathbf{a}}=\mathbf{M}(\mathbf{a})}$
	$\nabla f_A := \frac{\hat{\delta} f(\mathbf{a},\mathbf{b}(\mathbf{a}))}{\hat{\delta}\mathbf{a}}$
		`a⊢SMSReal[a$$]`
		`b⊢SMSFreeze[fb[a],"Dependency"→{a,M}]`
		`∇fA⊨SMSD[f[a,b],a]`

B	$\mathbf{b} := \mathbf{f}_b(\mathbf{a})\vert_{\frac{D\mathbf{b}}{D\mathbf{a}}=\mathbf{0}}$
	$\nabla f_B := \frac{\hat{\delta} f(\mathbf{a},\mathbf{b}(\mathbf{a}))}{\hat{\delta}\mathbf{a}}$
		`a⊢SMSReal[a$$]`
		`b⊢SMSFreeze[fb[a],"Dependency"→{a,0}]`
		`∇fB⊨SMSD[f[a,b],a]`

C	$\mathbf{b} := \mathbf{f}_b\vert_{\frac{D\mathbf{b}}{D\mathbf{a}}=\mathbf{M}}$
	$\nabla f_C := \frac{\hat{\delta} f(\mathbf{a},\mathbf{b})}{\hat{\delta}\mathbf{a}}$
		`a⊢SMSReal[a$$]`
		`b⊢SMSReal[b$$,"Dependency"→{a,M}]`
		`∇fC⊨SMSD[f[a,b],a]`

D	$\mathbf{c} := \mathbf{f}_c(\mathbf{a})$
	$\mathbf{b} := \mathbf{f}_b(\mathbf{c})\vert_{\frac{D\mathbf{b}}{D\mathbf{c}}=\mathbf{M}}$
	$\nabla f_D := \frac{\hat{\delta} f(\mathbf{a},\mathbf{b}(\mathbf{c}(\mathbf{a})))}{\hat{\delta}\mathbf{a}}$
		`a⊢SMSReal[a$$]`
		`c⊢SMSFreeze[fc[a]]`
		`b⊢SMSFreeze[fb[c],"Dependency"→{c,M}]`
		`∇fD⊨SMSD[f[a,b],a]`

A.3 Finite Elements Template Constants

The *AceGen* uses a set of global constants that, at the code generation phase, define the major characteristics of the finite element. They are called *the finite element template constants*. In most cases the element topology and the number of nodal degrees of freedom are sufficient to generate a proper interface code. Some of the FE environments do not support all the possibilities given here. The *AceGen* tries to accommodate the differences and always generates the code. The template constants are initialized by the `SMSTemplate` command.

`"SMSTopology"` constant specifies a keyword that defines element topology. For example "H1" keyword defines three-dimensional, 8 noded, hexahedral element with 3 degrees of freedom per node. Setting the topology constant also set other constants that directly depend on topology, e.g. number of nodes (`"SMSNoNodes"`), spatial dimension (`"SMSNoDimensions"`), etc. Their values are available after the `SMSTemplate` command and they can be used to make the *AceGen* input more general and problem independent. Some typical values of the `"SMSTopology"` constant are:

- `V2` point in 2D,
- `L1` straight line segment in 2D,
- `L2` curve with three nodes in 2D,
- `LX` curve with arbitrary number of nodes and arbitrary numbering in 2D,
- `T1` triangle with 3 nodes in 2D,
- `T2` triangle with 6 nodes in 2D,
- `TX` triangle with arbitrary number of nodes and arbitrary numbering in 2D,
- `Q1` quadrilateral with 4 nodes in 2D,
- `Q2` quadrilateral with 9 nodes in 2D,
- `Q2S` quadrilateral with 8 nodes (serendipity family) in 2D,
- `QX` quadrilateral with arbitrary number of nodes and arbitrary numbering in 2D,
- `V3` point in 3D,
- `C1` straight line segment in 3D,
- `C2` curve with three nodes in 3D,
- `CX` curve with arbitrary number of nodes and arbitrary numbering in 3D,
- `P1` triangle with 3 nodes in 3D,
- `P2` triangle with 6 nodes in 3D,
- `PX` triangle with arbitrary number of nodes and arbitrary numbering in 3D,
- `S1` quadrilateral with 4 nodes in 3D,
- `S2` quadrilateral with 9 nodes in 3D,
- `S2S` quadrilateral with 8 nodes (serendipity family) in 3D,
- `SX` quadrilateral with arbitrary number of nodes and arbitrary numbering in 3D,
- `O1` tetrahedron with 4 nodes (numerical integration rules and post-processing routines assume “AREA CCORDINATES” of the reference element!),

`O2` tetrahedron with 10 nodes,
`OX` tetrahedron with arbitrary number of nodes and arbitrary numbering,
`H1` hexahedron with 8 nodes,
`H2` hexahedron with 27 nodes,
`H2S` hexahedron with 20 nodes (serendipity family),
`HX` hexahedron with arbitrary number of nodes and arbitrary numbering.

`"SMSDomainDataNames"` constant specifies a list of keywords that characterize the material constants ($\mathbf{d}_B$). Material constants appear as an input data for finite element simulation.

`"SMSDefaultData"` constant specifies the default (most common) values for material constants $\mathbf{d}_B$.

`"SMSSymmetricTangent"` constant specifies weather the tangent matrix is symmetric or not. Only the upper triangular matrix has to be explicitly formed when the tangent matrix is symmetric. The option also effects the selection of algorithm used to solve the resulting system of linear equations.

`"SMSNoNodes"` $\equiv n_{en}$ constants has to be given when the element has more nodes than it is the default number of nodes for the elements with the chosen topology (as defined by `"SMSTopology"` constant).

`"SMSAdditionalNodes"` constant specifies a function that returns the position of nodes not already defined by the chosen topology of the element. Arguments of the function are the coordinates of nodes given as mesh input data. For example, the

```
Function[{X1, X2}, {(X1 + X2)/2}]
```

function adds one additional node in the middle of the nodes 1 and 2. Position of an additional nodes have to be specified when the element has more nodes than specified by the `"SMSTopology"` constant.

`"SMSNoTimeStorage"` $\equiv l_{dhe}$ constant specifies the length of history data vector per element (see also Sect. 5.2.6).

`"SMSPostIterationCall"` option specifies that after the convergence of the global system of equations has been achieved, all the local problem have to be solved again in order to have the global problem and the local problems solved with the same accuracy.

`"SMSDOFGlobal"` contains the total number of nodal unknowns per node for all nodes. By default the number of nodal unknowns is the same as the number of spatial dimensions.

`SMSNoDOFGlobal` $\equiv n_p$ is the number of global unknowns of the element. It is automatically determined from the `"SMSNoNodes"` and `SMSDOFGlobal` constants.

`"SMSNodeID"` contains for all nodes the node identification or *NodeID*. The *NodeID* is an arbitrary keyword that is used for identification of the nodes accordingly to the physical meaning of the nodal degree's of freedom. In order to have a consistent set of elements one has to use the same *NodeID* for the nodes with the same physical meaning (see also Sect. 6.4.1).

`"SMSNoDOFCondense"` $\equiv n_{he}$ number of unknowns local to the element that have to be eliminated before the element quantities are assembled (see also Sect. 6.5).

`"SMSSegments"` constant is used for visualization and defines the sequence of the element nodes that define the edge of the segment.

`"SMSSegmentsTriangulation"` constant is also used for visualization and specifies how to split an arbitrary shaped elements into a set of triangular or quadrilateral segments.

`"SMSSensitivityNames"` template constant that specifies the ordering of input data fields $\boldsymbol{\psi}_e$ for which sensitivity is defined within the element. (see also Sect. 8.4.1)

A.4 Finite Elements User Subroutines Interface

A general way of how to pass data from the main program into the automatically generated subroutine and how to get the results back to the main program is through a list of input/output parameter of the generated user subroutine. Unfortunately, different finite element environments are written in a different program languages and in general use different organization of data related to nodes, elements, material constants etc. Consequently, it would be awkward to use directly the names and the data structure of the input/output parameters of a specific finite element environment. Instead, the *AceGen* uses a generalized form of the input/output parameters that is automatically translated by *AceGen* to the form of input/output parameters required by a specific finite element environment. A selected list of generalized forms of input/output parameters is given here.

`nd$$[I, "X", j]` $\equiv X_{Ij}$ is the *j*th component of the initial coordinates of the *I*th node.

`nd$$[I, "at", j]` $\equiv p_{Ij}$ is the *j*th unknown in *I*th element node.

`es$$["Data", i]` $\equiv d_{Bi}$ is the *i*th material constant (as previously defined by the `"SMSDomainDataNames"` template constant).

`es$$["IntPoints", i, j]` $\equiv \Xi_i$ is the *i*th coordinate of *j*th Gauss point.

`es$$["IntPoints", 4, j]` $\equiv w_{gp}$ is the Gauss point weight in *j*th Gauss point.

`es$$["id", "NoIntPoints"]` $\equiv n_g$ is the number of Gauss points.

`ed$$["ht" , i]` $\equiv d_{hei}$ is the *i*th component of the element history data vector at time t_{n+1}.

`ed$$["hp", i]` $\equiv d_{heni}$ is the *i*th component of the element history data vector at time t_n.

`rdata$$["SubIterationTolerance"]` variable is a global variable and is specific for the chosen finite element environment. It specifies the tolerance with which the local problem has to be solved.

`idata$$["Iteration"]` $\equiv i_{NR}$ global variable specifies the index of the current global Newton iteration.

`nd$$[J, "SDVF", Iϕ, k]` is sensitivity of the kth sensitivity depended input parameter (ordering is defined by `"SMSSensitivityNames"` template constant) with respect to the I_ϕth sensitivity parameter in Jth element node. Data is used to construct the general design velocity field $D_{\phi_I}\boldsymbol{\psi}_e \equiv$ `DψeDϕIO` as defined by Eq. (8.3.1).

`nd$$[J,"st",Iϕ, k]` is sensitivity of the kth nodal unknown with respect to the I_ϕth sensitivity parameter in Jth element node in time t_{n+1}. Data is used to construct the $D_{\phi_I}\mathbf{p}_e \equiv$ `DpeDϕIO` data structure as defined by Eq. (8.36).

`nd$$[J,"sp",Iϕ, k]` is sensitivity of the kth nodal unknown with respect to the I_ϕth sensitivity parameter in Jth element node in time t_n. Data is used to construct the $D_{\phi_I}\mathbf{p}_{en} \equiv$ `DpenDϕIO` data structure.

`idata$$["SensIndexStart"]` $\equiv \phi_{L\text{start}}$ is an index of the first sensitivity parameter within the currently calculated subset of sensitivity parameters $\boldsymbol{\phi}_L$.

`idata$$["SensIndexEnd"]` $\equiv \phi_{L\text{end}}$ is an index of the last sensitivity parameter within $\boldsymbol{\phi}_L$.

`p$$` $\equiv \mathbf{R}_e$ element residual vector

`s$$` is depending on the user subroutine been generated:
element tangent matrix in `"Tangent and residual"` subroutine (`s$$` $\equiv \mathbf{K}_e$) and an element matrix of sensitivity pseudo-load vectors in `"Sensitivity pseudo-load"` subroutine ($\tilde{\mathbf{R}}_e^T \equiv$ `s$$`) with dimension $(\phi_{L\text{end}} - \phi_{L\text{start}} + 1) \times (n_p + n_{he})$.

Appendix B
Vectors and Tensors

This section summarizes sum rules of tensor algebra and tensor analysis which can be found in e.g. Eringen (1967) and Marsden and Hughes (1983). This summary is not meant to be complete, but it should help in understanding some mathematical derivations and results provided in the previous chapters. It is also aimed to introduce some of the notation used in *Mathematica* and *AceGen* for the operations discussed in this appendix.

B.1 Tensor Algebra

B.1.1 Definition of a Tensor

A tensor id defined as linear map between two vector spaces $\mathcal{V}$ and $\mathcal{W}$. This yields

$$\begin{aligned} \boldsymbol{T} \, : \, \mathcal{V} &\mapsto \mathcal{W} \\ \boldsymbol{v} &\mapsto \boldsymbol{w} = \boldsymbol{T}\,\boldsymbol{v} \qquad \boldsymbol{v} \in \mathcal{V}, \boldsymbol{w} \in \mathcal{W}, \\ \boldsymbol{T}(\boldsymbol{u} + \boldsymbol{v}) &= \boldsymbol{T}\,\boldsymbol{u} + \boldsymbol{T}\,\boldsymbol{v}, \\ \boldsymbol{T}(\alpha\,\boldsymbol{u}) &= \alpha\,\boldsymbol{T}\,\boldsymbol{u}. \end{aligned}$$

A special tensor is the dyad which consists of vectors defined in the spaces $\mathcal{V}$ and $\mathcal{W}$

$$\boldsymbol{T} = \boldsymbol{a} \otimes \boldsymbol{b}, \qquad \boldsymbol{a} \in \mathcal{W}, \boldsymbol{b} \in \mathcal{V}.$$

A linear map of a vector $\boldsymbol{c} \in \mathcal{V}$ to the space $\mathcal{W}$ is obtained with the dyad following the rule

$$(\boldsymbol{a} \otimes \boldsymbol{b})\,\boldsymbol{c} = (\boldsymbol{b} \cdot \boldsymbol{c})\,\boldsymbol{a} \qquad \boldsymbol{c}, \boldsymbol{b} \in \mathcal{V} \quad \boldsymbol{a} \in \mathcal{W}.$$

J. Korelc and P. Wriggers, *Automation of Finite Element Methods*,
DOI 10.1007/978-3-319-39005-5

Note that the vector spaces have to be chosen such that the scalar product $\boldsymbol{b} \cdot \boldsymbol{c}$ is defined which means that $\boldsymbol{b}$ and $\boldsymbol{c}$ are in the same vector space $\mathcal{V}$.

B.1.2 Vectors and Tensors in a Base System

The vectors and tensors defined Sect. B.1.1 have to be written with respect to a basis. This can be a cartesian coordinate system, defined by orthogonal base vectors $\boldsymbol{E}_1, \boldsymbol{E}_2, \boldsymbol{E}_3$ or $\boldsymbol{e}_1, \boldsymbol{e}_2, \boldsymbol{e}_3$. A more general basis is provided by a convective coordinate systems that yields co-variant, $\boldsymbol{g}_i$, and contra-variant, $\boldsymbol{g}^i$, base vectors. The associated convective coordinates are denoted by $\{\Theta^j\}$. They are inscribed in the bodies, see Fig. B.1. Hence the convective coordinates are deformed when the body is deformed under the action of loads. Let us assume that the cartesian coordinates of the initial and current configuration $\{X_A\}$ and $\{x_i\}$, see Chap. 1, can be written as a function of the convective coordinates $\{\Theta^j\}$

$$X_A = \hat{X}_A\,(\Theta^1, \Theta^2, \Theta^3), \qquad x_i = \hat{x}_i\,(\Theta^1, \Theta^2, \Theta^3).$$

In short the transformation reads: $\boldsymbol{X} = \hat{\boldsymbol{X}}\,(\Theta^j)$ and $\boldsymbol{x} = \hat{\boldsymbol{x}}\,(\Theta^j)$.

When convective coordinates are used, then the base vectors are different at each point of a body in the initial and current configuration. The so-called covariant base vectors are given for a point $\boldsymbol{X}$ in the initial configuration B of a body by

$$\boldsymbol{G}_j = \frac{\partial \boldsymbol{X}}{\partial \Theta^j} = \boldsymbol{X}_{,j}.$$

In an analogous way the co-variant base vector for a point $\boldsymbol{\varphi}\,(\boldsymbol{X}, t)$ in the current configuration $\varphi(B)$ is obtained

$$\boldsymbol{g}_j = \frac{\partial \boldsymbol{\varphi}\,(\boldsymbol{X}, t)}{\partial \Theta^j} = \boldsymbol{\varphi}_{,j}.$$

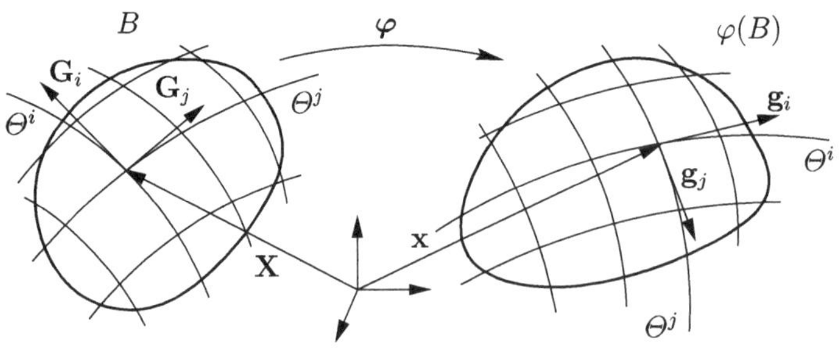

Fig. B.1 Convective coordinates of configurations B and $\varphi(B)$

The co-variant base vectors are tangents to the convective coordinates, see Fig. B.1.

An example how the co-variant base vectors are computed within the finite element context is provided in Sect. 6.2.3, see also the *AceGen* input in Box 6.4.

Vectors and tensors can be described with respect to the convective coordinate system. For the following details the basis $\{\boldsymbol{g}_i\}$ is selected. When using convective coordinates additional base vectors, so-called contra-variant base vectors, have to be defined. These are given by

$$\boldsymbol{g}_i \cdot \boldsymbol{g}^k = \delta_i^k$$

with the Kronecker symbol

$$\delta_k^i = \begin{cases} 1 \text{ for } i = k \\ 0 \text{ for } i \neq k \end{cases}.$$

From this definition it is observed that the contra-variant base vectors are orthogonal to the co-variant base vectors (e.g. $\boldsymbol{g}^1 \perp \boldsymbol{g}_2,\ \boldsymbol{g}_3$). Now vectors and tensors can be defined with respect to this base system:

- A vector $\boldsymbol{u}$ is given with respect to co- and contra-variant basis by

  $$\boldsymbol{u} = u^i\, \boldsymbol{g}_i \qquad \boldsymbol{v} = v_i\, \boldsymbol{g}^i$$

  where contra-variant u^i and the co-variant v_i components are defined by

  $$u^i = \boldsymbol{u} \cdot \boldsymbol{g}^i, \quad v_i = \boldsymbol{v} \cdot \boldsymbol{g}_i.$$

  Note further that the co- and contra-variant basis vectors can be transformed by the metric tensor **metric tensor**

  $$\begin{aligned} \boldsymbol{g} &= g_{ik}\, \boldsymbol{g}^i \otimes \boldsymbol{g}^k, \\ \boldsymbol{g}^{-1} &= g^{ik}\, \boldsymbol{g}_i \otimes \boldsymbol{g}_k, \\ \boldsymbol{g}^{-1}\, \boldsymbol{g} &= \mathbf{1}. \end{aligned}$$

  This yields

  $$\boldsymbol{g}^i = g^{ik}\, \boldsymbol{g}_k, \quad \boldsymbol{g}_i = g_{ik}\, \boldsymbol{g}^k.$$

  Note that in the special case of orthogonal cartesian coordinates X_1, X_2, X_3 condition $\boldsymbol{E}_i \cdot \boldsymbol{E}_k = \delta_{ik}$ holds and hence $g_{ik} = \delta_{ik}$ with the Kronecker symbol of the cartesian basis

  $$\delta_{ik} = \begin{cases} 1 \text{ for } i = k \\ 0 \text{ for } i \neq k \end{cases}$$

  Thus the metric tensor is equal to the unit tensor.
- Different forms can be found for the representation of a second order tensor with respect to a co- and contra-variant basis

$$S = S^{ik}\, \boldsymbol{g}_i \otimes \boldsymbol{g}_k, \quad \boldsymbol{T} = T^{i}_{.k}\, \boldsymbol{g}_i \otimes \boldsymbol{g}^k, \quad \boldsymbol{U} = U_{ik}\, \boldsymbol{g}^i \otimes \boldsymbol{g}^k.$$

The related components follow from

$$S^{ik} = \boldsymbol{g}^i \cdot \boldsymbol{S}\, \boldsymbol{g}^k, \quad T^{i}_{.k} = \boldsymbol{g}^i \cdot \boldsymbol{T}\, \boldsymbol{g}_k, \quad U_{ik} = \boldsymbol{g}_i \cdot \boldsymbol{U}\, \boldsymbol{g}_k$$

with the co-variant $\boldsymbol{g}_i$ and the contra-variant $\boldsymbol{g}^i$ base vectors.

In the special case of a cartesian coordinate system co- and contra-variant bases are identical. Hence the component form of a vector is given by

$$\boldsymbol{u} = u_i\, \boldsymbol{E}_i$$

while a tensor has the form

$$\boldsymbol{T} = T_{ik}\, \boldsymbol{E}_i \otimes \boldsymbol{E}_k,$$

where $\boldsymbol{E}_i$ denotes the cartesian basis. Basically all formulae presented in the following, which are written in either co- or contra-variant form, can be reduced to cartesian basis by using indices as subscripts.

The linear map of a vector by a tensor has for arbitrary tensors of second order and for a dyadic the following representation in components

$$\begin{aligned} \boldsymbol{T}\, \boldsymbol{u} &= (T^{ik}\, \boldsymbol{g}_i \otimes \boldsymbol{g}_k)\, u_l\, \boldsymbol{g}^l = T^{ik}\, u_l (\boldsymbol{g}_k \cdot \boldsymbol{g}^l)\, \boldsymbol{g}_i \\ &= T^{ik}\, u_l\, \delta^l_k\, \boldsymbol{g}_i = T^{ik}\, u_k\, \boldsymbol{g}_i, \\ (\boldsymbol{a} \otimes \boldsymbol{b})\, \boldsymbol{c} &= (\boldsymbol{b} \cdot \boldsymbol{c})\, \boldsymbol{a} = (b^i\, c_i)\, a^m\, \boldsymbol{g}_m \end{aligned}$$

B.1.3 Operations with Vectors and Tensors

Vectors and tensors can be combined using different operations. Some important possibilities are summarized below

1. Scalar product of vectors and tensors:

$$\begin{aligned} \boldsymbol{a} \cdot \boldsymbol{b} &= a^i\, b_i, \\ (\boldsymbol{a} \otimes \boldsymbol{b}) \cdot (\boldsymbol{c} \otimes \boldsymbol{d}) &= (\boldsymbol{a} \cdot \boldsymbol{c})\, (\boldsymbol{b} \cdot \boldsymbol{d}), \\ \boldsymbol{S} \cdot \boldsymbol{T} &= (S^{ik} \boldsymbol{g}_i \otimes \boldsymbol{g}_k) \cdot (T^{lm} \boldsymbol{g}_l \otimes \boldsymbol{g}_m) \\ &= S^{ik}\, T^{lm}\, (\boldsymbol{g}_i \cdot \boldsymbol{g}_l)\, (\boldsymbol{g}_k \cdot \boldsymbol{g}_m) = S^{ik}\, T^{lm}\, g_{il}\, g_{km} \\ &= S^{ik}\, T_{ik}. \end{aligned}$$

The scalar product can be defined between two vectors or tensors of same order. It always yields a single number, the scalar.

In *AceGen* a vector **a** is defined by the *Mathematica* command

$$\texttt{a} = \{\texttt{a1}, \texttt{a2}, \texttt{a3}\}$$

and a component of vector by

$$a_i \equiv a[[i]]$$

which means that the base vectors are omitted and only the components are introduced. A dyadic tensor $\boldsymbol{T} = \boldsymbol{a} \otimes \boldsymbol{b}$ can be defined in *Mathematica* by the command

$$\begin{aligned} \boldsymbol{a} \otimes \boldsymbol{b} &\equiv \texttt{Outer[Times, \{a1, a2, a3\}, \{b1, b2, b3\}]} \\ &\equiv \texttt{Table[a[[i, k]] b[[i, k]], \{i, 3\}, \{k, 3\}]} \end{aligned}$$

Only the components enter the definitions of the vectors and matrices. Thus it is clear that the user of *AceGen* has to make sure that the coordinate systems used for a theoretical formulation match the rules for vectors and tensors.

Box B.1 *AceGen* input for the definition of vectors and tensors

When using *AceGen* the scalar product between two vectors is just defined as above

$$\boldsymbol{a} \cdot \boldsymbol{b} \equiv \texttt{a.b}$$

However the scalar product of two tensors has to be computed via the trace command, see below in Sect. B.1.5.

$$\begin{aligned} \boldsymbol{S} \cdot \boldsymbol{T} &\equiv \texttt{Tr[S.T}^{\texttt{T}}\texttt{]} \\ &\equiv \texttt{Sum[S[[i, k]] T[[i, k]], \{i, 3\}, \{k, 3\}]} \end{aligned}$$

Box B.2 *AceGen* input for the scalar product and trace operation of vectors and tensors

2. Cross product of vectors in a cartesian basis:

$$\boldsymbol{a} \times \boldsymbol{b} = e_{ikl}\, a_i\, b_k \boldsymbol{E}_l,$$

with the permutation symbol

$$e_{ikl} = \begin{cases} \;\;0 & \text{for } i = k, i = l, k = l, i = k = l \\ +1 & \text{for } ikl = 123, = 312, = 231 \\ -1 & \text{for } ikl = 321, = 213, = 132 \end{cases}$$

The cross product of two vectors yields a vector perpendicular to both vectors.

When using *AceGen* the cross product between two vectors is just defined as above

$$\boldsymbol{a} \times \boldsymbol{b} \equiv \texttt{a} \times \texttt{b} \equiv \texttt{Cross[a, b]}.$$

Box B.3 *AceGen* input for the cross product

3. Product of two second order tensors leads again to a second order tensor

$$
\begin{aligned}
(\boldsymbol{a} \otimes \boldsymbol{b})\,(\boldsymbol{c} \otimes \boldsymbol{d}) &= (\boldsymbol{b} \cdot \boldsymbol{c})\,\boldsymbol{a} \otimes \boldsymbol{d}, \\
\boldsymbol{T}\,\boldsymbol{S} &= (\,T^{ik}\,\boldsymbol{g}_i \otimes \boldsymbol{g}_k)(\,S^{lm}\,\boldsymbol{g}_l \otimes \boldsymbol{g}_m) \\
&= T^{ik}\,S_{lm}\,(\boldsymbol{g}_k \cdot \boldsymbol{g}_l)\,\boldsymbol{g}_i \otimes \boldsymbol{g}_m \\
&= T^{ik}\,S_{lm}\,g_{kl}\,\boldsymbol{g}_i \otimes \boldsymbol{g}_m = T^{ik}\,S^l_k\,\boldsymbol{g}_i \otimes \boldsymbol{g}_l.
\end{aligned}
$$

When using *AceGen* the product between two tensors leads to

$$
\begin{aligned}
\boldsymbol{T}\ \boldsymbol{S} &\equiv \texttt{T.S} \\
&\equiv \texttt{Table[Sum[T[[i, k]] S[[k, l]], \{k, 3\}], \{i, 3\}, \{l, 3\}]}
\end{aligned}
$$

Box B.4 *AceGen* input for the tensor product

B.1.4 Special Forms of Tensors

Transposed tensor

$$
\begin{aligned}
(\,\boldsymbol{a} \otimes \boldsymbol{b}\,)^T &= \boldsymbol{b} \otimes \boldsymbol{a}, \\
\boldsymbol{T}^T &= T^{ik}\,\boldsymbol{g}_k \otimes \boldsymbol{g}_i, \\
\boldsymbol{u} \cdot \boldsymbol{T}\,\boldsymbol{v} &= \boldsymbol{v} \cdot \boldsymbol{T}^T\,\boldsymbol{u} \\
\boldsymbol{a} \cdot (\boldsymbol{b} \otimes \boldsymbol{c})\,\boldsymbol{d} &= \boldsymbol{d} \cdot (\boldsymbol{c} \otimes \boldsymbol{b})\,\boldsymbol{a}.
\end{aligned}
$$

Inverse tensor

$$
\boldsymbol{T}\,\boldsymbol{T}^{-1} = \mathbf{1}.
$$

A tensor multiplied by its inverse tensor yields the unit tensor.

AceGen input for the transposed tensor

$$
\boldsymbol{T}^T \equiv \texttt{Transpose[T]} \equiv \texttt{T}^{\texttt{T}}.
$$

and the inverse tensor

$$
\boldsymbol{T}^{-1} \equiv \texttt{Inverse[T]}.
$$

Box B.5 *AceGen* input for the transposed and inverse tensor

Sherman–Morrison formula for the inverse of the sum of an arbitrary second order tensor and a dyadic product

$$
[\,\boldsymbol{T} + \boldsymbol{a} \otimes \boldsymbol{b}\,]^{-1} = \boldsymbol{T}^{-1} - \frac{\boldsymbol{T}^{-1}\,\boldsymbol{a} \otimes \boldsymbol{b}\,\boldsymbol{T}^{-1}}{1 + \boldsymbol{b} \cdot \boldsymbol{T}^{-1}\,\boldsymbol{a}}.
$$

Unit tensor

$$\mathbf{1} = \delta^i_k\, \boldsymbol{g}_i \otimes \boldsymbol{g}^k = \boldsymbol{g}_i \otimes \boldsymbol{g}^i .$$

The unit tensor related to a cartesian basis is given by

$$\mathbf{1} = \delta_{ik}\, \boldsymbol{E}_i \otimes \boldsymbol{E}_k = \boldsymbol{E}_i \otimes \boldsymbol{E}_i .$$

Axial vector t_A

$$\boldsymbol{T}_A\, \boldsymbol{v} = \boldsymbol{t}_A \times \boldsymbol{v},$$

The axial vector represents a vector obtained by the linear map of an arbitrary skew symmetric Tensor $\boldsymbol{T}_A$ and an arbitrary vector $\boldsymbol{v}$.

$$\text{special case: } \boldsymbol{T}_A = \frac{1}{2}\,(\boldsymbol{a} \otimes \boldsymbol{b} - \boldsymbol{b} \otimes \boldsymbol{a}) \Rightarrow \boldsymbol{t}_A = \frac{1}{2}\,(\boldsymbol{b} \times \boldsymbol{a}).$$

Orthogonal tensor preserves the scalar product

$$(\boldsymbol{Q}\,\boldsymbol{a}) \cdot (\boldsymbol{Q}\,\boldsymbol{b}) = \boldsymbol{a} \cdot \boldsymbol{b} = \boldsymbol{b} \cdot (\boldsymbol{Q}^T\, \boldsymbol{Q}\, \boldsymbol{a}) \longrightarrow \boldsymbol{Q}^T\, \boldsymbol{Q} - \mathbf{1}.$$

Thus an orthogonal tensor has the properties

$$\boldsymbol{Q}^{-1} = \boldsymbol{Q}^T \iff \boldsymbol{Q}\,\boldsymbol{Q}^T = \mathbf{1}.$$

The orthogonal tensor $\boldsymbol{Q}$ has in an orthogonal basis $\{\boldsymbol{r}, \boldsymbol{s}, \boldsymbol{t}\}$ the representation

$$\boldsymbol{Q} = \boldsymbol{r} \otimes \boldsymbol{r} + (\,\boldsymbol{s} \otimes \boldsymbol{s} + \boldsymbol{t} \otimes \boldsymbol{t}\,)\, \cos\theta - (\,\boldsymbol{s} \otimes \boldsymbol{t} - \boldsymbol{t} \otimes \boldsymbol{s}\,)\, \sin\theta .$$

Hence the orthogonal tensor represents a rotation about axis $\boldsymbol{r}$.

B.1.5 Eigenvalues and Invariants of Tensors

Before the invariants and eigenvalues of second order tensors are discussed it is useful to define the trace and determinant of a second order tensor.

Trace of a tensor

$$\begin{aligned} \operatorname{tr} \boldsymbol{T} &= \mathbf{1} \cdot \boldsymbol{T} = (\boldsymbol{g}_i \otimes \boldsymbol{g}^i) \cdot (\,T^{lm}\, \boldsymbol{g}_l \otimes \boldsymbol{g}_m) \\ &= (\boldsymbol{g}_i \cdot \boldsymbol{g}_l)\,(\boldsymbol{g}^i \cdot \boldsymbol{g}_m)\, T^{lm} = g_{il}\, \delta^i_m\, T^{lm} = T^l_l, \\ \operatorname{tr}(\boldsymbol{a} \otimes \boldsymbol{b}) &= \boldsymbol{a} \cdot \boldsymbol{b}, \end{aligned}$$

with the properties

$$\operatorname{tr} \boldsymbol{T}^T = \operatorname{tr} \boldsymbol{T} \quad \text{and} \quad \operatorname{tr}(\boldsymbol{S} + \boldsymbol{T}) = \operatorname{tr} \boldsymbol{S} + \operatorname{tr} \boldsymbol{T}$$
$$\operatorname{tr}(\boldsymbol{S}\,\boldsymbol{T}) = \operatorname{tr}(\boldsymbol{T}\,\boldsymbol{S}) = \boldsymbol{S} \cdot \boldsymbol{T} = S_{ik}\, T^{ik} = \operatorname{tr}(\boldsymbol{S}\,\boldsymbol{T}^T) = \boldsymbol{T}^T \cdot \boldsymbol{S}^T.$$

Determinant

$$\det(\alpha\, \boldsymbol{T}) = \alpha^3 \det \boldsymbol{T},$$
$$\det(\boldsymbol{S}\,\boldsymbol{T}) = \det \boldsymbol{S} \det \boldsymbol{T} \quad \text{and} \quad \det(\boldsymbol{T}^{-1}) = \frac{1}{\det \boldsymbol{T}}.$$

Cofactor of a tensor[1]

$$\operatorname{Cof} \boldsymbol{T} = \det \boldsymbol{T}\, \boldsymbol{T}^{-T}.$$

Eigenvalues of a tensor
From the special eigenvalue problem

$$(\boldsymbol{T} - \lambda\, \mathbf{1})\, \boldsymbol{\varphi} = \mathbf{0}$$

follows a cubic equation for the eigenvalues, known as characteristic polynomial,

$$\det(\boldsymbol{T} - \lambda\, \mathbf{1}) = \lambda^3 - I_T\, \lambda^2 + II_T\, \lambda - III_T = 0$$

with the three invariants of the tensor $\boldsymbol{T}$:

$$I_T = \operatorname{tr} \boldsymbol{T},$$
$$II_T = \operatorname{tr} \operatorname{Cof} \boldsymbol{T},$$
$$III_T = \det \boldsymbol{T}.$$

When knowing the eigenvalues the invariants

$$I_T = \lambda_1 + \lambda_2 + \lambda_3,$$
$$II_T = \lambda_1\, \lambda_2 + \lambda_2\, \lambda_3 + \lambda_3\, \lambda_1,$$
$$III_T = \lambda_1\, \lambda_2\, \lambda_3,$$

can be computed in a simpler way.

---

[1] The abbreviation for the inverse of the transposed tensor $\boldsymbol{T}^{-T}$ can be introduced since $(\boldsymbol{T}^T)^{-1} = (\boldsymbol{T}^{-1})^T$.

When using *AceGen* the tensor invariants are given by

$$I_T = \operatorname{tr} \boldsymbol{T} \equiv \texttt{Tr[T]},$$
$$II_T = \operatorname{tr} \operatorname{Cof} \boldsymbol{T} \equiv \texttt{Simplify[Tr[Det[T] Inverse[T]}^{\mathrm{T}}\texttt{]]},$$
$$III_T = \det \boldsymbol{T} \equiv \texttt{Det[T]}.$$

Box B.6 *AceGen* input for the computation of the tensor invariants

B.1.6 Tensors of Higher Order

Tensors of higher order will be defined in terms of dyadic products. The derived rules can then be applied to the bases of arbitrary tensors. Here only tensors of third and fourth order are considered.

1. Third order dyadic product

$$(\boldsymbol{a} \otimes \boldsymbol{b}) \otimes \boldsymbol{c} = \boldsymbol{a} \otimes \boldsymbol{b} \otimes \boldsymbol{c}$$

$$(\boldsymbol{a} \otimes \boldsymbol{b} \otimes \boldsymbol{c})\, (\boldsymbol{d} \otimes \boldsymbol{e}) = (\boldsymbol{b} \cdot \boldsymbol{d})\, (\boldsymbol{c} \cdot \boldsymbol{e})\, \boldsymbol{a}$$

2. Fourth order dyadic product

$$(\boldsymbol{a} \otimes \boldsymbol{b}) \otimes (\boldsymbol{c} \otimes \boldsymbol{d}) = \boldsymbol{a} \otimes \boldsymbol{b} \otimes \boldsymbol{c} \otimes \boldsymbol{d}$$

$$(\boldsymbol{a} \otimes \boldsymbol{b} \otimes \boldsymbol{c} \otimes \boldsymbol{d})\, (\boldsymbol{f} \otimes \boldsymbol{g}) = (\boldsymbol{c} \cdot \boldsymbol{f})\, (\boldsymbol{d} \cdot \boldsymbol{g})\, (\boldsymbol{a} \otimes \boldsymbol{b})$$

Rules:

$$\begin{aligned}
(\boldsymbol{T} \otimes \boldsymbol{c})\, \boldsymbol{v} &= (\boldsymbol{c} \cdot \boldsymbol{v})\, \boldsymbol{T}, \\
(\boldsymbol{a} \otimes \boldsymbol{T})\, \boldsymbol{R} &= (\boldsymbol{T} \cdot \boldsymbol{R})\, \boldsymbol{a}, \\
(\boldsymbol{a} \otimes \boldsymbol{b} \otimes \boldsymbol{c})\, \mathbf{1} &= (\boldsymbol{b} \cdot \boldsymbol{c})\, \boldsymbol{a}, \\
(\boldsymbol{T} \otimes \boldsymbol{v})\, (\boldsymbol{a} \otimes \boldsymbol{b}) &= (\boldsymbol{b} \cdot \boldsymbol{v})\, \boldsymbol{T}\, \boldsymbol{a}, \\
(\boldsymbol{T} \otimes \boldsymbol{v})\, \boldsymbol{R} &= (\boldsymbol{T}\, \boldsymbol{R})\, \boldsymbol{v}, \\
(\boldsymbol{T} \otimes \boldsymbol{v})\, \mathbf{1} &= \boldsymbol{T}\, \boldsymbol{v}, \\
(\boldsymbol{T} \otimes \boldsymbol{R})\, \boldsymbol{S} &= (\boldsymbol{R} \cdot \boldsymbol{S})\, \boldsymbol{T}, \\
(\boldsymbol{T} \otimes \boldsymbol{R})\, \boldsymbol{v} &= \boldsymbol{T} \otimes \boldsymbol{R}\, \boldsymbol{v}, \\
(\boldsymbol{T} \otimes \boldsymbol{R})\, \mathbf{1} &= (\operatorname{tr} \boldsymbol{R})\, \boldsymbol{T}.
\end{aligned}$$

In general the representation of a fourth order tensor is given by

$$\mathbb{C} = C^{ijkl}\, \boldsymbol{g}_i \otimes \boldsymbol{g}_j \otimes \boldsymbol{g}_k \otimes \boldsymbol{g}_l .$$

A tensor of fourth order can be applied to define a linear mapping between two tensors of second order

$$
\begin{aligned}
\boldsymbol{U} &= \mathbb{C}[\, \boldsymbol{V} \,] \\
U^{ij}\, \boldsymbol{g}_i \otimes \boldsymbol{g}_j &= (\, C^{ijkl}\, \boldsymbol{g}_i \otimes \boldsymbol{g}_j \otimes \boldsymbol{g}_k \otimes \boldsymbol{g}_l)(\, V_{mn}\, \boldsymbol{g}^m \otimes \boldsymbol{g}^n \,) \\
&= C^{ijkl}\, V_{mn}\, \delta_k^m\, \delta_l^n\, \boldsymbol{g}_i \otimes \boldsymbol{g}_j = C^{ijkl}\, V_{kl}\, \boldsymbol{g}_i \otimes \boldsymbol{g}_j
\end{aligned}
$$

A general tensor transformation can be formulated by combining the `Table` and `Sum` *Mathematica* commands. Note that the `Table` and `Sum` commands are executed symbolically within *Mathematica* and consequently, the generated source code contains no loops.

$\mathbb{C}[\, \boldsymbol{V} \,] \equiv$ `Table[Sum[Cm[[i, j, k, l]] V[[k, l]], {k, 3}, {l, 3}], {i, 3}, {j, 3}]`

Box B.7 *AceGen* input to define a linear mapping between two tensors of second order

B.2 Tensor Analysis

In this section scalars, vectors and tensors are discussed which are functions of a position vector $\boldsymbol{X}$ and time t. For that the following fields are defined:

- Scalar field: $\alpha(\boldsymbol{X}, t)$
- Vector field: $\boldsymbol{v}(\boldsymbol{X}, t)$
- Tensor field: $\boldsymbol{T}(\boldsymbol{X}, t)$

Examples for scalar fields are density, pressure or temperature. Displacements, velocities or momentum can be described by vector fields. Stresses or strains are represented by tensor fields.

B.2.1 Differentiation with Respect to a Real Variable

The differentiation with respect to a real variable, e.g. the time, is based on the following definitions

$$
\text{Definition:} \qquad \dot{\boldsymbol{v}}(\boldsymbol{X}, t) = \frac{\partial \boldsymbol{v}(\boldsymbol{X}, t)}{\partial t}
$$

When using *AceGen* the time derivative of a vector or tensor is given by

$$\dot{\boldsymbol{v}}(\boldsymbol{X},t) \equiv \texttt{SMSD[v, t]} \quad \text{and} \quad \dot{\boldsymbol{T}}(\boldsymbol{X},t) \equiv \texttt{SMSD[T, t]}$$

or

$$(\boldsymbol{T}^{-1})^{\cdot} \equiv \texttt{SMSD[Inverse[T], t]}$$

Box B.8 *AceGen* input for the computation of the time derivative

B.2.2 Gradient of a Field

The gradient of a field yields always a fields which is one order higher. Hence a gradient of a scalar field yields a vector field and so forth.

$$\begin{aligned}\boldsymbol{v} &= \operatorname{Grad}\alpha(\boldsymbol{X},t) = \frac{\partial\alpha}{\partial\boldsymbol{X}} = \frac{\partial\alpha}{\partial X_i}\boldsymbol{G}^i,\\ \boldsymbol{T} &= \operatorname{Grad}\boldsymbol{v}(\boldsymbol{X},t) = \frac{\partial\boldsymbol{v}}{\partial\boldsymbol{X}} = \frac{\partial\boldsymbol{v}}{\partial X_i}\otimes\boldsymbol{G}^i.\end{aligned}$$

Often the Nabla operator ∇ is used instead of the gradient operator Grad. Then

$$\begin{aligned}\operatorname{grad}\alpha &= \nabla\,\alpha,\\ \operatorname{grad}\boldsymbol{u} &= \nabla\,\boldsymbol{u}.\end{aligned}$$

is written.

When using *AceGen* the gradient of a scalar, vector or tensor is given by

$$\text{scalar: } \operatorname{Grad}\alpha(\boldsymbol{X},t) \equiv \texttt{SMSD[, X]}$$

$$\text{vector: } \operatorname{Grad}\boldsymbol{v}(\boldsymbol{X},t) \equiv \texttt{SMSD[v, X]}$$

Box B.9 *AceGen* input for the computation of the gradient of a scalar and a vector

B.2.3 Divergence of a Field

The computation of the divergence of a field reduces the order of the field by one. Thus the divergence of e.g. a tensor field yields a vector field.

$$\begin{aligned}\operatorname{Div}\boldsymbol{v}(\boldsymbol{X},t) &= \operatorname{Grad}\boldsymbol{v}(\boldsymbol{X},t)\cdot\mathbf{1} = \operatorname{tr}(\operatorname{Grad}\boldsymbol{v}(\boldsymbol{X},t)),\\ \operatorname{Div}\boldsymbol{T}(\boldsymbol{X},t) &= \operatorname{Grad}\boldsymbol{T}(\boldsymbol{X},t)\,\mathbf{1},\end{aligned}$$

B.2.4 Pull Back and Push Forward Operations

In this section the *pull back* (φ^*) and *push forward* (φ_*) operations will be discussed. The co-variant basis vectors $\{\boldsymbol{g}_i\}$, see Fig. B.1 are related to the tangent space while the contra-variant basis vectors $\{\boldsymbol{g}^i\}$ can be denoted as one forms. These bases have different behaviour when pulled back from the current to the initial configuration and when pushed forward from the initial configuration to the current one. The following table depicts the different behaviour during transformation

$$\boldsymbol{g}_i = \boldsymbol{F}\,\boldsymbol{G}_i, \quad \boldsymbol{g}^i = \boldsymbol{F}^{-T}\,\boldsymbol{G}^i,$$

$$\boldsymbol{G}_i = \boldsymbol{F}^{-1}\,\boldsymbol{g}_i, \quad \boldsymbol{G}^i = \boldsymbol{F}^T\,\boldsymbol{g}^i.$$

As in Chap. 1 small letters are associated with quantities measured in the current configuration and capital letters are related to quantities in the initial configuration. Thus basis vectors $\boldsymbol{G}_i$ refer to the initial configuration B and basis vectors $\boldsymbol{g}_i$ refer to the current configuration $\varphi(B)$.

The transformation behaviour of the divergence operator is given by

$$\operatorname{div}\boldsymbol{v} = \tfrac{1}{J_F}\operatorname{Div}\boldsymbol{v}, \ \operatorname{Div}\boldsymbol{v} = J_F\operatorname{div}\boldsymbol{v}.$$

In analogous way the gradients can be transferred:

$$\operatorname{grad}\alpha = \boldsymbol{F}^{-T}\operatorname{Grad}\alpha, \operatorname{Grad}\alpha = \boldsymbol{F}^T\operatorname{grad}\alpha,$$
$$\operatorname{grad}\boldsymbol{v} = \operatorname{Grad}\boldsymbol{v}\,\boldsymbol{F}, \operatorname{Grad}\boldsymbol{v} = \operatorname{grad}\boldsymbol{v}\,\boldsymbol{F}^{-1}.$$

Tensors can, as well as gradients, be referred to base systems of the initial or current configuration. Here this will be exemplarily performed for the Cauchy stress tensor $\boldsymbol{\sigma}$, which is defined with respect to the current configuration. With the representation of $\boldsymbol{\sigma}$ with respect to co- and contra variant bases

$$\boldsymbol{\sigma}^\flat = \sigma_{ik}\,\boldsymbol{g}^i \otimes \boldsymbol{g}^k,$$

$$\boldsymbol{\sigma}^\sharp = \sigma^{ik}\,\boldsymbol{g}_i \otimes \boldsymbol{g}_k,$$

$$\boldsymbol{\sigma}_1 = \sigma^i_{\cdot k}\,\boldsymbol{g}_i \otimes \boldsymbol{g}^k,$$

the pull back and push forward operations are

push forward	*pull back*
$\boldsymbol{\sigma}^\flat = \boldsymbol{F}^{-T}\,\boldsymbol{\Sigma}^\flat\,\boldsymbol{F}^{-1}$	$\boldsymbol{\Sigma}^\flat = \boldsymbol{F}^T\,\boldsymbol{\sigma}^\flat\,\boldsymbol{F}$

$$\boldsymbol{\sigma}^{\sharp} = \boldsymbol{F}\,\boldsymbol{\Sigma}^{\sharp}\,\boldsymbol{F}^{T} \quad \boldsymbol{\Sigma}^{\sharp} = \boldsymbol{F}^{-1}\,\boldsymbol{\sigma}^{\sharp}\,\boldsymbol{F}^{-T}$$

$$\boldsymbol{\sigma}_{1} = \boldsymbol{F}\,\boldsymbol{\Sigma}_{1}\,\boldsymbol{F}^{-1} \quad \boldsymbol{\Sigma}_{1} = \boldsymbol{F}^{-1}\,\boldsymbol{\sigma}_{1}\,\boldsymbol{F}.$$

Here $\boldsymbol{\Sigma}$ denotes the stress tensor referred to the initial configuration B.

B.2.5 Lie-Derivative of Stress Tensors

The Lie derivative is a derivative of a spatial tensor with respect to time. It is defined by

$$L_{v}\,(\boldsymbol{t}) = \Phi_{t*}\left[\frac{d}{dt}\,\Phi_{t}^{*}\,(\boldsymbol{t})\right].$$

This means that one has first to pull the tensor back to initial configuration, then perform the time derivative and after that push the result forward to the spatial or current configuration.

It can be applied to tensors which are defined with respect to the current configuration, see e.g. Sect. 1.1.4. The Lie derivative yields the flux related to the used stress tensor. From the definition of the pull back and push forward operations it is clear that the selected tensor basis (co- or contra-variant) has influence on the result of the Lie derivative. Since stress tensors are mostly written with respect to a co-variant basis this basis is used for the following considerations.

Application of the Lie derivative to the Kirchhoff stress tensor $\boldsymbol{\tau}$ yields the so called Oldroyd stress flux or stress rate

$$L_{v}\,(\boldsymbol{\tau}^{\sharp}) = \boldsymbol{F}\,\frac{d}{dt}\,\left[\boldsymbol{F}^{-1}\,\boldsymbol{\tau}\,\boldsymbol{F}^{-t}\right]\boldsymbol{F}^{T}.$$

With (1.37) the result $\boldsymbol{F}\,\dot{\boldsymbol{F}}^{-1} = -\boldsymbol{l}$ is obtained from $\frac{d}{dt}(\boldsymbol{F}\,\boldsymbol{F}^{-1}) = \boldsymbol{0}$ and hence the final expression for the Oldroyd stress rate is

$$L_{v}\,(\boldsymbol{\tau}^{\sharp}) = \dot{\boldsymbol{\tau}}^{\sharp} - \boldsymbol{l}\,\boldsymbol{\tau}^{\sharp} - \boldsymbol{\tau}^{\sharp}\,\boldsymbol{l}^{T}.$$

The Truesdell stress rate $L_{v}^{J}\,(\boldsymbol{\sigma}^{\sharp})$ follows from the relation between the 2nd Piola–Kirchhoff stress tensor and the Cauchy stress tensor

$$\begin{aligned} L_{v}^{J}\,(\boldsymbol{\sigma}^{\sharp}) &= J_{F}^{-1}\,\boldsymbol{F}\,\frac{d}{dt}[\,J_{F}\boldsymbol{F}^{-1}\,\boldsymbol{\sigma}^{\sharp}\,\boldsymbol{F}^{-T}\,]\,\boldsymbol{F}^{T} \\ &= \dot{\boldsymbol{\sigma}}^{\sharp} - \boldsymbol{l}\,\boldsymbol{\sigma}^{\sharp} - \boldsymbol{\sigma}^{\sharp}\,\boldsymbol{l}^{T} + \boldsymbol{\sigma}^{\sharp}\,tr(\boldsymbol{d}\,). \end{aligned}$$

These stress rates are objective, see Sect. 1.2.4.

Further known stress rates can be derived in an analogous way. Let us remark that the addition of two objective stress rates yield again an objective stress rate.

References

Eringen, A. 1967. *Mechanics of continua*. New York: Wiley.
Marsden, J.E., and T.J.R. Hughes. 1983. *Mathematical foundations of elasticity*. Englewood Cliffs: Prentice-Hall.

Appendix C
Tables for Gauss Integration

In the following tables the Gauss integration rules for one-, two- and three-dimensional integration are summarized.

C.1 One-Dimensional Integration

$$\int_{-1}^{+1} f(\xi)\, J_e(\xi)\, \delta\xi \approx \sum_{g=1}^{n_g} f(\xi_g)\, J_e(\xi_g)\, w_{gp}.$$

These integration formulae integrate polynomials in ξ^i exact up to the order $i \leq m$ (Table C.1).

Table C.1 One-dimensional Gauss-Integration

AceFEM	m	n_g	g	ξ_g	w_{gp}	
20	1	1	1	0	2	
21	3	2	1	$1/\sqrt{3}$	1	
			2	$1/\sqrt{3}$	1	
22	5	3	1	$-\sqrt{3/5}$	5/9	
			2	0	8/9	
			3	$+\sqrt{3/5}$	5/9	

J. Korelc and P. Wriggers, *Automation of Finite Element Methods*,
DOI 10.1007/978-3-319-39005-5

C.2 Two-Dimensional Integration

C.2.1 Quadrilateral Elements

$$\int_{-1}^{+1}\int_{-1}^{+1} f(\xi,\eta)\,\det \boldsymbol{J}_e\, d\xi\, d\eta \approx \sum_{g=1}^{n_g} f(\xi_g,\eta_g)\,\det \boldsymbol{J}_e(\xi_g,\eta_g)\, w_{gp}.$$

These integration formulae integrate polynomials in $\xi^i\,\eta^k$ exact up to the order $i+k \leq m$.

C.2.2 Triangular Elements

$$\int_{0}^{1}\int_{0}^{1-\xi} f(\xi,\eta)\,\det \boldsymbol{J}_e\, d\eta\, d\xi \approx \sum_{g=1}^{n_g} f(\xi_g,\eta_g)\,\det \boldsymbol{J}_e(\xi_g,\eta_g)\, w_{gp}$$

C.3 Three-Dimensional Integration

C.3.1 Hexahedral Elements

$$\int_{-1}^{+1}\int_{-1}^{+1}\int_{-1}^{+1} f(\xi,\eta\,\zeta)\,\det \boldsymbol{J}_e\, d\xi\, d\eta\, d\zeta \approx \sum_{g=1}^{n_g} f(\boldsymbol{\Xi}_g)\,\det \boldsymbol{J}_e(\boldsymbol{\Xi}_g)\, w_{gp}$$

The coordinates of the Gauss point $\boldsymbol{\Xi}_g = (\xi_g,\eta_g,\zeta_g)$ can easily be derived from the two-dimensional case, see Table C.2, by expanding these coordinates into the third coordinate direction.

These integration formulae integrate polynomials in $\xi^i\,\eta^k\,\zeta^l$ exact up to the order $i+k+l \leq m$.

Table C.2 Two-dimensional Gauss integration for quadrilateral elements

AceFEM	m	n_g	g	ξ_g	η_g	w_{gp}	Position of points
1	1	1	1	0	0	4	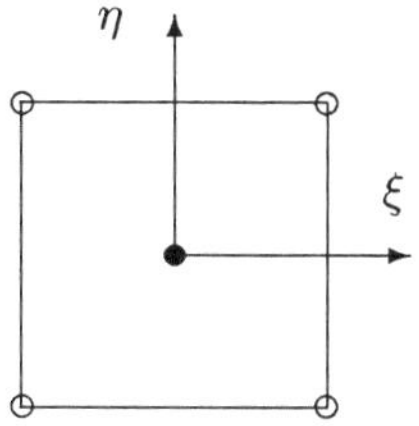
2	3	4	1	$-1/\sqrt{3}$	$-1/\sqrt{3}$	1	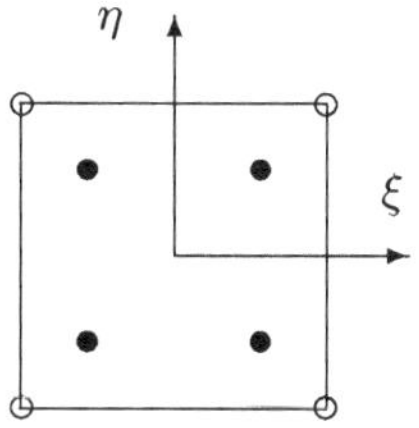
			2	$+1/\sqrt{3}$	$-1/\sqrt{3}$	1	
			3	$-1/\sqrt{3}$	$+1/\sqrt{3}$	1	
			4	$+1/\sqrt{3}$	$+1/\sqrt{3}$	1	
3	5	9	1	$-\sqrt{3/5}$	$-\sqrt{3/5}$	25/81	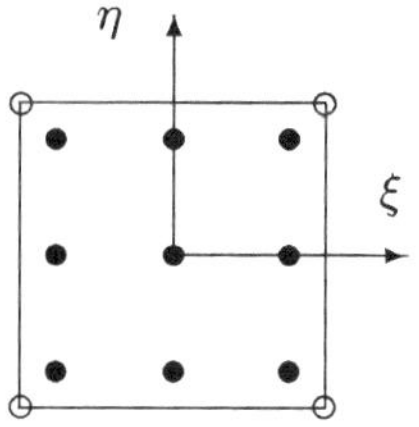
			2	0	$-\sqrt{3/5}$	40/81	
			3	$+\sqrt{3/5}$	$-\sqrt{3/5}$	25/81	
			4	$-\sqrt{3/5}$	0	40/81	
			5	0	0	64/81	
			6	$+\sqrt{3/5}$	0	40/81	
			7	$-\sqrt{3/5}$	$+\sqrt{3/5}$	25/81	
			8	0	$+\sqrt{3/5}$	40/81	
			9	$+\sqrt{3/5}$	$+\sqrt{3/5}$	25/81	

C.3.2 Tetrahedral Elements

There exist special integration rules for three-dimensional finite elements. These rules need less integration points for a given accuracy and thus are more efficient to use for large scale computations. The following table summarizes three different rules (Tables C.3, C.4, C.5 and C.6).

Table C.3 Two-dimensional integration for triangular elements

AceFEM	m	n_g	g	ξ_g	η_g	w_{gp}	Position of points
12	1	1	1	1/3	1/3	1/2	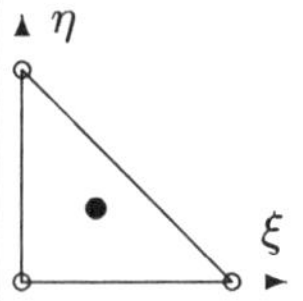
14	2	3	1	1/2	1/2	1/6	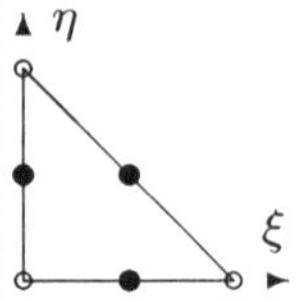
			2	0	1/2	1/6	
			3	1/2	0	1/6	
35	2	3	1	1/6	1/6	1/6	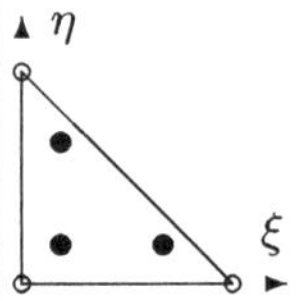
			2	2/3	1/6	1/6	
			3	1/6	2/3	1/6	
16	3	4	1	1/3	1/3	−27/96	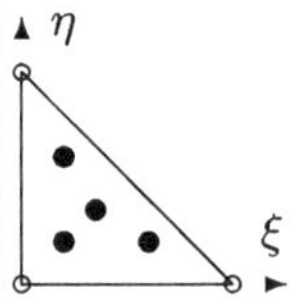
			2	1/5	1/5	25/96	
			3	3/5	1/5	25/96	
			4	1/5	3/5	25/96	

Table C.4 Three-dimensional Gauss integration for hexahedral elements

AceFEM	m	n_g
6	1	$1 \times 1 \times 1$
7	3	$2 \times 2 \times 2$
8	5	$3 \times 3 \times 3$

Table C.5 Special integration for hexahedral elements

AceFEM	m	n_g	g	ξ_g	η_g	ζ_g	w_{gp}
36	2	4	1	0	$\sqrt{2/3}$	$-1/\sqrt{3}$	2
			2	0	$-\sqrt{2/3}$	$-1/\sqrt{3}$	2
			3	$\sqrt{2/3}$	0	$1/\sqrt{3}$	2
			4	$-\sqrt{2/3}$	0	$1/\sqrt{3}$	2
37	3	6	1	1	0	0	4/3
			2	-1	0	0	4/3
			3	0	1	0	4/3
			4	0	-1	0	4/3
			5	0	0	1	4/3
			6	0	0	-1	4/3
38	5	14	1	$\sqrt{19/30}$	0	0	320/361
			2	$-\sqrt{19/30}$	0	0	320/361
			3	0	$\sqrt{19/30}$	0	320/361
			4	0	$-\sqrt{19/30}$	0	320/361
			5	0	0	$\sqrt{19/30}$	320/361
			6	0	0	$-\sqrt{19/30}$	320/361
			7	$\sqrt{19/33}$	$\sqrt{19/33}$	$\sqrt{19/33}$	121/361
			8	$-\sqrt{19/33}$	$\sqrt{19/33}$	$\sqrt{19/33}$	121/361
			9	$\sqrt{19/33}$	-$\sqrt{19/33}$	$\sqrt{19/33}$	121/361
			10	$-\sqrt{19/33}$	$-\sqrt{19/33}$	$\sqrt{19/33}$	121/361
			11	$\sqrt{19/33}$	$\sqrt{19/33}$	$-\sqrt{19/33}$	121/361
			12	$\sqrt{19/33}$	-$\sqrt{19/33}$	$-\sqrt{19/33}$	121/361
			13	$-\sqrt{19/33}$	$-\sqrt{19/33}$	$-\sqrt{19/33}$	121/361
			14	-$\sqrt{19/33}$	$\sqrt{19/33}$	$-\sqrt{19/33}$	121/361

Table C.6 Three-dimensional integration for tetrahedral elements

AceFEM	m	n_g	g	ξ_g	η_g	ζ_g	w_{gp}
15	1	1	1	1/4	1/4	1/4	1/6
19	3	5	1	1/4	1/4	1/4	−2/15
			2	1/6	1/6	1/6	3/40
			3	1/6	1/6	1/2	3/40
			4	1/6	1/2	1/6	3/40
			5	1/2	1/6	1/6	3/40

Index

J. Korelc and P. Wriggers, *Automation of Finite Element Methods*,
DOI 10.1007/978-3-319-39005-5